Amplitudes, Hodge Theory and Ramification

From Periods and Motives to Feynman Amplitudes

Clay Mathematics Proceedings
Volume 21

Amplitudes, Hodge Theory and Ramification

From Periods and Motives to Feynman Amplitudes

2014 Clay Mathematics Institute Summer School
Periods and Motives: Feynman
Amplitudes and the 21st Century
Instituto de Ciencias Matemáticas
Madrid, Spain
June 30–July 25, 2014

K. Ebrahimi-Fard
J. I. Burgos Gil
D. Manchon
Editors

American Mathematical Society
Clay Mathematics Institute

2020 *Mathematics Subject Classification.* Primary 81Q30, 14C15, 11S15.

Library of Congress Cataloging-in-Publication Data

Names: Clay Mathematics Institute. Summer School (2014 : Madrid, Spain), author. | Ebrahimi-Fard, Kurusch, 1973- editor. — Burgos Gil, José I. (José Ignacio), 1962- editor.| Manchon, Dominique, 1962- editor.

Title: Amplitudes, Hodge theory and ramification : from periods and motives to Feynman amplitudes : 2014 Clay Institute Summer School : periods and motives : Feynman amplitudes in the 21st century, June 30-July 25, 2014, Instituto de Ciencias Matemáticas, Madrid, Spain / K. Ebrahimi-Fard, J.I. Burgos Gil, D. Manchon, editors.

Description: Providence, RI : Published by the American Mathematical Society for the Clay Mathematics Institute, [2020] | Series: Clay mathematics proceedings, 1534-6455 ; volume 21 | Includes bibliographical references.

Identifiers: LCCN 2020001922 | ISBN 9781470443290 (softcover) | ISBN 9781470456986 (ebook)

Subjects: LCSH: Mathematical physics–Congresses. | Quantum field theory–Congresses. | Feynman integrals–Congresses. | Hodge theory–Congresses. | Motives (Mathematics)–Congresses. | Geometry, Algebraic–Congresses. | Algebraic number theory–Congresses. | AMS: Quantum theory – General mathematical topics and methods in quantum theory – Feynman integrals and graphs; applications of algebraic topology and algebraic geometry. — Algebraic geometry – Cycles and subschemes – (Equivariant) Chow groups and rings; motives. | Number theory – Algebraic number theory: local and p-adic fields – Ramification and extension theory.

Classification: LCC QC19.2 .C586 2020 | DDC 530.15–dc23

LC record available at https://lccn.loc.gov/2020001922

Published by the American Mathematical Society, Providence, RI,
for the Clay Mathematics Institute, Cambridge, MA.
Printed in the United States of America.

∞ The paper used in this book is acid-free and falls within the guidelines
established to ensure permanence and durability.
Visit the AMS home page at `https://www.ams.org/`
Visit the Clay Mathematics Institute home page at `https://www.claymath.org/`

10 9 8 7 6 5 4 3 2 1 25 24 23 22 21 20

Contents

Preface

The Clay Mathematics Institute 2014 Summer School, "Periods and Motives: Feynman amplitudes in the 21st century" took place from June 30 to July 25, 2014, at the Instituto de Ciencias Matemáticas – ICMAT (Institute of Mathematical Sciences) in Madrid, Spain.

The central theme of the summer school was, as the title indicates, periods and motives and their interplay with Feynman amplitudes in perturbative quantum field theory. The school brought together more than 80 selected students from around the world to attend to several courses presenting background as well as advanced tools and techniques necessary to play an active role in future mathematics research on periods and motives.

The first three weeks were organized into four grand lecture series supplemented by exercise and problem sessions. The four series were

- **Spencer Bloch** (University of Chicago, USA): *Lectures on the Mathematics of Feynman Amplitudes*,
- **José Ignacio Burgos Gil** (ICMAT-CSIC, Spain): *Motivic multiple zeta values*,
- **Lars Kindler** and **Kay Rülling** (Freie Universität Berlin, Germany): *ℓ-adic Galois representations of function fields over finite fields I*,
- **Matilde Marcolli** (Caltech, USA): *Feynman integrals, periods and motives.*

The fourth week of the school consisted of five mini-courses which were more advanced and focused on specific topics. They also included presentations of some open problems which are under current investigation, ranging from pure mathematics to theoretical physics:

- **Joseph Ayoub** (University of Zürich, Switzerland): *Motives, motivic Galois groups and periods*,
- **Francis Brown** (Oxford University, UK and CNRS, France): *Multiple modular values*,
- **James Drummond** (CNRS, France): *Bootstraps for scattering amplitudes in $N = 4$ super Yang–Mills theory*,
- **Claude Duhr** (Durham University, UK): *Feynman integrals, scattering amplitudes and the Hopf algebra of multiple polylogarithms*,
- **Hélène Esnault** (Freie Universität Berlin, Germany): *ℓ-adic Galois representations of function fields over finite fields II.*

This is the first of two volumes containing articles that grew out of the lectures presented at the summer school. The volume at hand includes the following three lecture notes:

- **Feynman Integrals in Mathematics and Physics,**
  by Spencer Bloch,
- **Feynman integrals and periods in configuration spaces,**
  by Özgür Ceyhan and Matilde Marcolli,
- **Introductory course on ℓ-adic sheaves and their ramification theory on curves,**
  by Lars Kindler and Kay Rülling

Spencer Bloch's contribution is based on his summer school lectures. The focus lies on some mathematical questions associated to Feynman integrals, including Hodge structures, relations with string theory, and monodromy (Cutkosky rules).

The notes by Lars Kindler and Kay Rülling resulted from their joint introductory lecture series on ℓ-adic Galois representations of function fields and ramification theory over curves. They also provided the necessary foundations for the advanced course of Prof. Esnault's in the last week of the school.

The contribution by Özgür Ceyhan and Matilde Marcolli is based on the lectures given by Matilde Marcolli. In her course she explained the occurrence of motives, in particular mixed Tate motives, and periods in the computation of Feynman integrals in two different settings: in momentum space and in configuration space.

The second volume contains the single manuscript

- **Multiple zeta values: from numbers to motives,**
  by José Ignacio Burgos Gil and Javier Fresán. First chapter in collaboration with Ulf Kühn.

It is based on the series of lectures given by José Ignacio Burgos Gil, which aimed at explaining the results of Francis Brown, Pierre Deligne, Alexander Goncharov and Tomohide Terasoma on the algebra of multiple zeta values, as well as presenting some of the required prerequisites needed to get familiar with the fundamental works of the aforementioned authors.

The term "period", understood as the integral of an algebraic differential form over a topological cycle on an algebraic variety, goes back to the study of elliptic integrals. The term originated from the period of a periodic elliptic function, which can be computed as an elliptic integral. Since then, periods have played an important role in algebraic geometry. Kontsevich and Zagier have formalized the notion of "ring of periods", that is, the set of complex numbers whose real as well as imaginary part are given in terms of absolutely convergent multiple integrals of a rational differential form with rational coefficients, over a domain in $\mathbb{R}^n$, which is defined by polynomial equations and inequalities with rational coefficients. They conjectured, roughly speaking, that the only relations among periods are the "obvious" ones coming from geometry, that is linearity, change of variables formula and Stoke's theorem. This conjecture encompasses many transcendence conjectures.

The theory of motives was introduced by Grothendieck as a "universal cohomology theory" to explain the common properties of different cohomology theories. One may describe motives as an intermediate step between algebraic varieties and their linear invariants (cohomology). The motives associated to smooth projective varieties are called "pure", while the motives associated to quasi-projective and singular varieties are called "mixed" (they consist of different pure pieces). A complete and satisfactory theory of mixed motives is not yet available, although some pieces of it have already been developed. For instance the category of mixed Tate motives over number fields (the simplest mixed motives) is well understood. Motives form an active area of current research in pure mathematics. As an example we mention a recent letter of Deligne to Drinfeld, in which he proved the finiteness of the number of irreducible lisse $\mathbb{Q}_\ell$-sheaves with bounded ramification, up to isomorphism and up to twist, on a smooth variety defined over a finite field. This result is inspired by motivic considerations, and its proof relies on Lafforgue's Langlands correspondence over curves. This was the topic of the series of lectures by Esnault, Kindler and Rülling.

Being intermediate between algebraic varieties and cohomology theories, the theory of motives provides a conceptual framework for the study of periods. For instance the periods that arise from mixed Tate motives over $\mathbb{Z}$ are combinations of multiple zeta values.

Multiple zeta values (MZVs) are defined by iterated sums. Their analysis can be traced back to the work of Euler in the 18th century, but it was not until the last decades that the theory of MZVs has been intensively developed. Indeed, a systematic exploration of these objects only started with the seminal work of Zagier in the early 1990s. MZVs are special values of two related families of special functions, i.e., multiple zeta functions and multiple polylogarithms. These two families of functions stand at the intersection of several mathematical areas, including algebraic geometry, Lie group theory (Kashiwara–Vergne conjecture, recently proved by Alekseev and Meinrenken following an approach initiated by Torossian), algebra and combinatorics.

Kontsevich discovered a representation of MZVs in terms of iterated integrals, which shows that they appear as periods of mixed Tate motives. Moreover, the fact that they originate from iterated integrals as well as from iterated sums gives rise to one of the most fundamental properties of them, i.e., they satisfy three families of nontrivial polynomial relations, regrouped under the terminology "double shuffle and regularization". One of the prominent open problems (roughly speaking, the conjectural algebraic independence of MZVs over the rational numbers, modulo double-shuffle and regularization relations) is still far from being solved, and is related to Kontsevich's and Zagier's conjecture on the structure of the ring of periods.

The study of MZV's through the theory of motives is the topic of the course by Burgos Gil and will be the main theme of the second volume.

Recently, the notion of Feynman diagrams entered the picture of motives and periods. In perturbative quantum field theory (QFT) Feynman diagrams appear as coefficients of power expansions in coupling parameters. These diagrams encode highly intricate integrals over a large number of variables. The correspondence between diagrams and integrals is established via the Feynman rules of a QFT, which associate to any Feynman diagram a Feynman amplitude. Their efficient calculation is of foremost importance for theoretical predictions, however the integrals thus obtained are complicated and hard to compute. Moreover, the occurrence of ultraviolet and infrared singularities demands the introduction of regularization and renormalization procedures in order to obtain meaningful results. The former procedure replaces a divergent integral with a function of an additional—non-physical—parameter, which has a pole or singularity at the particular value of the parameter that corresponds to the original integral. It is otherwise well defined and finite at nearby values of the parameter. Renormalization provides a mean for extracting finite values from the regularized expressions in a way that is consistent with physics as well as the combinatorics of nested diagrams. Over the years, several regularization methods and renormalization schemes have been developed.

Thanks to the pathbreaking work of Bloch, Esnault, Kreimer, Connes, Marcolli and others, a beautiful relation between Feynman integrals and periods has emerged. In particular Bloch, Esnault, and Kreimer showed that Feynman integrals (modulo divergences) can be written as integrals of algebraic differential forms on an algebraic variety, integrated over a cycle with boundary contained on a divisor in the variety. It turns out that, in many simple situations, this period corresponds to a mixed Tate motive, and thus to multiple zeta values. The interplay between Feynman integrals and motives, and some consequences of this relation are the topic of the courses by Bloch and Marcolli.

Two remarks are in order. First, recent computations involving more complicated Feynman diagrams clearly indicate that we are starting to uncover a much richer interplay between the theory of motives and QFT, one that will involve other kinds of motives and special functions. Second, we should emphasize that the relation between QFT and the motivic point of view is not only of "theoretical" interest, but has already given rise to practical consequences. As an example we mention the recent work of Goncharov et al. on the amplitude associated to a particular type of Feynman diagram, whose original expansion spanned an expression covering several pages. Thanks to the motivic interpretation and the development of MPL functional equations, this expansion has been simplified to just two lines, which significantly reduced the time needed to compute it. This result underlines the importance that these developments are having in theoretical physics and is a prelude to future progress.

It goes without saying that we are most thankful for the help and expertise we received from the referees in preparing this volume. A large summer school like this one can only succeed in a place centered around the idea of bringing together established and aspiring mathematicians to fruitfully discuss and do research. ICMAT is such a place and we greatly acknowledge the availability of its facilities that were key in making this a fruitful event. We are as well thankful for the warm hospitality and professionality that we have experienced during the summer school.

K. Ebrahimi-Fard
J. I. Burgos Gil
D. Manchon

Clay Mathematics Proceedings
Volume **21**, 2020
https://doi.org/10.1090/cmip/21/475

FOREWORD

Yuri I. Manin

The most overlooked symbol in any equation is the humble equals sign.
Robbert Dijkgraaf
IAS Bulletin, Dec. 2017

This collection is dedicated to one of the most active research area bridging theoretical physics (quantum field theory) and pure mathematics (algebraic geometry and number theory).

In the articles by Spencer Bloch, and by Özgür Ceyhan with Matilde Marcolli, the readers will be introduced to the study of a remarkable class of complex numbers: "periods" (Kontsevich–Zagier, [**2**]). They will find information about expressions of periods as integrals of rational functions over the domains bounded by locally algebraic varieties, and the problem of finding as many as possible (ideally, "all") relations among periods.

Among these numbers, many are, or are expected to be, transcendental ones, but nevertheless, there is an amazing machinery for discovering hidden symmetries among them, a generalisation of the classical Galois group acting upon all algebraic numbers. This generalisation is achieved by passing from varieties to their motives, a conjectural universal cohomology theory, or at least to various approximations to it given by developed cohomology theories such as De Rham and Hodge ones: see S. Bloch [**1**].

Finally, it is only natural to expect that periods have also hidden arithmetic properties. They must be encoded in l–adic cohomologies of motives, and some points of this theory are studied in the paper by Lars Kindler and Kay Rülling.

In physics, periods appear as terms of asymptotic expansions of Feynman integrals, and I will dedicate the remaining part of this Foreword to some explanations regarding this powerful and mysterious toolbox.

***

In the early years of quantum physics, it was understood that *measurements*, at least in microworld, are not the same as classical *observations*, say, of positions of planets. Results of quantum measurements give random answers, mathematical models of which are based upon the notion of *probability amplitudes*.

The state of a quantum system is mathematically described as a vector φ on the unit sphere of complex Hilbert space $\mathcal{H}$ (more precisely, on the image of this unit sphere in the projective space of complex lines in $\mathcal{H}$ passing through the origin), and the probability amplitude to observe the state ψ of the system which is in the state φ, is the scalar product $< \psi|\varphi >$. In this notation due to Paul Dirac, the scalar product is **C**–linear, resp. antilinear, with respect to the ket–vector φ, resp. bra–vector ψ.

The modern machinery for calculating these probability amplitudes in quantum field theory was designed by Richard Feynman under the name of *path integrals* and their *perturbative expansions.*

Such an expansion is generally a formal series, whose terms are labelled by graphs, and each term is a classical integral of a sophisticated but very beautiful structure.

For a mathematician, the passage from a generic expression for amplitude to Feynman's integral to the terms of its perturbative expansion is a path consisting of a chain of formal computations each equals sign in which embodies a rather mysterious "equality", thus illustrating Robbert Dijkgraaf's wise aphorism.

Here are some steps that can be made mathematically precise in the context of "toy models", and thus make physical intuition somewhat more accessible to mathematicians.

Start with considering a finite–dimensional Hilbert state $\mathcal{H}$ endowed with an orthonormal basis $\{e_i\}$, $i = 1, \dots, n$. One easily sees that

$$< \psi|\varphi > = \sum_{k,l} < \psi|e_l >< e_l|e_k >< e_k|\varphi >$$

that is, the initial probability amplitude is the sum over all "paths"

$$\varphi \to e_k \to e_l \to \psi$$

of respective amplitudes. To make this picture even closer to Feynman's summation, one can fix some N, as value of "discrete time t", consider all paths of length N, and insert exponentiated Hamiltonian operators of time evolution $\exp(iHt/N)$ at appropriate places. We leave this as an exercise for the reader and pass to a more sophisticated toy, which reminds me of the Christmas train that Spencer Bloch bought for his 4 years old grandson.

Using this toy we show, how a perturbation series over Feynman diagrams arises, and how the structure of its terms depends on the action functional S embodying our former $\exp(iH\tau/N)$.

Now the *Feynman path integral* for us will be an heuristic expression of the form

$$\frac{\int_{\mathcal{P}} e^{S(\varphi)} D(\varphi)}{\int_{\mathcal{P}} e^{S_0(\varphi)} D(\varphi)} \tag{1}$$

(we omit the slightly more complicated expression for correlation functions).

In the expression (1), $\mathcal{P}$ is imagined as a functional space of *classical fields* φ on a *space–time manifold* M. Space–time may be endowed with Minkowski or Euclidean metric, but not necessarily: in models of quantum gravity, metric is one of the fields. Fields in general may include scalar functions, tensors of various ranks, sections of vector bundles, connections etc.

$S : \mathcal{P} \to \mathbf{C}$ is a functional of *classical action*: usually $S(\varphi)$ is expressed as an integral over M of a local density on M which is called *Lagrangian.* In our notation (1)

$$S(\varphi) = -\int_M L(\varphi(x))dx,$$

Lagrangian density may depend on derivatives, include distributions etc.

Usually $S(\varphi)$ is represented as the sum of *a quadratic part* $S_0(\varphi)$ (Lagrangian of free fields) and remaining terms which are interpreted as interaction and treated perturbatively.

Finally, the integration measure $D(\varphi)$ and the integral itself $\int_{\mathcal{P}}$ should be considered as simply a part of the total expression (1.1) expressing the idea of "summing over trajectories".

In our toy model, we will replace $\mathcal{P}$ by a finite–dimensional real space. We endow it with a basis indexed by a finite set of "colors" A, and an Euclidean metric g encoded by the symmetric tensor (g^{ab}), $a, b \in A$. We put $(g^{ab}) = (g_{ab})^{-1}$.

The action functional $S(\varphi)$ will be a formal series in linear coordinates on $\mathcal{P}$, (φ^a), of the form

$$S(\varphi) = S_0(\varphi) + S_1(\varphi), \quad S_0(\varphi) := -\frac{1}{2}\sum_{a,b} g_{ab}\varphi^a\varphi^b,$$

$$S_1(\varphi) := \sum_{k=1}^{\infty}\frac{1}{k!}\sum_{a_1,\dots,a_k\in A} C_{a_1,\dots,a_k}\varphi^{a_1}\dots\varphi^{a_k} \tag{2}$$

where $(C_{a_1,\dots,a_n})$ are certain symmetric tensors. If these tensors vanish for all sufficiently large ranks n, $S(\varphi)$ becomes a polynomial and can be considered as a genuine function on $\mathcal{P}$.

On the other hand, below we will mostly consider (g_{ab}) and $(C_{a_1,\dots,a_n})$ as independent formal variables as well, "formal coordinates on the space of theories".

We can now state our first theorem, expressing the toy version of (1) as a series over (isomorphism classes of) graphs.

To be precise, a graph τ for us consists of two finite sets, edges E_τ and vertices V_τ, and the incidence map sending E_τ to the set of unordered pairs of vertices. Each vertex is supposed to be incident to at least one edge. There is one *empty graph.*

THEOREM. *We have, for a formal parameter* λ

$$\frac{\int_{\mathcal{P}} e^{\lambda^{-1}S(\varphi)}D(\varphi)}{\int_{\mathcal{P}} e^{\lambda^{-1}S_0(\varphi)}D(\varphi)} \quad \text{“}=\text{''} \quad \sum_{\tau\in\Gamma}\frac{\lambda^{-\chi(\tau)}}{|\mathrm{Aut}\,\tau|}w(\tau) \tag{3}$$

In the r.h.s. of (3) the summation is taken over (representatives of) all isomorphism classes of all finite graphs τ*. The weight* $w(\tau)$ *of such a graph is determined by the action functional (2) as follows:*

$$w(\tau) := \sum_{u:\,F_\tau\to A}\ \prod_{e\in E_\tau} g^{u(\partial e)} \prod_{v\in V_\tau} C_{u(F_\tau(v))}\,.$$

Here F_τ *is the set of flags, or "half–edges" of* τ*. Each edge* e *consists of a pair of flags denoted* ∂e*, and each vertex* v *determines the set of flags incident to it denoted* $F_\tau(v)$*. Finally,* $\chi(\tau)$ *is the Euler characteristic of* τ*.*

The equals sign in (3) is Dijkgraaf's "most overlooked symbol". For its meaning here and deduction of (2), see [**3**]. Graphs τ in our toy model imitate Feyman graphs of physicists, and the whole right hand side of (3) corresponds to Feynman's asymptotic expansion.

References

[1] Spencer Bloch, *Motives associated to graphs*, Jpn. J. Math. **2** (2007), no. 1, 165–196, DOI 10.1007/s11537-007-0648-9. MR2295619

[2] Maxim Kontsevich and Don Zagier, *Periods*, Mathematics unlimited—2001 and beyond, Springer, Berlin, 2001, pp. 771–808. MR1852188

[3] Yuri I. Manin, *Renormalization and computation I: motivation and background* (English, with English and French summaries), OPERADS 2009, Sémin. Congr., vol. 26, Soc. Math. France, Paris, 2013, pp. 181–222. MR3203373

Max–Planck–Institut für Mathematik, Bonn, Germany

Clay Mathematics Proceedings
Volume **21**, 2020
https://doi.org/10.1090/cmip/21/476

Feynman integrals in mathematics and physics

Spencer Bloch

ABSTRACT. These are notes of lectures given at the CMI conference in August, 2014 at ICMAT in Madrid. The focus is on some mathematical questions associated to Feynman integrals, including Hodge structures, relations with string theory, and monodromy (Cutkosky rules).

CONTENTS

1. Introduction

What follows are notes from some lectures I gave at a Clay Math conference at ICMAT in Madrid in August, 2014. I am endebted to the organizers for inviting me, and most particularly to José Burgos Gil for giving me his notes.

The subject matter here is (a slice of) modern physics as viewed by a mathematician. To understand what this means, I told the audience about my grandson who, when he was 4 years old, was very much into trains. (He is 5 now, and more into dinosaurs.) For Christmas, I bought him an elaborate train set. At first, the little boy was at a complete loss. The train set was much too complicated. Amazingly, everything worked out! Even though the train itself was hopelessly technical and difficult, the box the train came in was fantastic; with wonderful, imaginative pictures of engines and freight cars. Christmas passed happily in fantasy play inspired by the images of the trains on the box. Modern physics is much too complicated for anyone but a "trained" and dedicated physicist to follow. However, "the box it comes in", the superstructure of mathematical metaphor and analogy which surrounds it, can be a delightful inspiration for mathematical fantasy play.

The particular focus of these notes is the Feynman integral. Essentially, to a physical theory the physicist associates a lagrangian, and to the lagrangian a collection of graphs, and to each graph a function called the integral on a space of external momenta. Section 2 develops the basic algebra of the first and second Symanzik polynomials in the context of configurations, i.e. invariants associated to a finite dimensional based vector space $\mathbb{Q}^E$ with a given subspace $H \subset \mathbb{Q}^E$. Section 3 considers the first Symanzik (aka Kirchoff polynomial) ψ_G in the case $E = \text{edges}(G)$ and $H = H_1(G)$ for a graph G. $\psi_G = \det(M_G)$ is a determinant, so the *graph hypersurface* $X_G : \psi_G = 0$ admits a birational cover $\pi : \Lambda_G \to X_G$ with fibre over $x \in X_G$ the projective space associated to $\ker M_G(x)$. The main result in this section is theorem 3.7. A classical theorem of Riemann considers singularities of the theta divisor $\Theta \subset J^{g-1}(C)$ where C is a Riemann surface of genus g. Here $J^{g-1}(C)$ is the space of rational equivalence classes of divisors of degree $g-1$ on C and $\Theta \subset J^{g-1}$ is the subspace of effective divisors. One shows that Θ is (locally) defined by the vanishing of a determinant, and the evident map $\text{Sym}^{g-1}(C) \to \Theta$ has fibres the projectivized kernels. Riemann's result is that the dimension of the fibre over $x \in \Theta$ is one less than the multiplicity of x on Θ. Theorem 3.7 is the analogous result for X_G, viz. $\dim(\pi^{-1}(x)) = \text{Mult}_x(X_G) - 1$.

Section 4 considers in more detail the Hodge structure of Λ_G. It turns out that whereas the Hodge structure for X_G is subtle and complicated, the Hodge structure on Λ_G is fairly simple. In particular, Λ_G is mixed Tate while X_G is known not to be mixed Tate in general [**3**].

Sections 5-9 focus on the integral A_G, viewed as a multi-valued function of external momenta. Various standard formulas for A_G are given in section 5. Sections 6-9 develop from a Hodge-theoretic perspective a result relating A_G to a limit in string theory when the string tension $\alpha' \to 0$. Associated to G is a family of stable rational curves which are unions of Riemann spheres with dual graph G. We view this family as lying on the boundary of a moduli space of pointed curves of genus $g = h_1(G)$. We show how external momenta are associated to limits at the boundary of marked points, and we explicit the second Symanzik as a limit of heights. The integral A_G then becomes an integral over the space of nilpotent orbits associated to the degeneration of the Hodge structure given by H^1_{Betti} of the curves. This is joint work with José Burgos Gil, Omid Amini, and Javier Fresan. We are greatly endebted to P. Vanhove and P. Tourkine for their insights, ([**23**], [**24**] and references cited there.) I met my co-authors at the conference, and this work grew out of an attempt to make sense of vague and imprecise suggestions I had made in the lectures.

The last topic (sections 10 - 12) concerns joint work with Dirk Kreimer on Cutkosky Rules. The integral A_G is a multi-valued function of external momenta. Cutkosky rules give a formula for the *variation* as external momenta winds around the *threshold divisor* which is the discriminant locus in the space of external momenta for the quadric propagators. Over the years there has been some uncertainty in the physics community as to the precise conditions for Cutkosky rules to apply. Using a modified notion of vanishing cycle due to F. Pham, we prove that Cutkosky rules do apply in the case of what are called *physical singularities*.

Perhaps a word about what is *not* in these notes. In recent years, there has been enormous progress in calculating scattering amplitudes. The work I should have liked to talk about is due to Francis Brown and Oliver Schnetz [**3**] which exhibits an algorithm for calculating many Feynman integrals and gives a beautiful example where the algorithm fails. In this case the integral is the period of a modular form. The whole subject of polylogarithms, which is intimately linked to integrals, is not addressed at all. I apologize; there simply wasn't time.

Finally, a shout-out to the organizers, most particularly Kurusch Ebrahimi-Fard who worked very very hard on both the practical and scientific aspects of the conference.

2. Configuration polynomials

Let H be a finite dimensional vector space of dimension g over a field k, and suppose we are given a finite set E and an embedding $\iota : H \hookrightarrow k^E$, where $k^E := \{\sum_{e\in E} \kappa_e e \mid \kappa_e \in k\}$. Write $W = k^E/H$. Let $e^\vee : k^E \to k$ be the evident functional that simply takes the e-th coordinate of a vector, and write $e^\vee$ as well for the composition $H \hookrightarrow k^E \to k$. The function $h \mapsto (e^\vee(h))^2$ defines a rank 1 quadratic form $e^{\vee,2}$ on H. If we fix a basis of H, we can identify the quadratic form with a rank 1 $g \times g$ symmetric matrix M_e in such a way that, thinking of elements of H as column vectors, we have

$$e^{\vee,2}(h) = h^t M_e h. \tag{2.1}$$

Let A_e, $e \in E$ be variables, and consider $M := \sum_{e\in E} A_e M_e$. We can interpret M as a $g \times g$ symmetric matrix with entries that are linear forms in the A_e. More canonically, $M : H \to H^\vee$ is independent of the choice of basis of H.

DEFINITION 2.1. The first Symanzik polynomial $\psi(H, \{A_e\})$ associated to $H \hookrightarrow k^E$ is the determinant

$$\psi(H, \{A_e\}) := \det(M). \tag{2.2}$$

REMARK 2.1. A different choice of basis multiplies $\psi(H, \{A_e\})$ by an element in $k^{\times,2}$. Indeed, M is replaced by $N^t M N$ where N is a $g \times g$ invertible matrix with entries in k. When the configuration comes from a graph, which is the case which interests us, it is possible to scale so all the monomials appear with coefficient $+1$. We will refer (abusively) to the resulting polynomial as $\psi(H, \{A_e\})$.

For $w \in W$, define H_w to be the inverse image in k^E of $k \cdot w \subset W = k^E/H$. We have $H \subset H_w \subset k^E$, and we can calculate the first Symanzik polynomial of $\psi(H_w, \{A_e\})$.

DEFINITION 2.2. The second Symanzik polynomial

$$\phi(H, w, \{A_e\}) := \psi(H_w, \{A_e\}). \tag{2.3}$$

REMARK 2.2. There are various definitions of the second Symanzik polynomial in the literature. That the definition given here coincides with the usual physics definition in the case when external momenta have dimension 1 is a result of Patterson [**19**]. We will need the more general case when external momenta have dimension $D \geq 1$. In this case, everything that follows can be found in the classical physics literature, e.g. [**8**].

LEMMA 2.3. *(i) The first Symanzik $\psi(H,\{A_e\})$ is homogeneous of degree $g = \dim H$ in the A_e.*

(ii) The second Symanzik $\phi(H,w,\{A_e\})$ is homogeneous of degree $g+1$ in the A_e and is quadratic in w.

PROOF. The assertions about homogeneity in A_e are clear. To check that ϕ is quadratic in w, it suffices to note that

$$\phi(H,w,\{A_e\}) = \det\Big(\sum A_e \begin{pmatrix} M_e & W_e \\ W_e^t & Q(W_e)\end{pmatrix}\Big) \tag{2.4}$$

is quadratic in W_e, where $W_e = (w_{e,1},\ldots,w_{e,g})$ is a column vector and $Q(W_e)$ is a quadratic form in the $w_{e,i}$. This is straightforward, expanding the determinant by the last row for example. □

REMARK 2.3. In fact, one can do a bit better. Let V be a vector space and let q be a quadratic form on V. We can take the $W_e = (v_{e,1},\ldots,v_{e,g})$ to have entries in V and define $Q(W_e) := \sum_1^g q(v_{e,i})$. Using the quadratic form q, one can make sense of the determinant expression on the right in (2.4) and define $\phi(H,w,\{A_e\})$ for $w \in W\otimes V$. Typically, in physics $V = \mathbb{R}^D$ is space-time and q is the Minkowski metric.

REMARK 2.4. To relate the second Symanzik to the height (see sections 5-9), it will be convenient to rewrite the above determinant expression as a bilinear form. Suppose we are given a symmetric $(g+1)\times(g+1)$ matrix of the form

$$A := \begin{pmatrix} M & W \\ W^t & S\end{pmatrix} \tag{2.5}$$

where M is $g\times g$ and invertible, W is a column vector of length g, and S is a scalar. Recall the classical formula $(\det M)M^{-1} = \operatorname{adj}(M)$ where $\operatorname{adj}(M)$ is the matrix of minors. Then

$$\det A = -W^t \operatorname{adj}(M) W + S\det M. \tag{2.6}$$

In the context of the previous remark, if W has entries in a vector space V with a quadratic form Q, the determinant in (2.4) can be written

$$\phi(H,w,\{A_e\})/\psi(H,\{A_e\}) = -W^t M^{-1} W + Q(W). \tag{2.7}$$

where the product $W^tM^{-1}W$ is interpreted via the quadratic form. (The referee points out that this formula can be interpreted geometrically via the Green's function associated to a metric on the graph (world line formalism). See [**24**], (3.19).)

3. The first Symanzik polynomial for graphs

Of primary importance for us will be configurations associated to graphs.

DEFINITION 3.1. A graph G is determined by sets $E = E(G)$ (edges) and $V = V(G)$ (vertices) together with a set Λ which we can think of as the set of half edges of G. We are given a diagram of projections

$$E \overset{p}{\leftarrow} \Lambda \overset{q}{\rightarrow} V. \tag{3.1}$$

We assume p is surjective and $p^{-1}(e)$ has 2 elements for all $e\in E$.

(This definition may seem a bit fussy, but half edges are very useful when one wants to talk about automorphisms of a graph. For example, there is a unique graph G with one edge and one vertex, and $\mathrm{Aut}(G) = \mathbb{Z}/2\mathbb{Z}$.) An *orientation* of G is an ordering of $p^{-1}(e)$ for all edges e, i.e. a map $\Lambda \to E \times \{-1, 1\}$ satisfying obvious conditions. For an oriented graph G there is a boundary map $\partial : \mathbb{Z}^E \to \mathbb{Z}^V$, $e \mapsto \partial^+(e) - \partial^-(e)$. We define the homology of G via the exact sequence

$$0 \to H_1(G, \mathbb{Z}) \to \mathbb{Z}^E \xrightarrow{\partial} \mathbb{Z}^V \to H_0(G, \mathbb{Z}) \to 0. \tag{3.2}$$

The homology of G with coefficients in an abelian group A is defined similarly.

DEFINITION 3.2. The graph polynomial $\psi_G(\{A_e\}_{e \in E(G)})$ (also sometimes called the Kirchoff polynomial or the first Symanzik polynomial of G) for G is the first Symanzik polynomial for the configuration $H_1(G, \mathbb{Z}) \subset \mathbb{Z}^E$.

Let G be a graph and let $e \in G$ be an edge. The graph $G - e$ (resp. G/e) is obtained by cutting (resp. contracting) the edge e in G. In words, cutting an edge leaves a graph $G - e$ with one fewer edges and the same set of vertices. (A vertex may be left isolated.) Shrinking an edge with two distinct endpoints brings the endpoints together to yield a graph with one fewer vertices and one fewer edges. If the edge being contracted is a self-loop, it simply removes the loop. The general notion of cutting or contracting in a configuration is explained by the diagram

$$\begin{array}{ccccc}
0 & \longrightarrow & H(\text{cut } e) & \longrightarrow & k^{E-\{e\}} \\
 & & \big\downarrow \text{inject} & & \big\downarrow \text{inject} \\
0 & \longrightarrow & H & \longrightarrow & k^E \\
 & & \big\downarrow & & \big\downarrow \text{surject} \\
0 & \longrightarrow & H(\text{shrink } e) & \longrightarrow & k^{E-\{e\}}
\end{array} \tag{3.3}$$

PROPOSITION 3.3. *(i)* $\psi(H(\text{cut } e)) = \frac{\partial}{\partial A_e}\psi(H)$.
(ii) $\psi(H(\text{shrink } e)) = \psi(H)|_{A_e=0}$ *unless e is a tadpole (i.e. the edge e has only one vertex). If e is a tadpole,* $\psi(H)|_{A_e=0} = 0$.
(iii) $\psi(H)$ *has degree* ≤ 1 *in every* A_e.

PROOF. (i) and (ii) are straightforward from (3.3). For (iii), note ψ depends only up to scale on the choice of basis of H. We choose a basis $h_1, \ldots, h_g$ such that $e^\vee(h_i) = 0$ for $2 \leq i \leq g$. The matrix M_e is then $g \times g$ diagonal with $g - 1$ zeroes and a single 1 in position $(1, 1)$. The matrix $\sum_{\varepsilon \in E} A_\varepsilon M_\varepsilon$ then involves A_e only in position $(1, 1)$, and (iii) follows. □

In the case of the graph polynomial of a graph G, one can be more precise. A *spanning tree* $T \subset G$ is a connected subgraph of G with $V(T) = V(G)$ and $H_1(T, \mathbb{Z}) = (0)$.

PROPOSITION 3.4. *Let G be a graph. Then the graph polynomial can be written*

$$\psi_G = \sum_{T \subset G} \prod_{e \notin T} A_e. \tag{3.4}$$

(Here T runs through all spanning trees of G.)

PROOF. We know by proposition 3.3 (iii) that every monomial in ψ_G is a product of $g = h_1(G)$ distinct edge variables A_e. For a given set $\{e_1, \ldots, e_g\}$ of distinct edges, we know by proposition 3.3 (ii) that this term is exactly the graph polynomial of G with all the edges $\varepsilon \notin \{e_1, \ldots, e_g\}$ shrunk. If the edges not in $\{e_1, \ldots, e_g\}$ form a spanning tree for the graph, shrinking them will yield a *rose* with g loops, i.e. a union of g tadpoles. The graph polynomial of such a graph is simply $\prod_1^g A_{e_i}$ so the coefficient of this product in ψ_G is 1. On the other hand, if the edges A_ε not in $\{e_1, \ldots, e_g\}$ do not form a spanning tree, they must necessarily contain a loop and so setting them to 0 kills ψ_G and there is no term in $A_{e_1}, \ldots, A_{e_g}$. □

We continue to assume G is a graph. We write $n = \#E(G)$ and $g = \dim H_1(G)$. The hypersurface $X_G : \psi_G = 0$ in $\mathbb{P}^{n-1}$ is the *graph hypersurface*. The symmetric matrix $M := \sum_e A_e M_e$ defines a linear map $\bigoplus_g \mathcal{O}_{\mathbb{P}^{n-1}} \to \bigoplus_g \mathcal{O}_{\mathbb{P}^{n-1}}(1)$. Define

$$\Lambda_G := \{(a, \beta) \in \mathbb{P}^{n-1} \times \mathbb{P}^{g-1} \mid M_a(\beta) = 0\} \tag{3.5}$$

Note $\Lambda_G \subset X_G \times \mathbb{P}^{g-1} \subset \mathbb{P}^{n-1} \times \mathbb{P}^{g-1}$.

PROPOSITION 3.5. *(i) There exist coherent sheaves $\mathcal{E}$ on X_G and $\mathcal{F}$ on $\mathbb{P}^{g-1}$ such that $\Lambda_G \cong Proj(Sym(\mathcal{E})) \cong Proj(Sym(\mathcal{F}))$.*

(ii) Λ_G is a reduced, irreducible variety of dimension $n-2$ which is a complete intersection of codimension g in $\mathbb{P}^{n-1} \times \mathbb{P}^{g-1}$. The projection $\pi : \Lambda_G \to X_G$ is birational.

PROOF. Define $\mathcal{E}$ by the presentation

$$H_1(G) \otimes \mathcal{O}_{X_G} \xrightarrow{M} H^1(G) \otimes \mathcal{O}_{X_G}(1) \to \mathcal{E} \to 0. \tag{3.6}$$

Here for $a \in X_G$, $M_a = \sum a_e M_e$.

For $\mathcal{F}$, the map which is given over $\beta \in \mathbb{P}^{g-1}$ by $a \mapsto \sum_E a_e (e^\vee(\beta)) e^\vee$ dualizes to a presentation

$$\mathcal{O}^g_{\mathbb{P}^{g-1}} \to \mathcal{O}^n_{\mathbb{P}^{g-1}}(1) \to \mathcal{F} \to 0. \tag{3.7}$$

The fibre $\mathcal{F}_\beta$ is the quotient of $k^{n,\vee}$ by the space of functionals of the form $a \mapsto \sum_e a_e e^\vee(\beta) e^\vee$. We have $\dim \mathcal{F}_\beta = n - g + \varepsilon(\beta)$ where $\varepsilon(\beta)$ is the codimension in $k^{g,\vee}$ of the span of $\{e^\vee \mid e^\vee(\beta) \neq 0\}$. Since the $e^\vee$ span $H_1(G)^\vee$ it follows that for β general we have $\varepsilon(\beta) = 0$ so $\Lambda_G = \mathrm{Proj}(\mathrm{Sym}(\mathcal{F}))$ has dimension $n-2$. Since the fibre of $\mathcal{E}$ over a point a is the kernel of M_a it is non-zero for $a \in X_G$, whence $\Lambda_G \twoheadrightarrow X_G$ with fibres projective spaces. Since the two varieties have the same dimension, it follows that $\Lambda_G \to X_G$ is birational.

Finally, to realize Λ_G as a complete intersection, let A_e (resp. B_j) be a basis for the homogeneous coordinates on $\mathbb{P}^{n-1}$ (resp. $\mathbb{P}^{g-1}$). We can think of $B_j \in H_1(G)$. Write $w_{e,j} = e^\vee(B_j)$. Then the defining equations are

$$0 = \sum_{j=1}^{g} \sum_E A_e w_{e,j} B_j w_{e,i} = 0; \quad 1 \leq i \leq g. \tag{3.8}$$

□

We will further investigate the motives of X_G and Λ_G in the sequel. There is an interesting analogy between X_G and Θ, the theta divisor of a genus g Riemann surface C. Both are determinantal varieties, and $\Lambda_G \to X_G$ corresponds to

$\mathrm{Sym}^{g-1}C \xrightarrow{\pi} \Theta$. One has that Θ is a divisor on the jacobian $J(C)$, and a classical theorem of Riemann states that the multiplicity of Θ at a point x is equal to $1+\dim \pi^{-1}(x)$. We will show below that the same result holds for graph hypersurfaces X_G.

EXAMPLE 3.6. Consider a determinantal variety defined by the determinant of a diagonal variety

$$X : \det \begin{pmatrix} f_1 & 0 & \dots & 0 \\ 0 & f_2 & \dots & 0 \\ \vdots & \vdots & \dots & \vdots \\ 0 & 0 & \dots & f_g \end{pmatrix} = 0$$

Then X satisfies Riemann's theorem iff the zero sets of the f_i are smooth and transverse. The fibre over a point $x \in X$ is a projective space of dimension one less than the number of f_i vanishing at x.

THEOREM 3.7 (E. Patterson [**18**]). *Let G be a graph, and let $\pi : \Lambda_G \to X_G$ be the birational cover of X_G defined above. For $x \in X_G$ the fibre $\pi^{-1}(x)$ has dimension equal to $\mathrm{mult}_x(X_G) - 1$.*

PROOF. Define $X_p := \{x \in X_G \mid \dim(\pi^{-1}(x)) \geq p\}$ and $X(p) := \{x \in X_G \mid \mathrm{mult}_x(X_G) \geq p+1\}$. We want to show $X_p = X(p)$.

LEMMA 3.8. $X_p \subset X(p)$.

PROOF OF LEMMA. We have

$$x \in X_p \Leftrightarrow \dim(\ker(M_x = \sum x_e M_e : H_1(G) \to H_1(G))) \geq p+1.$$

On the other hand, $X(p)$ is defined by the vanishing of all p-fold derivatives $\frac{\partial^p}{\partial A_{e_1} \cdots \partial A_{e_p}} \psi_G = \psi_{G-\{e_1,\dots,e_p\}}$. If we associate to G the quadratic form $\sum x_e e^{\vee,2}$ on $H_1(G)$, then X_p is the set of x for which the null space of this form has dimension $\geq p+1$. The quadratic form associated to $G - \{e_1,\dots,e_p\}$ is simply $\sum x_e e^{\vee,2} | H_1(G) \cap \bigcap_{i=1}^p \{e_i^\vee = 0\}$. This restricted form cannot be nondegenerate if the null space of the form on $H_1(G)$ has dimension $\geq p+1$. □

LEMMA 3.9. *Let Q be a quadratic form on a vector space H. Let $N \subset H$ be the null space of Q. We assume $\dim N = s > 0$. Suppose we are given an embedding $H \hookrightarrow k^E$ for a finite set E. For $e \in E$ write $e^\vee : H \to k$ for the corresponding functional. Then there exists a subset $\{e_1,\dots,e_s\} \subset E$ such that $Q|\{e_1^\vee = \cdots = e_s^\vee = 0\}$ is non-degenerate.*

PROOF OF LEMMA. It suffices to take $e_1,\dots,e_s$ such that $N \cap \{e_1^\vee = \cdots = e_s^\vee = 0\} = (0)$. Indeed, if L is any codimension s subspace with $L \cap N = (0)$ we will necessarily have $Q|L$ nondegenerate. Since $H = L \oplus N$, any ℓ in the null space of $Q|L$ will necessarily be orthogonal to $L \oplus N = H$. □

We return to the proof of the theorem. By the first lemma we have $X_i \subset X(i)$. As a consequence of the last lemma we see that

$$X_i - X_{i+1} \subset X(i) - X(i+1). \tag{3.9}$$

In other words, if the nullspace has dimension exactly $i+1$, then there exists some $(i+1)$-st order partial which doesn't vanish. Taking the disjoint union we get $X_0 - X_j \subset X(0) - X(j)$ for any j. Since $X_0 = X(0)$ it follows that $X(j) \subset X_j$. This completes the proof. □

4. X_G and Λ_G

It turns out that the hypersurface $X_G : \psi_G = 0$ in $\mathbb{P}^{n-1}$ is quite subtle and complicated, while the variety Λ_G introduced above is rather simple. Recall we have a birational map $\pi : \Lambda_G \to X_G$ and the fibre $\pi^{-1}(a)$ is the projectivized kernel of $M_a = \sum a_e M_e$.

Our arguments at this point are completely geometric but for simplicity we focus on the the case of Betti cohomology and varieties over the complex numbers. The key point is

PROPOSITION 4.1. *The Hodge structure on Betti cohomology $H^*(\Lambda_G, \mathbb{Q})$ is mixed Tate.*

Recall Betti cohomology of a variety over $\mathbb{C}$ carries a Hodge structure.

DEFINITION 4.2. A Hodge structure on a finite dimensional $\mathbb{Q}$-vector space H is a pair of filtrations (W_*, F^*) with $W_*H_{\mathbb{Q}}$ a finite increasing filtration (separated and exhaustive) and $F^* = F^*H_{\mathbb{C}}$ a finite (separated and exhaustive) decreasing filtration. The filtration induced by F on $gr_p^W H_{\mathbb{C}}$ should be p-opposite to its complex conjugate, meaning that

$$gr_p^W H_{\mathbb{C}} = \bigoplus F^q gr_p^W H_{\mathbb{C}} \cap \overline{F}^{p-q} gr_p^W H_{\mathbb{C}} \tag{4.1}$$

A Hodge structure is *pure* if its weight filtration has a single non-trivial weight.

EXAMPLE 4.3. The Tate Hodge structures $\mathbb{Q}(n)$ are one dimensional $\mathbb{Q}$-vector spaces with weight $\mathbb{Q}(n) = -2n$ and Hodge filtration $F^{-n}\mathbb{Q}(n)_{\mathbb{C}} = \mathbb{Q}(n)_{\mathbb{C}} \supset F^{-n+1} = (0)$.

DEFINITION 4.4. A Hodge structure H is mixed Tate if $gr_{-2n}^W H = \bigoplus \mathbb{Q}(n)$ for all n.

LEMMA 4.5. *Let V be a variety over $\mathbb{C}$. Assume V admits a finite stratification $V = \amalg V_i$ by Zariski locally closed sets such that $H_c^*(V_i)$ is mixed Tate for all i. (Here H_c is cohomology with compact supports.) Then $H_c^*(V)$ is mixed Tate.*

PROOF. The functor $H^* \mapsto gr_p^W H^*$ is exact on the category of Hodge structures. We apply this functor to the spectral sequence which relates $H_c^*(V_i)$ to $H_c^*(V)$ and deduce a spectral sequence converging to $gr_p^W H^*(V)$ with initial terms direct sums of $\mathbb{Q}(p)$. Since extensions of $\mathbb{Q}(p)$ are all split, it follows that $gr_p^W H^*(V)$ is a direct sum of $\mathbb{Q}(p)$ so $H_c^*(V)$ is mixed Tate. □

Note of course that $H_c^*(V) = H^*(V)$ if the variety V is proper.

PROOF OF PROPOSITION 4.1. Let $\varepsilon : \mathbb{P}^{g-1} \to \mathbb{N}$ be as in the proof of Proposition 3.5. Define

$$T^m = \{\beta \mid \varepsilon(\beta) \geq m\}. \tag{4.2}$$

It is clear that $T^m \subset \mathbb{P}^{g-1}$ is closed, and $T^{m+1} \subset T^m$. The sets $S^m := T^m - T^{m+1}$ form a locally closed stratification on $\mathbb{P}^{g-1}$. Let $\mathcal{F}$ be the constructible sheaf on $\mathbb{P}^{g-1}$ defined in Proposition 3.5. The fibres of $\mathcal{F}$ over S^m have constant rank, so $\mathcal{F}|_{S^m}$ is a vector bundle and $\Lambda|_{S^m}$ is a projective bundle. It will suffice by the lemma to show $H_c^*(\Lambda|_{S^m})$ is mixed Tate, and by the projective bundle theorem this will follow if we show $H_c^*(S^m)$ is mixed Tate.

The set T^m can be described as follows. Let $Z \subset 2^E$ be the set of all subsets $z \subset E$ such that the span of $e^\vee|H_1(G), e \in z$ has codimension $< m$ in $H^1(G)$. Then T^m is the set of β such that $e^\vee(\beta) = 0$ for at least one $e \in z$. Said another way, for any subset $W \subset E$ containing at least one edge from each $z \in Z$, let $L_W \subset \mathbb{P}^{g-1}$ be the set of those β such that $e^\vee(\beta) = 0$ for all $e \in W$. Then $T^m = \bigcup L_W$ is the union of the L_W. Since the cohomology of a union of linear spaces is mixed Tate, we see that $H^*(T^m)$ is mixed Tate. Finally, from the long exact sequence relating the cohomologies of T^m, T^{m+1} to the compactly supported cohomology of S^m we deduce that $H_c^*(S^m)$ is mixed Tate as well. □

We had mentioned the analogy between X_G and the theta divisor $\Theta \subset J(C)$ of an algebraic curve C of genus $n-1$. From this point of view, the birational map $\pi : \Lambda_G \to X_G$ is analogous to the map $\mathrm{Sym}^{n-2}C \to \Theta$. Unlike the curve case, however, Λ_G is not usually smooth. To understand this, we consider partitions $E(G) = E' \amalg E''$. Let $G', G'' \subset G$ be the unions of the corresponding edge sets. We say our partition is *non-trivial on loops* if neither $\{e^\vee\}_{e\in E'}$ nor $\{e^\vee\}_{e\in E''}$ span $H^1(G)$.

We have seen that $\Lambda_G = \mathrm{Proj}(\mathcal{F})$ is a projective fibre space over $\mathbb{P}^{g-1}$. The general fibres have dimension $n-g-1$. Of course, over the open set with fibres of dimension exactly $n-g-1$, Λ_G is a projective bundle, hence smooth. Singularities can occur only when the fibre dimension jumps.

PROPOSITION 4.6. *(i) The fibre of Λ_G over $\beta \in \mathbb{P}^{g-1}$ has dimension $> n-g-1$ iff there exists a partition $E = E' \amalg E''$ which is non-trivial on loops such that $e^\vee(\beta) = 0$ for all $e \in E'$.*

(ii) Λ_G is singular iff there exists a partition $E = E' \amalg E''$ which is non-trivial on loops.

PROOF. For (i), we may take $E'' = \{e \mid e^\vee(\beta) \neq 0\}$. Assertion (i) is now straightforward from the definition of $\varepsilon(\beta)$ in the proof of Proposition 3.5.

For (ii), note that if no such partition exists, then Λ_G is a projective bundle over $\mathbb{P}^{g-1}$, hence smooth, so the existence of a partition is certainly necessary. Assume such a partition exists. The equation for Λ_G can be written in vector form $\sum_{e\in E} a_e e^\vee(\beta) e^\vee|_{H_1(G)}$ (compare (3.8)). We take β so $e^\vee(\beta) = 0$ for all $e \in E'$ and we take $a_e = 0$ for all $e \in E'$. (The value of the equation for such β does not depend on the choice of a_e, $e \in E'$, so this is a free choice.) It is clear that the partial derivatives at such a point with respect to the a_e and also with respect to coordinates on $H_1(G)$ all vanish, so this is a singular point. □

REMARK 4.1. The singular structure of Λ_G is more complicated than the above argument suggests because it may happen that there exists $\emptyset \neq F \subsetneq E'$ such that $e^\vee, e \in F \amalg E''$ still does not span $H^1(G)$. In such a case, it suffices to take $a_e = 0$ for $e \in E' - F$.

EXAMPLE 4.7 (Wheel and spoke graphs with 3 and 4 edges.). (i) The wheel with 3 spokes graph G_3 has vertices $1, 2, 3, 4$ and edges

$$\{1,2\}, \{2,3\}, \{3,1\}, \{1,4\}, \{2,4\}, \{3,4\}. \tag{4.3}$$

It has 3 loops, but it is easy to check there are no partitions of the edges which are non-trivial on the loops. It follows from Prop. 4.6 that Λ_{G_3} is a $\mathbb{P}^2$-bundle over $\mathbb{P}^2$ and hence non-singular.

(ii) The wheel with 4 spokes G_4 has 5 vertices, 4 loops, and 8 edges:

$$\{1,2\},\{2,3\},\{3,4\},\{4,1\},\{1,5\},\{2,5\},\{3,5\},\{4,5\}. \tag{4.4}$$

With the aid of a computer, one can show that $\Lambda_{G_4} \to \mathbb{P}^3$ is a $\mathbb{P}^3$-bundle over $\mathbb{P}^3 - 4$ points. Over the 4 points, the fibre jumps to $\mathbb{P}^4$.

We next calculate $gr^W H^*(\Lambda_G)$ for a general graph G. Let $f : \Lambda_G \to \mathbb{P}^{n-1}$ be the projection.

LEMMA 4.8. *The sheaves $R^a f_* \mathbb{Q}_\Lambda$ are zero for a odd. For $a = 2b$, we have*

$$R^a f_* \mathbb{Q}_\Lambda \cong \mathbb{Q}(-b)|_{S_b} \tag{4.5}$$

where $S_b \subset \mathbb{P}^{g-1}$ is the closed set where the fibre dimension of f is $\geq b$. In particular, $S_b = \mathbb{P}^{g-1}$ for $b \leq n-g-1$.

PROOF. Let $p : \Lambda_G \to \mathbb{P}^{n-1}$ be the other projection (recall $\Lambda_G \subset \mathbb{P}^{g-1} \times \mathbb{P}^{n-1}$.) We have a map of sheaves on Λ_G

$$p^* : H^a(\mathbb{P}^{n-1}, \mathbb{Q})_{\mathbb{P}^{g-1}} \to R^a f_* \mathbb{Q}_{\Lambda_G}. \tag{4.6}$$

The lemma follows from the fact that p^* is surjective with support on S_b. (Both assertions are checked fibrewise.) □

Consider the Leray spectral sequence

$$E_2^{pq} = H^p(\mathbb{P}^{g-1}, R^q f_* \mathbb{Q}_{\Lambda_G}) \Rightarrow H^{p+q}(\Lambda_G, \mathbb{Q}). \tag{4.7}$$

It follows from the lemma that $E_2^{pq} = H^p(S_{q/2}, \mathbb{Q}(-q/2))$ (zero for q odd) has weights $\leq p+q$ with equality if either $p = 0$ or $q \leq 2(n-g-1)$. Since E_s, $s \geq 2$ is a subquotient of E_2 we get the same assertion for E_s. From the complex

$$E_s^{p-s,q+s-1} \to E_s^{pq} \to E_s^{p+s,q-s+1} \tag{4.8}$$

we deduce

PROPOSITION 4.9. *For the spectral sequence (4.7) we find in the range $q \leq 2(n-g-1)$ or $p = 0$, $q \leq 2(n-g)$ that $E_\infty^{pq} = \mathbb{Q}(-(p+q)/2)$ if both p, q are even, and $E_\infty^{pq} = (0)$ otherwise. In particular, the pullback $H^s(\mathbb{P}^{n-1} \times \mathbb{P}^{g-1}, \mathbb{Q}) \to H^s(\Lambda_G, \mathbb{Q})$ is an isomorphism for $s \leq 2(n-g)$.*

An interesting special case is that of *log divergent graphs*. By definition, G is log divergent if $n = 2g$.

COROLLARY 4.10. *For G log divergent, $W_{n-3}H^{n-2}(X_G, \mathbb{Q})$ dies in $H^{n-2}(\Lambda_G, \mathbb{Q})$.*

PROOF. Indeed, as a consequence of the proposition, $H^{n-2}(\Lambda_G, \mathbb{Q})$ is pure of weight $n-2$. Since pullback respects the Hodge structure (and in particular the weights), the corollary follows. □

Concerning the motive of the graph hypersurface X_G for an arbitrary graph G we deduce

THEOREM 4.11. *$gr_a^W H^a(X_G, \mathbb{Q})$ is pure Tate for any $a \in \mathbb{Z}$.*

PROOF. Consider the maps $\widetilde{\Lambda}_G \xrightarrow{\rho} \Lambda_G \xrightarrow{\pi} X_G$, where $\widetilde{\Lambda}_G$ is a resolution of singularities of Λ_G. By [**7**], Prop. 8.2.5, the image $\rho^*\pi^* H^a(X_G) \subset H^a(\widetilde{\Lambda}_G)$ is isomorphic to $gr_a^W H^a(X_G, \mathbb{Q})$. This image is a subquotient of $H^a(\Lambda_G, \mathbb{Q})$ which is mixed Tate by Prop. 4.1. Since the image is pure of weight a, the theorem follows. □

5. The second Symanzik polynomial for graphs

The second Symanzik polynomial for a graph G depends on masses and external momenta. More precisely, to each vertex v one associates $p_v \in \mathbb{R}^D$ where D is the dimension of space-time. The conservation of momentum condition is

$$\sum_v p_v = 0. \tag{5.1}$$

In addition, to each edge e is attached a mass $m_e \in \mathbb{R}$. The *propagator* f_e associated to an edge e is defined via the diagram

$$\begin{array}{ccccc} H_1(G,\mathbb{R}^D) & \longrightarrow & (\mathbb{R}^D)^E & \xrightarrow{e^{\vee,2}-m_e^2} & \mathbb{R} \\ & & \big\downarrow \partial & & \\ & & p \in (\mathbb{R}^D)^{V,0} & & \end{array} \tag{5.2}$$

By definition $f_e = q_e^2 - m_e^2$ is the function $e^{\vee,2} - m_e^2$ restricted to $\partial^{-1}(p) \cong H_1(G,\mathbb{R}^D)$. The integral is

$$A_G(p,m) := \int_{\partial^{-1}(p)} \frac{d^{Dg}x}{\prod_{e\in E} f_e} \tag{5.3}$$

Of course, $\partial^{-1}(p)$ can be identified (non-canonically) with $H_1(G,\mathbb{R}^D)$ and the integral can be viewed as an integral over $H_1(G,\mathbb{R}^D)$.

Let $n = \#E(G)$. We consider edge variables A_e as homogeneous coordinates on $\mathbb{P}^{n-1}$. (Sometimes it is convenient to order the variables and write A_i rather than A_e.) Write

$$\Omega := \sum (-1)^{i-1} A_i dA_1 \wedge \cdots \wedge \widehat{A_i} \wedge \cdots \wedge dA_n \tag{5.4}$$

For $F(A_e)$ homogeneous of degree n, the ratio Ω/F is a meromorphic form of top degree $n-1$ on $\mathbb{P}^{n-1}$. A chain of integration σ is defined by

$$\sigma = \{(\ldots, A_e, \ldots) \mid A_e \geq 0\} \subset \mathbb{P}^{n-1}(\mathbb{R}) \tag{5.5}$$

A choice of ordering of the edges orients σ.

EXAMPLE 5.1. Suppose G is a tree, i.e. $g = 0$. Then $\partial^{-1}(p)$ is a point, and the integral simply becomes evaluation of $\frac{1}{\prod_e f_e}$ at this point. If, for example, G is just a string with vertices $\{1,2,\ldots,n\}$ and edges $(i,i+1)$, $1 \leq i \leq n-1$, then with evident orientation we have $\partial(i,i+1) = (i+1) - (i)$. Write p_i for the external momentum at the vertex i. We must find $q_{i,i+1} \in \mathbb{R}^D$, $1 \leq i \leq n$ such that $q_{i-1,i} - q_{i,i+1} = p_i, 1 \leq i \leq n$ where $q_{0,1} = q_{n,n+1} = 0$. This yields $q_{i,i+1} = -p_1 - p_2 - \cdots - p_i$ and the integral is

$$A_G(p,m) = \frac{1}{(p_1^2 - m_1^2)((p_1+p_2)^2 - m_2^2)\cdots((p_1+\ldots+p_{n-1})^2 - m_{n-1}^2)}. \tag{5.6}$$

LEMMA 5.2 (Schwinger parameters). *Viewing the f_i as independent coordinates, we have*

$$\frac{1}{\prod_{i=1}^n f_i} = (n-1)! \int_\sigma \frac{\Omega}{(\sum A_i f_i)^n}. \tag{5.7}$$

PROOF. In affine coordinates $a_i = A_i/A_n$ the assertion becomes

$$\frac{1}{\prod_{i=1}^n f_i} = (n-1)! \int_{0^{n-1}}^{\infty^{n-1}} \frac{da_1 \cdots da_{n-1}}{(a_1 f_1 + \cdots + a_{n-1} f_{n-1} + f_n)^n}. \tag{5.8}$$

We have

$$d\Big(\frac{da_2 \cdots da_{n-1}}{(a_1 f_1 + \cdots + a_{n-1} f_{n-1} + f_n)^{n-1}}\Big) = \\ - (n-1) f_1 \frac{da_1 \cdots da_{n-1}}{(a_1 f_1 + \cdots + a_{n-1} f_{n-1} + f_n)^n}, \tag{5.9}$$

and the result follows by induction. □

The integral (5.3) can thus be rewritten

$$A_G = \frac{1}{(n-1)!} \int_{\mathbb{R}^{Dg}} d^{Dg}x \int_\sigma \frac{\Omega}{(\sum A_e f_e)^n}. \tag{5.10}$$

We would like to interchange the two integration operations. Following the physicists, we take the metric on $\mathbb{R}^D$ to be Euclidean. There is still an issue of convergence because the quadratic form is only positive semi-definite, but as mathematicians we are looking for interesting motives to study. The issue of convergence of a particular period integral is of secondary concern. (Added in proof: The referee mentions the practice in physics to add a small imaginary term $i\varepsilon$ to each factor in the denominator in equation (5.6) so the integrand has no pole on the chain of integration. Sadly, this doesn't work. The chain of integration extends to ∞. Projectivizing, one finds that adding $i\varepsilon$ simply moves the pole to infinity. Physicists know this, but they choose to ignore it. This is not the place to discuss renormalization in detail, but in work with D. Kreimer [**2**], the author is looking at "Cutkosky rules" which calculate the variation in (5.3) as momenta loop around a degenerate value. This variation is given by an integral over a compact chain which is well-defined for $i\varepsilon \neq 0$. Thus, one has an ill-defined multi-valued function whose variations are well-defined. Could physicists work with such a thing?)

We must evaluate

$$\int_{\mathbb{R}^{Dg}} \frac{d^{Dg}x}{(\sum A_e f_e)^n}. \tag{5.11}$$

To this end, we first complete the square for the quadratic form $\sum A_e f_e$. We identify $H_1(G, \mathbb{R}^D) = (\mathbb{R}^D)^g$ and write

$$x_i = (x_i^1, \ldots, x_i^D) : H_1(G, \mathbb{R}^D) \to \mathbb{R}^D; \quad 1 \le i \le g; \quad \mathbf{x} = (x_1^t, \ldots, x_g^t).$$

Similarly, $\mathbf{p} = (\ldots, p_v^t, \ldots)_{v \neq v_0}$ where $p_v \in \mathbb{R}^D$ are the external momenta and we omit one external vertex v_0. We write

$$\sum A_e f_e = \mathbf{x} M \mathbf{x}^t - 2\mathbf{x} B \mathbf{p} + \mathbf{p} \Gamma \mathbf{p}^t - \mu \tag{5.12}$$

Here M (resp. B, resp. Γ) is a $g \times g$ (resp. $g \times (\#V-1)$, resp. $(\#V-1) \times (\#V-1)$) matrix with entries which are linear forms in the A_e; and $\mu = \sum_e m_e^2 A_e$. Note that M is the symmetric $g \times g$ matrix associated to the configuration $H_1(G, \mathbb{R}) \subset \mathbb{R}^E$ as in (2.2). In particular, $\psi_G = \det(M)$. Note also that the matrix operations in (5.12) are a bit exotic. Whenever column D-vectors in $\mathbf{x}$ and $\mathbf{p}$ are to be multiplied, the multiplication is given by the quadratic form on $\mathbb{R}^D$.

To complete the square write $\mathbf{x} = \mathbf{x}' + M^{-1}B\mathbf{p}$ and $\mathbf{x}' = \mathbf{x}''R$ where R is orthogonal with $RMR^t = \mathbb{D}$ diagonal. We get

$$\sum A_e f_e = \mathbf{x}''\mathbb{D}\mathbf{x}''^t - (B\mathbf{p})(M^{-1})(B\mathbf{p})^t + \mathbf{p}\Gamma\mathbf{p}^t - \mu \tag{5.13}$$

By definition, the second Symanzik polynomial is

$$\phi_G(,\{A_e\},\mathbf{p},\{m_e\}) := (B\mathbf{p})(\mathrm{adj}(M))(B\mathbf{p})^t + (\mathbf{p}\Gamma\mathbf{p}^t - \mu)\psi_G(\{A_e\}). \tag{5.14}$$

Here $\mathrm{adj}(M)$ is the adjoint matrix, so $M^{-1} = \mathrm{adj}(M)\det(M)^{-1}$. We can rewrite (5.13)

$$\sum A_e f_e = \mathbf{x}''\mathbb{D}\mathbf{x}''^t - \frac{\phi_G(A,p,m)}{\psi_G(A)} \tag{5.15}$$

Using the elementary identity

(5.16)
$$\int_{-\infty^N}^{\infty^N} \frac{du_1\cdots du_N}{((C_1u_1^2+\cdots+C_Nu_N^2)+L)^n} = \pi^{N/2}\frac{\Gamma(n-N/2)}{\Gamma(n)}\prod_{i=1}^N C_i^{-1/2}L^{-n+(N/2)},$$

we now find (taking $N = gD$)

$$\int_{\mathbb{R}^{Dg}} \frac{d^{Dg}x}{(\sum A_e f_e)^n} = \frac{\pi^{Dg}\psi_G^{n-(g+1)D/2}}{\phi_G^{n-gD/2}} \tag{5.17}$$

$$A_G = \frac{\pi^{Dg}}{(n-1)!}\int_{\sigma\subset\mathbb{P}^{n-1}(\mathbb{R})} \frac{\psi_G^{n-(g+1)D/2}\Omega}{\phi_G^{n-gD/2}}. \tag{5.18}$$

REMARK 5.1. Of particular interest is the *log divergent* case $D = 2n/g$, when $A_G = \frac{\pi^{Dg}}{(n-1)!}\int_\sigma \frac{\Omega}{\psi_G^{D/2}}$ is independent of masses and external momenta.

To summarize, we now have three formulas for the integral

$$A_G = \int_{\mathbb{R}^{Dg}} \frac{d^{Dg}x}{\prod_{e\in E} f_e} \tag{5.19}$$

$$A_G = \frac{1}{(n-1)!}\int_{\mathbb{R}^{Dg}\times\sigma} \frac{d^{Dg}x\Omega}{(\sum A_e f_e)^n} \tag{5.20}$$

$$A_G = \frac{\pi^{Dg}}{(n-1)!}\int_{\sigma\subset\mathbb{P}^{n-1}(\mathbb{R})} \frac{\psi_G^{n-(g+1)D/2}\Omega}{\phi_G^{n-gD/2}}. \tag{5.21}$$

To these, we add without proof a fourth ([**10**], formula (6-89))

$$A_G = \frac{1}{(i(4\pi)^2)^g}\int_{\widetilde{\sigma}} \frac{\exp(i\phi_G/\psi_G)\prod_E dA_e}{\psi_G^{D/2}}. \tag{5.22}$$

Here $\widetilde{\sigma} = [0,\infty]^E$ so $\sigma = \widetilde{\sigma} - \{0\}/\mathbb{R}_+^\times$. Philosophically, we can think of $\widetilde{\sigma}$ as the space of metrics (i.e. lengths of edges) on G. The integral then looks like a path integral on a space of metrics, with the action ϕ_G/ψ_G. We will see in what sense this action is a limit of string theory action.

Here is a useful way to think about the second Symanzik polynomial when the metric on space-time is euclidean. Let G be a connected graph, and assume the metric on $\mathbb{R}^D$ is positive definite (i.e. euclidean). Then for $a_e > 0$ the metric

$\sum_e a_e e^{\vee,2}$ on $(\mathbb{R}^D)^E$ is positive definite as well, so there are induced metrics on $H_1(G, \mathbb{R}^D)$ and $(\mathbb{R}^D)^{V,0}$. For $p \in (\mathbb{R}^D)^{V,0}$ let $m_a(p)$ be the value of the metric.

PROPOSITION 5.3. *We have* $m_a(p) = \phi_G(a,p,0)/\psi_G(a)$.

PROOF. The symmetric matrix M above is positive definite when the edge coordinates $A_e > 0$. It is then clear from (5.15) that the minimum of the metric in the fibre $\partial^{-1}(p)$ is given by $-\frac{\phi_G(A,p,0)}{\psi_G(A)}$. This is how the metric on the quotient is defined. (In general, if V has a positive definite metric, the metric on a quotient V/W is defined by identifying $V/W \cong W^\perp \subset V$.) $\square$

6. Riemann surfaces

In sections 6 - 9 we will reinterpret formula (5.22) for the integral. The graph G becomes the dual graph of a singular rational curve C_0 of arithmetic genus g which we view as lying at infinity on a moduli space of pointed curves of genus g. Vertices of G correspond to irreducible components of C_0, and the external momentum associated to a vertex is interpreted in terms of families of points meeting the given irreducible component of C_0. The chain of integration in (5.22) is identified with the nilpotent orbit associated to the degenerating Hodge structures, and the action $\exp(i\phi_G/\psi_G)$ is shown to be a limit of actions involving heights. This is joint work with José Burgos Gil, Javier Fresan, and Omid Amini [**1**].

As a first step, in this section we will interpret the rank 1 symmetric matrices M_e on $H_1(G)$, (2.1), in terms of the monodromy of the degenerating family of genus g curves. We continue to assume G is a connected graph with g loops and n edges. Stable rational curves C_0 associated to G arise taking quotients of $\coprod_{V(G)} \mathbb{P}^1$ identifying a chosen point of $\mathbb{P}^1_v$ with a chosen point of $\mathbb{P}^1_w$ whenever there exists an edge e with $\partial e = \{v, w\}$. For the moment we assume that every vertex of G meets at least 3 edges, so every $\mathbb{P}^1 \subset C_0$ has at least three "distinguished" points which are singularities of C_0. Note that if there is a vertex meeting ≥ 4 edges, the corresponding $\mathbb{P}^1$ will have ≥ 4 distinguished points and C_0 will have moduli.

DEFINITION 6.1. Let $C = \bigcup \mathbb{P}^1$ be a curve obtained by identifying a finite set of pairs of points in $\coprod_V \mathbb{P}^1$ for some finite set V. The dual graph of C is the graph with vertex set V and edge set E the set of pairs of points being identified. If $e \in E$ corresponds to $\{p_1, p_2\}$ with $p_i \in \mathbb{P}^1_{v_i}$ then the edge e is taken to connect the vertices v_1, v_2.

EXAMPLE 6.2. The dual graph of the curve C_0 constructed above is G.

PROPOSITION 6.3. *There is a canonical identification* $H^1(C_0, \mathcal{O}_{C_0}) \cong H^1(G, \mathbb{Q}) \cong H^1(C_0, \mathbb{Q})$. *In particular, the arithmetic genus of* C_0 *is equal to* g, *the loop number of* G.

PROOF. Let $p : \coprod_{V(G)} \mathbb{P}^1 \to C_0$ be the identification map. We have an exact sequence of sheaves

$$0 \to \mathcal{O}_{C_0} \to p_*\mathcal{O}_{\coprod \mathbb{P}^1} \to \mathcal{S} \to 0 \tag{6.1}$$

where $\mathcal{S}$ is a skyscraper sheaf with stalk k over each singular point. Since p is a finite map, taking cohomology commutes with p_* and we find

$$0 \to k \to \bigoplus_{V(G)} k \xrightarrow{\delta} \bigoplus_{E(G)} k \to H^1(C_0, \mathcal{O}_{C_0}) \to 0 \tag{6.2}$$

It is straightforward to check that δ in the above can be identified with the dual to the boundary map calculating $H_1(G)$, so $\mathrm{coker}(\delta) \cong H^1(G,k)$. The proof that $H^1(G,\mathbb{Q}) \cong H^1(C_0,\mathbb{Q})$ is similar. One simply replaces the exact sequence of coherent sheaves (6.1) with an analogous sequence of constructible sheaves calculating Betti cohomology. $\square$

We recall some basic results about deformation theory for C_0, **[9]**. There exists a smooth formal scheme $\widehat{S} = \mathrm{Spf}\ k[[t_1,\ldots,t_p]]$ and a (formal) family of curves $\widehat{\mathcal{C}} \xrightarrow{\pi} \widehat{S}$ such that the fibre $\mathcal{C}_0$ over $0 \in \widehat{S}$ is identified with the curve C_0 above, and such that the family is in some sense maximal. In particular, $\widehat{\mathcal{C}}$ is formally smooth over k and the tangent space T to $\widehat{S}$ at 0 is identified with $\mathrm{Ext}^1(\Omega^1_{C_0}, \mathcal{O}_{C_0})$. Étale locally at the singular points $C_0 \cong \mathrm{Spec}\ k[x,y]/(xy) =: \mathrm{Spec}\ R$ so $\Omega^1_{C_0} \cong Rdx \oplus Rdy/(xdy+ydx)$. Thus $xdy \in \Omega^1_{C_0}$ is killed by both x and y and so Ω^1 has a non-trivial torsion subsheaf supported at the singular points. The 5-term exact sequence of low degree terms for the local to global Ext spectral sequence yields in this case a short exact sequence

$$0 \to H^1(C_0, \underline{Hom}(\Omega^1_{C_0}, \mathcal{O}_{C_0})) \to \mathrm{Ext}^1(\Omega^1_{C_0}, \mathcal{O}_{C_0}) \to \Gamma(C_0, \underline{Ext}^1(\Omega^1_{C_0}, \mathcal{O}_{C_0})) \to 0. \tag{6.3}$$

The local ext sheaf on the right is easily calculated using the local presentation at the singular point as above

$$0 \to R \xrightarrow{1 \mapsto xdy+ydx} Rdx \oplus Rdy \to \Omega^1_R \to 0. \tag{6.4}$$

One identifies in this way $\underline{Ext}^1(\Omega^1_{C_0}, \mathcal{O}_{C_0})$ with the skyscraper sheaf having one copy of k supported at each singular point. The subspace $H^1(C_0, \underline{Hom}(\Omega^1_{C_0}, \mathcal{O}_{C_0})) \subset T$ corresponds to deformations which keep all the double points, i.e. only the chosen points on the $\mathbb{P}^1_v$ move. In general, r points on $\mathbb{P}^1$ have $r-3$ moduli, so

$$\dim H^1(C_0, \underline{Hom}(\Omega^1_{C_0}, \mathcal{O}_{C_0})) = \sum_{v \in V(G)} (\#\text{edges through } v - 3) \tag{6.5}$$

$$\dim T = \sum_{v \in V(G)} (\#\text{edges through } v - 3) + \#E(G) = \tag{6.6}$$
$$-3\#V(G) + 3\#E(G) = 3h_1(G) - 3 = 3g_a(C_0) - 3.$$

Here $g_a(C_0)$ is the arithmetic genus which coincides with the usual genus of a smooth deformation of C_0. The last identity follows from proposition 6.3. Note $3g-3$ is the dimension of the moduli space of genus g curves.

The map $T = Tan_{\widehat{S},0} \twoheadrightarrow \bigoplus_E k$ arises as follows. For each $e \in E$ there is a singular point $p_e \in C_0$. The deformations of C_0 which preserve the singularity at p_e give a divisor $\widehat{D}_e \subset \widehat{S}$. Let $g_e \in \mathcal{O}_{\widehat{S}}$ define $\widehat{D}_e$. The functional $T \to \bigoplus_E k \xrightarrow{pr_e} k$ is defined by dg_e. The geometric picture is then a collection of principal divisors $\widehat{D}_e \subset \widehat{S}$ meeting transversally. The subvariety cut out by the divisors is the locus of equisingular deformations of C_0 given by moving the singular points. If G is trivalent, i.e. if every vertex of G has exactly three adjacent edges, then $\{0\} = \bigcap \widehat{D}_e$.

Deformation theory leads to a formal *versal* deformation $\widehat{\mathcal{C}} \to \widehat{S}$, but these formal schemes can be spread out to yield an analytic deformation $\mathcal{C} \to S$. Here S is a polydisk of dimension $3g-3$. The divisors lift to analytic divisors $D_e : g_e = 0$ on S. We fix a basepoint $s_0 \in S - \bigcup D_e$, and we wish to study the monodromy action on $H_1(C_{s_0}, \mathbb{Q})$. We choose simple loops $\ell_e \subset S - \bigcup_E D_e$ based at s_0 looping around D_e. We assume ℓ_e is contractible in $S - \bigcup_{\varepsilon \neq e} D_\varepsilon$. The *Picard Lefschetz formula* gives the monodromy for the action of ℓ_* on $H_1(C_{s_0}, \mathbb{Q})$

$$b \mapsto b + \langle b, a_e \rangle a_e \tag{6.7}$$

where $a_e \in H_1(C_{s_0}, \mathbb{Q})$ is the vanishing cycle associated to the double point on the curve which remains as we deform along D_e.

A classical result in differential topology says that, possibly shrinking the polydisk S, the inclusion $C_0 \hookrightarrow \mathcal{C}$ admits a homotopy retraction $\mathcal{C} \to C_0$ in such a way that the composition $\mathcal{C} \to C_0 \to \mathcal{C}$ is homotopic to the identity. It follows that $C_0 \hookrightarrow \mathcal{C}$ is a homotopy equivalence. In this way, one defines the specialization map

$$sp : H_1(C_{s_0}, \mathbb{Q}) \to H_1(\mathcal{C}, \mathbb{Q}) \cong H_1(C_0, \mathbb{Q}). \tag{6.8}$$

LEMMA 6.4. *The specialization map sp above is surjective.*

PROOF. Intuitively, a loop in $H_1(C_0, \mathbb{Q})$ can be broken up into segments connecting double points of the curve. These double points arise from shrinking vanishing cycles on C_{S_0} so the segments can be modeled by segments in C_{s_0} connecting the vanishing cycles. These segments connect to yield a loop in C_{s_0} which specializes to the given loop in C_0. (One can give a more formal proof based on the Clemens-Schmid exact sequence, [**20**].) □

LEMMA 6.5. *The subspace $A \subset H_1(C_{s_0}, \mathbb{Q})$ spanned by the vanishing cycles a_e is maximal isotropic.*

PROOF. As the base point s_0 approaches $0 \in S$, the various a_e approach the singular points $p_e \in C_0$. In particular, if s_0 is taken close to 0, the a_e are physically disjoint, so $\langle a_e, a_{e'} \rangle = 0$. Since the pairing on H_1 is symplectic, one has $\langle a_e, a_e \rangle = 0$, so the subspace A spanned by the vanishing cycles is isotropic. To see it is maximal, note we can express C_0 as a topological colimit $C_0 = C_{s_0} / \coprod S^1$ so we get an exact sequence $A \to H_1(C_{s_0}) \xrightarrow{sp} H_1(C_0)$. In particular,

$$g = \dim H_1(C_0) \geq \dim H_1(C_{s_0}) - \dim A = 2g - \dim A. \tag{6.9}$$

It follows that $\dim A \geq g$ so A is maximal isotropic. In terms of a symplectic basis $a_1, \ldots, a_g, b_1, \ldots, b_g$ we write

$$a_e = \sum_{i=1}^{g} c_{e,i} a_i. \tag{6.10}$$

□

The link between the combinatorics of the graph polynomial and the monodromy is given by the following proposition. Write $N_e = \ell_e - id$ so by (6.7) we have $N_e(b) = \langle b, a_e \rangle a_e$. By lemma 6.5 we get $N_e^2 = 0$ so $N_e = \log(\ell_e)$. We consider the composition

$$H_1(G) \cong H_1(C_0) \cong H_1(C_{s_0})/A \xrightarrow{N_e} A \cong (H_1(C_{s_0})/A)^\vee \cong H_1(G)^\vee \tag{6.11}$$

PROPOSITION 6.6. *The bilinear form on $H_1(G)$ given by (6.11) coincides with the bilinear form M_e in (2.1).*

PROOF. Let $b \in H_1(C_{s_0})$. We can identify $sp(b) \in H_1(C_0) \cong H_1(G)$ with a loop $\sum_e n_e e$. Here $n_e = \langle b, a_e\rangle$ is the multiplicity of intersection of b with the vanishing cycle a_e. The quadratic form on $H_1(C_{s_0})$ corresponding to N_e sends $b \mapsto \langle b, \langle b, a_e\rangle a_e\rangle = n_e^2$. The quadratic form on $H_1(G)$ corresponding to M_e maps the loop $\sum n_\varepsilon \varepsilon$ to n_e^2. □

REMARK 6.1. In terms of the basis b_i for $B \cong H_1(G)$, we can write $M_e = (c_{e,i}c_{e,j})$ using the notation of (6.10). We will generalize this to relate the monodromy for punctured curves to the combinatorics of the second Symanzik in (9.11) through (9.15).

7. Biextensions and heights

Our objective in this section will be to link the second Symanzik polynomial (definition 2.2) to geometry. Recall for a graph G with edges E and vertices V, the second Symanzik $\phi(H, w, \{A_e\})$ depends on $H = H_1(G) \subset \mathbb{Q}^E$ and on $w \in \mathbb{R}^{V,0}$. (We will see later how to extend the construction and take $w \in (\mathbb{R}^D)^{V,0}$ where D is the dimension of space-time.) Recall that ϕ is quadratic in w. We will work with the corresponding bilinear function

$$\begin{aligned}(7.1)\quad & \phi(H, w, w', \{A_e\}) := \\ & \qquad \phi(H, w + w', \{A_e\}) - \phi(H, w, \{A_e\}) - \phi(H, w', \{A_e\}).\end{aligned}$$

We change notation and assume $\mathcal{C} \xrightarrow{\pi} S$ is a family of pointed curves. More precisely, we suppose given two collections $\sigma_{v,i} : S \to \mathcal{C},\ v \in V,\ i = 1, 2$ of sections of π. We assume $\sigma_{v,i}(S) \cap C_0 \in \mathbb{P}^1_v$. Since $\mathcal{C}$ is taken to be regular over the ground field, the sections cannot pass through double points of C_0. We assume further that the $\sigma_{v,1}$ and $\sigma_{v,2}$ are disjoint. It follows after possibly shrinking S that the multi-sections σ_1 and σ_2 are disjoint as well. Let W be an $\mathbb{R}$-vector space with a symmetric quadratic form (x, y). (In fact, for us $W = \mathbb{R}$ or $W = \mathbb{R}^D$ with the Minkowski metric.) We fix W-divisors

$$(7.2)\qquad \mathfrak{A}_i := \sum r_{v,i}\sigma_{v,i}; \quad r_{v,i} \in W, \ \sum_v r_{v,i} = 0, \ i = 1, 2.$$

More generally, we should work with a diagram

$$(7.3)\qquad \begin{array}{ccc} T_1 \amalg T_2 & \xrightarrow{\text{closed immersion}} & \mathcal{C} \\ \Big\| & & \downarrow \\ T_1 \amalg T_2 & \xrightarrow{\tau_1 \amalg \tau_2} & S \end{array}$$

where τ_i are finite étale. The labels $r_{v,i}$ would be replaced by sections of local systems of $\mathbb{R}$-vector spaces R_i on T_i equipped with trace maps $R_i \twoheadrightarrow \tau_i^*\mathbb{R}_S$ so we can talk about sections of degree 0. We have no use for this generalization, but we mention it exists.

The W-divisors will play the role of external momenta. We need to define the action which we write

$$(7.4)\qquad \mathbf{S}[C_{s_0}, \mathfrak{A}_1, \mathfrak{A}_2]$$

and which will tend to $\phi(H_1(G), \sum r_{v,1}v, \sum r_{v,2}v)$ as $s_0 \to 0$. This action is given by the archimedean height pairing which is defined as follows. We identify the section σ_1 on C_{s_0} with its image $\sigma_1 \subset C_{s_0}$. We consider the exact sequence of Hodge structures

$$0 \to H^1(C_{s_0}, \mathbb{Z}(1)) \to H^1(C_{s_0} - \sigma_1, \mathbb{Z}(1)) \xrightarrow{res} \coprod_{\sigma_1}^{0} \mathbb{Z} \to 0. \tag{7.5}$$

This sequence is canonically split as an exact sequence of $\mathbb{R}$-Hodge structures. I.e. there exists a canonical splitting

$$\coprod_{\sigma_1}^{0} \mathbb{R} \to F^0(H^1(C_{s_0} - \sigma_1)(1)) \cap H^1(C_{s_0} - \sigma_1, \mathbb{R}(1)); \quad \mathfrak{A} \mapsto \omega_{\mathfrak{A}}. \tag{7.6}$$

Here $\omega_{\mathfrak{A}} \in \Gamma(C_{s_0}, \Omega^1(\log \sigma_1))$. For $\mathfrak{B}$ another $\mathbb{R}$-divisor of degree 0 on C_{s_0} with support away from the support of $\mathfrak{A}$, we can write $\mathfrak{B} = \partial\beta$ where β is a 1-chain on $C_{s_0} - \mathrm{Supp}(\mathfrak{A})$ and define the archimedean height

$$\langle \mathfrak{A}, \mathfrak{B} \rangle := \int_\beta \omega_{\mathfrak{A}} \in \mathbb{C}/i\mathbb{R} = \mathbb{R}. \tag{7.7}$$

Because $\omega_{\mathfrak{A}}$ is an $\mathbb{R}(1)$-class, changing β by a class in $H_1(C_0, \mathbb{R})$ does not change the real part of the integral so the height is well-defined.

Finally, we can couple this pairing to the quadratic form on W and define $\langle \mathfrak{A}_1, \mathfrak{A}_2 \rangle$. Indeed, if we choose basepoints $\sigma_{0,i} \in \mathcal{C}_{s_0}$ we can write $\sum_v r_{v,i}\sigma_{v,i} = \sum_v r_{v,i}(\sigma_{v,i} - \sigma_{0,i})$ and define

$$\mathbf{S}[C_{s_0}, \mathfrak{A}_1, \mathfrak{A}_2] = \langle \mathfrak{A}_1, \mathfrak{A}_2 \rangle = \sum_{v,v'} (r_{v,1}, r_{v',2})\langle \sigma_{v,1} - \sigma_{0,1}, \sigma_{v',2} - \sigma_{0,2} \rangle. \tag{7.8}$$

This is independent of the choice of $\sigma_{0,i}$. Changing e.g. $\sigma_{0,1}$ to $\sigma_{0,1}{}'$ changes the above by

$$\sum_{v,v'} (r_{v,1}, r_{v',2})\langle \sigma_{0,1}{}' - \sigma_{0,1}, \sigma_{v',2} - \sigma_{0,2} \rangle$$

which vanishes because $\sum_v r_{v,1} = 0$.

8. The Poincaré bundle

To understand the behavior of the height in a degenerating family, it is convenient to use the Poincaré bundle. We first recall the Poincaré bundle for a single compact complex torus $T := V/\Lambda$, (V finite dimensional $\mathbb{C}$-vector space, $\Lambda \subset \mathbb{V}$ a cocompact lattice.) Let $\widehat{V} := \mathrm{Hom}_{\overline{\mathbb{C}}}(V, \mathbb{C})$ be the $\mathbb{C}$-vector space of $\mathbb{C}$-antilinear functionals on V, and let $\widehat{\Lambda} := \{\phi \in \widehat{V} \mid \phi(\Lambda) \subset \mathbb{Z}\}$. By definition, the dual torus $\widehat{T} := \widehat{V}/\widehat{\Lambda}$.

The Poincaré bundle is a $\mathbb{C}^\times$-bundle $\mathcal{P}^\times$ on $T \times \widehat{T}$. (NB. It will be more convenient to work with the principal $\mathbb{G}_m$ or $\mathbb{C}^\times$-bundle rather than the corresponding line bundle.) It is characterized by two properties:

(i)$\mathcal{P}^\times|_{\{0\}\times\widehat{T}}$ is trivial with a given trivialization.

(ii) $\mathcal{P}^\times|_{T\times\{\phi\}} = L_\phi$; where L_ϕ is the $\mathbb{C}^\times$-bundle on T associated to the representation of the fundamental group

$$\pi_1(T) = \Lambda \subset V \xrightarrow{\phi} \mathbb{C} \xrightarrow{\exp(2\pi i \cdot)} \mathbb{C}^\times. \tag{8.1}$$

To construct the Poincaré bundle over the family of all principally polarized abelian varieties of dimension g, we recall the Siegel domain

$$\mathbb{H}_g := \{M\ g \times g \text{ complex symmetric matrix} \mid Im(M) > 0\}. \tag{8.2}$$

(The following results about the Poincaré bundle are given in detail in [**1**]) The group $Sp_{2g}(\mathbb{R})$ acts on $\mathbb{H}_g$ by $\begin{pmatrix} A & B \\ C & D \end{pmatrix} M = (AM+B)(CM+D)^{-1}$. The quotient $\mathcal{A}_g := \mathbb{H}_g/Sp_{2g}(\mathbb{Z})$ is the Siegel moduli space parametrizing principally polarized abelian varieties of dimension g. We can also think of $\mathcal{A}_g$ as parametrizing polarized Hodge structures of weight 1. To $\Omega \in \mathbb{H}_g$ we associate the map $H_\mathbb{Z} := \mathbb{Z}^{2g} \to \mathbb{C}^g$ given by the $2g \times g$ matrix $\begin{pmatrix} \Omega \\ I_g \end{pmatrix}$. The symplectic form on $H_\mathbb{Z}$ is given by $\begin{pmatrix} 0 & I_g \\ -I_g & 0 \end{pmatrix}$.

The space

$$\widetilde{X} := \mathbb{H}_g \times \mathrm{Row}_g(\mathbb{C}) \times \mathrm{Col}_g(\mathbb{C}) \times \mathbb{C} \tag{8.3}$$

is a homogeneous space for the group

$$\widetilde{G} = \left\{ \begin{pmatrix} 1 & \lambda_1 & \lambda_2 & \alpha \\ 0 & A & B & \mu_1 \\ 0 & C & D & \mu_2 \\ 0 & 0 & 0 & 1 \end{pmatrix} \middle| \lambda_i \in \mathrm{Row}_g(\mathbb{R}),\ \mu_j \in \mathrm{Col}_g(\mathbb{R}), \alpha \in \mathbb{C}, \right. \\ \left. \begin{pmatrix} A & B \\ C & D \end{pmatrix} \in Sp_{2g}(\mathbb{R}) \right\}. \tag{8.4}$$

The action is determined by formulas

$$\begin{pmatrix} 1 & 0 & 0 & 0 \\ 0 & A & B & 0 \\ 0 & C & D & 0 \\ 0 & 0 & 0 & 1 \end{pmatrix} (\Omega, W, Z, \rho) = \tag{8.5}$$
$$((A\Omega+B)(C\Omega+D)^{-1}, W(C\Omega+D)^{-1}, ((C\Omega+D)^t)^{-1}Z, \rho - WC^t(C\Omega+D)^{-t}Z)$$

$$\begin{pmatrix} 1 & \lambda_1 & \lambda_2 & 0 \\ 0 & I & 0 & 0 \\ 0 & 0 & I & 0 \\ 0 & 0 & 0 & 1 \end{pmatrix} (\Omega, W, Z, \rho) = (\Omega, W + \lambda_1\Omega + \lambda_2, Z, \rho + \lambda_1 Z) \tag{8.6}$$

$$\begin{pmatrix} 1 & 0 & 0 & 0 \\ 0 & I & 0 & \mu_1 \\ 0 & 0 & I & \mu_2 \\ 0 & 0 & 0 & 1 \end{pmatrix} (\Omega, W, Z, \rho) = (\Omega, W, Z + \mu_1 - \Omega\mu_2, \rho - W\mu_2) \tag{8.7}$$

$$\begin{pmatrix} 1 & 0 & 0 & \alpha \\ 0 & I & 0 & 0 \\ 0 & 0 & I & 0 \\ 0 & 0 & 0 & 1 \end{pmatrix} (\Omega, W, Z, \rho) = (\Omega, W, Z, \rho + \alpha). \tag{8.8}$$

Write $\widetilde{G}(\mathbb{Z}) \subset \widetilde{G}$ for the subgroup with entries in $\mathbb{Z}$.

THEOREM 8.1. *The quotient*

$$\widetilde{G}(\mathbb{Z})\backslash(\mathbb{H}_g \times Row_g(\mathbb{C}) \times Col_g(\mathbb{C}) \times \mathbb{C}) \tag{8.9}$$

is the dual of the Poincaré bundle $\mathcal{P}^\times$.

PROOF. See [**1**]. □

THEOREM 8.2. *The Poincaré bundle admits a translation-invariant metric* $\log ||\cdot|| : \mathcal{P}^\times \to \mathbb{R}$. *For* $(\Omega, W, Z, \alpha) \in \mathbb{H}_g \times Row_g(\mathbb{C}) \times Col_g(\mathbb{C}) \times \mathbb{C}$, *we have*

$$\log ||(\Omega, W, Z, \alpha)|| = 4\pi\Big(Im(\alpha) + Im(W)(Im\Omega)^{-1}Im(Z)\Big). \tag{8.10}$$

PROOF. See [**1**]. □

The Poincaré bundle has a Hodge-theoretic interpretation as the moduli space for biextensions, which are mixed Hodge structures M with weights $-2, -1, 0$. We assume

$$W_{-2}M = \mathbb{Z}(1),\ gr^W_{-1}M = H,\ gr^W_0 M = \mathbb{Z}(0), \tag{8.11}$$

where H is a rank g principally polarized Hodge structure of weight -1. To see this, we remark that $\Lambda_\mathbb{C} = H_1(T, \mathbb{C})$ has a Hodge structure of weight -1 with $F^0\Lambda_\mathbb{C} := \ker(\Lambda_\mathbb{C} \twoheadrightarrow V)$ and $gr_F^{-1}\Lambda_\mathbb{C} = V$. For $0 \neq t \in T$ the relative homology $H_1(T, \{0, t\}, \mathbb{Z})$ yields an extension of Hodge structures

$$0 \to H_1(T, \mathbb{Z}) \to H_1(T, \{t, 0\}) \to \mathbb{Z}(0) \to 0. \tag{8.12}$$

A point $\phi \in \widehat{T} = Hom_{\overline{\mathbb{C}}}(V, \mathbb{C})/\widehat{\Lambda}$ yields a character $\exp(2\pi i\phi) : \Lambda \to \mathbb{C}^\times$ and hence a principal $\mathbb{C}^\times$-bundle $L_\phi^\times$ over T. The corresponding sequence of homology groups looks like

$$0 \to \mathbb{Z}(1) \to H_1(L_\phi^\times, \mathbb{Z}) \to H_1(T, \mathbb{Z}) \to 0 \tag{8.13}$$

It follows from exactness of the Hodge filtration functor that $F^0H_1(L_\phi^\times, \mathbb{C}) = F^0H_1(T, \mathbb{C})$. Thus

$$L_\phi^\times \cong gr_F^{-1}H_1(L_\phi^\times, \mathbb{C})/H_1(L_\phi^\times, \mathbb{Z}) = \mathrm{Ext}^1_{MHS}(\mathbb{Z}(0), H_1(L_\phi^\times, \mathbb{Z})). \tag{8.14}$$

The projection $L_\phi^\times \twoheadrightarrow T$ yields a diagram of Hodge structures coming from $\ell \in L_\phi^\times$ lying over $t \in T$ (Here $M_t = H_1(T, \{t, 1\})$ as a Hodge structure)

$$\begin{array}{ccccccccc}
 & & & & 0 & & 0 & & \\
 & & & & \downarrow & & \downarrow & & \\
0 & \longrightarrow & \mathbb{Z}(1) & \longrightarrow & H_1(L_\phi^\times, \mathbb{Z}) & \longrightarrow & H_1(T, \mathbb{Z}) & \longrightarrow & 0 \\
 & & \| & & \downarrow & & \downarrow & & \\
0 & \longrightarrow & \mathbb{Z}(1) & \longrightarrow & M_\ell & \longrightarrow & M_t & \longrightarrow & 0 \\
 & & & & \downarrow & & \downarrow & & \\
 & & & & \mathbb{Z}(0) & = & \mathbb{Z}(0) & & \\
 & & & & \downarrow & & \downarrow & & \\
 & & & & 0 & & 0 & &
\end{array} \tag{8.15}$$

The biextension corresponding to $\ell \in L_\phi^\times = \mathcal{P}|_{T\times\{\phi\}}$ is then M_ℓ with $W_{-1}M_\ell = H_1(L_\phi^\times, \mathbb{Z})$ and $M_\ell/W_{-2} = M_t$.

Let $\mathfrak{A}, \mathfrak{B}$ be divisors of degree 0 on a smooth curve C. Assume the supports $|\mathfrak{A}|$ and $|\mathfrak{B}|$ are disjoint. We associate to $\mathfrak{A}, \mathfrak{B}$ a biextension Hodge structure $H^1(C - \mathfrak{A}, \mathfrak{B}; \mathbb{Z})$. (Notation like $H^1(C-\mathfrak{A}, \mathfrak{B}; \mathbb{Z})$ or $H^1(C, \mathfrak{B}; \mathbb{Z})$ is of course abusive. These are subquotients of $H^1(C - |\mathfrak{A}|, |\mathfrak{B}|; \mathbb{Z})$ and $H^1(C, |\mathfrak{B}|; \mathbb{Z})$ which fit into a diagram

$$
\begin{array}{ccccccccc}
 & & & & 0 & & 0 & & \\
 & & & & \downarrow & & \downarrow & & \\
0 & \longrightarrow & \mathbb{Z}(1) & \longrightarrow & H^1(C, \mathfrak{B}; \mathbb{Z}(1)) & \longrightarrow & H^1(C, \mathbb{Z}(1)) & \longrightarrow & 0 \\
 & & \| & & \downarrow & & \downarrow & & \\
0 & \longrightarrow & \mathbb{Z}(1) & \longrightarrow & H^1(C - \mathfrak{A}, \mathfrak{B}; \mathbb{Z}(1)) & \longrightarrow & H^1(C - \mathfrak{A}, \mathbb{Z}(1)) & \longrightarrow & 0 \\
 & & & & \downarrow & & \downarrow & & \\
 & & & & \mathbb{Z}(0) & = & \mathbb{Z}(0) & & \\
 & & & & \downarrow & & \downarrow & & \\
 & & & & 0 & & 0.) & &
\end{array}
\tag{8.16}
$$

Diagrams (8.15) and (8.16) are related as follows. Take $T = J(C)$ and let $t \in T$ be the image of a 0-cycle $\mathfrak{A}$ on C of degree 0. The bundle $L_\phi^\times$ is a group which is an extension of T by $\mathbb{C}^\times$. We view C as embedded in $T = J(C)$. There exists a 0-cycle $\mathfrak{B}$ on C of degree 0 such that $L_\phi^\times|_C \cong \mathcal{O}_C(\mathfrak{B}) - 0$-section. Then $L_\phi^\times|_{C-|\mathfrak{B}|} \cong \mathbb{C}^\times \times (C - |\mathfrak{B}|)$, and this trivialization is canonical up to $c \in \mathbb{C}^\times$. Fix such a trivialization $\mu : C - |\mathfrak{B}| \to L_\phi^\times$. Write $\mathfrak{A} = \sum n_i a_i$, and let $\ell = \sum \mu(a_i)^{n_i}$. Then $\ell \in L_\phi^\times$ is well-defined independent of the choice of μ, and diagram (8.15) for $\ell \in L_\phi^\times$ coincides with (8.16) for $\mathfrak{A}$ and $\mathfrak{B}$.

Note that $L_\phi^\times$ is an abelian Lie group. As such, it has a unique maximal compact subgroup K, and

$$L_\phi^\times / K \cong \mathbb{C}^\times / S^1 \cong \mathbb{R}. \tag{8.17}$$

Let $\rho_\phi : L_\phi^\times \to \mathbb{R}$ be the induced map.

PROPOSITION 8.3. *With notation as above, the following quantities are equal*
(i) $\langle \mathfrak{A}, \mathfrak{B} \rangle$.
(ii) $\log ||H^1(C - \mathfrak{A}, \mathfrak{B}; \mathbb{Z}(1))||$.
(iii) $\rho_\phi(\ell)$.

PROOF. Omitted. □

9. The main theorem; nilpotent orbit and passage to the limit

Write $U^* = (\Delta^*)^E \times \Delta^{3g-3-E+N}$, where N parameters are added to accommodate markings associated to external momenta. Consider our family $\mathcal{C}^* \xrightarrow{\pi} U^*$. Let $\widetilde{U} = \mathbb{H}^E \times \Delta^{3g-3-E+N} \to U^*$ be the universal cover with coordinate map $t_e = \exp(2\pi i z_e)$. We fix a basepoint $t_0 \in U^*$. We choose a symplectic basis $a_1, \ldots, a_g, b_1, \ldots, b_g$ in $H_1(C_{t_0}, \mathbb{Z}) = A \oplus B$ with the vanishing cycles $a_e \in A$. We fix divisors $\delta = \delta_t, \mu = \mu_t$ which have degree 0 on each C_t and are disjoint. We choose 1-chains $\sigma_{\delta,t}$ and $\sigma_{\mu,t}$ on C_t which have δ_t and μ_t as boundary. We fix a basis $\{\omega_{i,t}\}$ of the holomorphic 1-forms on C_t such that $\int_{a_i} \omega_j = \delta_{ij}$. The classical period matrix is then $(\int_{b_i} \omega_{j,t})$. Finally, we choose 1-forms of the third kind $\omega_{\delta,t}, \omega_{\mu,t}$ with log poles on δ_t and μ_t respectively.

We define the period map $\widetilde{\phi} : \widetilde{U} \to \mathbb{H}_g \times \mathbb{C}^g \times \mathbb{C}^g \times \mathbb{C}$

$$\widetilde{\phi}(z) = \Big(\big(\int_{b_i} \omega_{j,t})_{i,j}, (\int_{\sigma_{\delta,t}} \omega_{j,t})_j, (\int_{\sigma_{\mu,t}} \omega_{j,t})_j, \int_{\sigma_{\delta,t}} \omega_{\mu,t}\Big) \tag{9.1}$$

We check that this is the correct map as follows. It suffices to work pointwise, so we drop the t. We identify $J(C) = \mathrm{Ext}^1_{MHS}(H^1(C,\mathbb{Z}), \mathbb{Z}(0))$ so e.g. μ defines an extension

$$0 \to \mathbb{Z}(0) \to H_\mu \to H^1(C,\mathbb{Z}) \to 0. \tag{9.2}$$

Dually, we can think of δ as defining an extension of $\mathbb{Z}(0)$ by $H^1(C,\mathbb{Z}(1))$. We want to think of δ as defining an extension of $\mathbb{Z}(0)$ by $H_\mu(1)$ (or dually an extension of $H_\mu^\vee$ by $\mathbb{Z}(1)$) which "lifts" this extension. The diagram looks like

$$\begin{array}{ccccccccc}
 & & 0 & & 0 & & & & \\
 & & \downarrow & & \downarrow & & & & \\
 & & \mathbb{Z}(1) & = & \mathbb{Z}(1) & & & & \\
 & & \downarrow & & \downarrow & & & & \\
0 & \longrightarrow & (H_\delta)^\vee(1) & \longrightarrow & M_{\mu,\delta} & \longrightarrow & \mathbb{Z}(0) & \longrightarrow & 0 \\
 & & \downarrow & & \downarrow & & \| & & \\
0 & \longrightarrow & H^1(C,\mathbb{Z}(1)) & \longrightarrow & H_\mu^\vee & \longrightarrow & \mathbb{Z}(0) & \longrightarrow & 0 \\
 & & \downarrow & & \downarrow & & & & \\
 & & 0 & & 0. & & & &
\end{array} \tag{9.3}$$

The exact sequence of ext groups is

$$\begin{aligned} 0 \to \mathrm{Ext}^1_{MHS}(\mathbb{Z}(0), \mathbb{Z}(1)) \to \mathrm{Ext}^1_{MHS}(\mathbb{Z}(0), H_\delta^\vee(1)) \to \\ \mathrm{Ext}^1_{MHS}(\mathbb{Z}(0), H^1(C,\mathbb{Z}(1))) \to 0 \end{aligned} \tag{9.4}$$

More concretely, we can think of $\int_{\sigma_\mu}$ as a functional on $\Gamma(C,\Omega^1) = F^1H^1(C,\mathbb{C})$. Let $\Gamma(C,\Omega^1(\log\delta)) \supset \Gamma(C,\Omega^1)$ be spanned by the ω_j together with the form of the third kind ω_δ. We can define $H_1(C-\delta,\mathbb{Z})$ as a quotient of $H_1(C - \mathrm{Supp}(\delta),\mathbb{Z})$ in such a way that we get an exact sequence

$$\begin{aligned} 0 \to \mathbb{C}/\mathbb{Z} \to \Gamma(C,\Omega^1(\log\delta))^\vee/H_1(C-\delta,\mathbb{Z}) \to \\ \Gamma(C,\Omega^1)^\vee/H_1(C,\mathbb{Z}) \to 0 \end{aligned} \tag{9.5}$$

The exact sequences (9.4) and (9.5) coincide. The description of $\widetilde{\phi}$ in (9.1) follows from this; lifting $\int_{\sigma_\mu}$ to a functional on $\Gamma(C,\Omega^1(\log\delta))$. Indeed, $(\int_{\sigma_{\delta,t}} \omega_{j,t})_j$ and $(\int_{\sigma_{\mu,t}} \omega_{j,t})_j$ yield the points in $J(C)$ associated to δ_t and μ_t respectively. The term $\Gamma(C,\Omega^1(\log\delta))^\vee/H_1(C-\delta,\mathbb{Z})$ in (9.5) shows that the class of δ in the generalized jacobian extension corresponding to μ is obtained by taking $\int_{\sigma_{\delta,t}} \omega_{\mu,t}$ as entry in $\mathbb{C}$.

The idea will be to use the nilpotent orbit theorem to understand the limiting behavior of (9.1). This theorem in the case of variations of polarized pure Hodge structures was proven by W. Schmid, [**22**]. We need the more general case of a variation of mixed Hodge structures, [**19**],[**11**],[**12**],[**13**]. In our case, we have the diagram

$$\begin{array}{ccc} \widetilde{U} & \xrightarrow{\tilde{\psi}} & \mathbb{H}_g \times \mathrm{Row}_g(\mathbb{C}) \times \mathrm{Col}_g(\mathbb{C}) \times \mathbb{C} \\ \downarrow & & \downarrow \\ U^* & \longrightarrow & \widetilde{G}(\mathbb{Z})\backslash\Big(\mathbb{H}_g \times \mathrm{Row}_g(\mathbb{C}) \times \mathrm{Col}_g(\mathbb{C}) \times \mathbb{C}\Big) \end{array} \tag{9.6}$$

The action of the fundamental group $\mathbb{Z}^E$ of U^* is unipotent, and we write $\mathcal{N}_e$ for the logarithm of the generator $1_e \in \mathbb{Z}^E$. The expression $\tilde{\psi}(z) := \exp(-\sum_E z_e\mathcal{N}_e)\tilde{\phi}(z)$ takes values in a compact dual $\check{\mathcal{M}}$ which is essentially a flag variety parametrizing filtrations $F^*\mathbb{C}^{g+2}$ satisfying the conditions to be the Hodge filtration on a biextension of genus g. It is invariant under $z_e \mapsto z_e + 1$ and so descends to a map $\psi : U^* \to \check{\mathcal{M}}$.

THEOREM 9.1. *[Nilpotent orbit theorem] (i) The limit*

$$F_\infty := \lim_{s\to 0} \psi(s) \in \check{\mathcal{M}}$$

exists.

(ii) We have

$$\exp(\sum_E z_e\mathcal{N}_e)F_\infty \in \mathbb{H}_g \times Row_g(\mathbb{C}) \times Col_g(\mathbb{C}) \times \mathbb{C}$$

whenever $Im(z_e) >> 0, \forall e$.

(iii) For an invariant metric d *on* $\mathbb{H}_g \times Row_g(\mathbb{C}) \times Col_g(\mathbb{C}) \times \mathbb{C}$ *there exist constants* b, K *such that for* $\min_e(Im(z_e)) >> 0$ *we have*

$$d(\tilde{\phi}(z), \exp(\sum_E z_e\mathcal{N}_e)F_\infty) \le K(\min_e(Im(z_e))^b \exp(-2\pi \min_e(Im(z_e))).$$

We want to understand how the logarithm of monodromy maps N_e act on the entries in (9.1). We have using (6.10) and $\int_{a_i} \omega_j = \delta_{ij}$

$$N_e(\int_{b_i} \omega_{j,t}) = \langle b_i, a_e\rangle \int_{a_e} \omega_{j,t} = c_{e,i}c_{e,j}. \tag{9.7}$$

By remark 6.1 we conclude

$$N_e(\int_{b_i} \omega_{j,t})_{i,j} = M_e. \tag{9.8}$$

Said another way, if we view $b_i, b_j \in H_1(G)$, we have

$$N_e(\int_{b_i} \omega_{j,t})_{i,j} = (\ldots, b_iM_eb_j^t \ldots) \tag{9.9}$$

To define the Picard-Lefschetz transformation on the path $\sigma_\delta \in H_1(C, \delta)$ we define vanishing cycles $a_e^\delta \in H_1(C - |\delta|)$. With this refinement, the Picard-Lefschetz formula

$$N_e(\sigma_\delta) = \langle \sigma_\delta, a_e^\delta \rangle a_e \tag{9.10}$$

holds. Here N_e is viewed as an endomorphism of $H_1(C, \delta)$. Thus

$$N_e(\int_{\sigma_{\delta,t}} \omega_{j,t})_j = (\langle \sigma_\delta, a_e^\delta \rangle c_{e,j})_j \tag{9.11}$$

To understand this, notice that the specialization identification $sp : B \cong H_1(G) \subset \mathbb{R}^E$ maps $b_j \mapsto \sum_e \langle b_j, a_e \rangle e = \sum_e c_{e,j} e$. Note that $\langle b_j, a_e \rangle$ counts how many times (with orientation) the closed chain $sp(b_j)$ passes through the singular point $pt_e \in C_0$. Similarly we can think of $sp(\sigma_\delta)$ as a 1-chain on C_0. It is not closed, but it bounds the 0-chain corresponding to δ. Thus, it corresponds to $\sum_e \langle \sigma_\delta, a_e^\delta \rangle e \in \mathbb{R}^E$. In this way we can embed $H_1(C_s, \delta_s)/A \hookrightarrow \mathbb{R}^E$ with the path σ_δ mapping as indicated. Formula (9.11) then can be interpreted as the pairing applied to σ_δ and b_j.

Finally we have

$$N_e(\int_{\sigma_{\delta,t}} \omega_{\mu,t}) = \langle \sigma_{\delta,t}, a_e^\delta \rangle \int_{a_e^\mu} \omega_\mu \tag{9.12}$$

Notice that $\omega_\mu \in H^1(C_t - \mu_t) \cong H_1(C_t, \mu_t)$. Modifying by a linear combination of the ω_j we can assume under that identification that ω_μ corresponds to σ_μ. Thus, we have

$$N_e(\int_{\sigma_{\delta,t}} \omega_{\mu,t}) = \langle \sigma_{\delta,t}, a_e^\delta \rangle \langle \sigma_{\mu,t}, a_e^\mu \rangle \tag{9.13}$$

We can summarize our computations as follows. Consider the relative homology $H_1(C_t, \delta_t, \mu_t)$ which contains classes for the two paths $\sigma_{\delta,t}$ and $\sigma_{\mu,t}$. Let $A_{\sigma,\mu} \subset H_1(C_t, \delta_t, \mu_t)$ be the kernel of the specialization map

$$sp : H_1(C_t, \delta_t, \mu_t)/A_{\sigma,\mu} \to H_1(C_0, \delta_0, \mu_0) \subset \mathbb{R}^E. \tag{9.14}$$

We can now consider the composition $((\ldots, y_e, \ldots)$ means multiplication by the diagonal matrix with entries $y_e)$

$$\begin{aligned} H_1(C_t, \delta_t, \mu_t)/A_{\sigma,\mu} &\xrightarrow{sp} H_1(C_0, \delta_0, \mu_0) \subset \mathbb{R}^E \\ &\xrightarrow{(\ldots, y_e, \ldots)} \mathbb{R}^E \to (H_1(C_t, \delta_t, \mu_t)/A_{\sigma,\mu})^\vee \end{aligned} \tag{9.15}$$

Replacing N_e by $\sum y_e N_e$, the formulas (9.8), (9.11), and (9.13) give the pairings $\langle b_i, b_j \rangle, \langle \sigma_\delta, b_i \rangle, \langle \sigma_\delta, \sigma_\mu \rangle$ for the pairing defined by (9.15).

In our case, the nilpotent orbit theorem 9.1 says that the family (9.1) is well-approximated by

$$\begin{aligned} \Big(\Omega_\infty + \sum_E z_e M_e, W_\infty + \sum_E z_e(\ldots, \langle \sigma_\delta, a_e^\delta \rangle c_{ej}, \ldots), \\ Z_\infty + \sum_E z_e(\ldots, \langle \sigma_\mu, a_e^\mu \rangle c_{ej}, \ldots), \alpha_\infty + \sum_E z_e \langle \sigma_{\delta,t}, a_e^\delta \rangle \langle \sigma_{\mu,t}, a_e^\mu \rangle \Big). \end{aligned} \tag{9.16}$$

We write $z_e = x_e + iy_e$ and $t_e = \exp(2\pi i z_e)$ so $y_e = -(\log |t_e|)/2\pi$. We imagine

$$y_e = Y_e/\alpha' \tag{9.17}$$

where $Y_e > 0$ is constant and the *string tension* $\alpha' \to 0$. Substituting from (8.10) yields the following expression for the height

$$\begin{aligned}(9.18)\quad & \frac{1}{\alpha'}\Big[(\alpha' ImW_\infty + \sum_e Y_e(\ldots, \langle\sigma_\delta, a_e^\delta\rangle c_{ej}\ldots))\cdot \\ & \cdot(\alpha' Im\Omega_\infty + \sum_e Y_e M_e)^{-1}(\alpha' ImZ_\infty + \sum_e Y_e(\ldots, \langle\sigma_\mu, a_e^\mu\rangle c_{ej}\ldots)) \\ & \qquad - \alpha' Im\alpha_\infty - \sum_e Y_e\langle\sigma_\delta, a_e^\delta\rangle\langle\sigma_\mu, a_e^\mu\rangle\Big]\end{aligned}$$

Now, using the determinant formula for the second Symanzik polynomial (remark 2.4) we conclude

THEOREM 9.2. *Let $\tilde{\phi} : \widetilde{U} \to \mathcal{P}$ be as in* (9.1). *Then*

$$\lim_{\alpha'\to 0} \alpha' \log ||\tilde{\phi}(z)|| = \phi_G(p, p', \{Y_e\})/\psi_G(\{Y_e\}). \tag{9.19}$$

Here $z_e = X_e + i\frac{Y_e}{\alpha'}$, ψ and ϕ are the first and second Symanzik polynomials, and p, p' denote external momenta (see (7.1)*).*

PROOF. The proof is simply a question of pulling together the linear algebra. We split the exact sequence of homology of the graph to write $\mathbb{R}^E = H_1(G, \mathbb{R}) \oplus \mathbb{R}^{V,0}$. With respect to this basis, the symmetric matrix corresponding to the quadratic form $\sum y_e e^{\vee,2}$ is

$$\begin{pmatrix} M_y & B_y^{\delta,t} \\ B_y^\mu & p^\mu \Gamma_y p^{\delta,t}\end{pmatrix} \tag{9.20}$$

Here the pairing on the b_i is $M_y = \sum y_e M_e$ with M_e as in remark 6.1. One external momentum maps to $\sum_e \langle\sigma_\delta, a_e^\delta\rangle e$. Pairing this with the b_j yields the column vector $B_y^{\delta,t} = (\ldots, \sum_e y_e\langle\sigma_\delta, a_e^\delta\rangle c_{e,j}, \ldots)^t$. The other external momentum pairs with the b_j to yield a row vector $B_y^\mu = (\ldots, \sum_e y_e\langle\sigma_\mu, a_e^\delta\rangle c_{e,j}, \ldots)$. Finally, the entry in the bottom right corner is $\sum_e y_e\langle\sigma_\delta, a_e^\delta\rangle\langle\sigma_\mu, a_e^\mu\rangle$. The theorem follows from remark 2.4. □

To summarize, our recipe for computing the Feynman integral (5.22) is the following. Given the external momentum $p := \sum_{v\in V(G)} r_v v$ with $r_v \in \mathbb{R}^D$ and $\sum_v r_v = 0$, choose an analytic neighborhood S of C_0 in a suitable local moduli space of marked curves and two relative $\mathbb{R}^D$-divisors $\mathfrak{A}_i := \sum_v r_v\sigma_{v,i}$, $i = 1, 2$ supported on the markings. We assume the markings $\sigma_{v,i}$ are all disjoint. Further, we require that the intersection numbers with components of the rational curve C_0 coincide, and that $\sigma_{v,1} \cdot C_{0,v'} = \sigma_{v,2} \cdot C_{0,v'} = \delta_{v,v'}$. (Kronecker delta.) (The issue here is that the height $\langle\mathfrak{A}_1, \mathfrak{A}_2\rangle$ is only defined if the supports of $\mathfrak{A}_1$, $\mathfrak{A}_2$ are disjoint.) Recall from the beginning of this section we have

$$\widetilde{U} = \mathbb{H}^{E(G)} \times \Delta^{3g-3-E(G)+N} \to U^* = \Delta^{*,E(G)} \times \Delta^{3g-3-E(G)+N};$$
$$X_e + iY_e/\alpha' = z_e = \frac{\log(t_e)}{2\pi i}.$$

Then (compare (7.8). The factor 2 occurs because of (7.1).)

$$\lim_{\alpha'\to 0} \alpha' \mathbf{S}[C_{t_e}, \mathfrak{A}_1, \mathfrak{A}_2] = 2\phi_G(p, \{Y_e\})/\psi_G(\{Y_e\}). \tag{9.21}$$

The Feynman integral arises by plugging this limit into the exponential in (5.22) and integrating over the the space of nilpotent orbits which is parametrized by $Y_e \geq 0$. (The above should be compared with computations from [**23**] where the same scaling $X + iY/\alpha'$ occurs.)

Let me suggest two directions for further research.

1. The rational curve C_0 with dual graph G corresponds to a maximally degenerate boundary point on the moduli space of (marked) curves of genus g. Are there physically meaningful quantities arising when we specialize the height action (7.8) to less degenerate boundary points. The classical interpretation of Feynman integrals as coefficients of Green's functions arising from path integrals is really more of a metaphor than a mathematically rigorous calculus. Perhaps studying partial degenerations might throw more light.

2. The height $\langle \mathfrak{A}_1, \mathfrak{A}_2 \rangle$ is just one example of an $\mathbb{R}$-valued function associated to a variation of mixed Hodge structures. (In this case, we consider variations of biextensions.) Are there other interesting quantities arising from integration over the space of nilpotent orbits when a variation of Hodge structures degenerates?

10. Cutkosky rules

The final topic in the notes is a report on some aspects of joint work with Dirk Kreimer [**2**] on Cutkosky rules. The integral $A_G = A_G(q_{ext})$ is a multi-valued function of external momenta q_{ext}, as one sees e.g. from (5.19). (For simplicity the masses are kept fixed.) The following beautiful physical interpretation of the *variation* $\mathrm{Var}(A_G)$ as q_{ext} winds around the threshold locus was given in [**6**], [**8**]:

$$\mathrm{Var}(A_G) = (-2\pi i)^r \int \frac{\delta^+(q_1^2 - m_1^2) \cdots \delta^+(q_r^2 - m_r^2) d^{Dg}x}{(q_{r+1}^2 - m_{r+1}^2) \cdots (q_n^2 - m_n^2)}. \tag{10.1}$$

Here A_G is the integral associated to a graph G with n edges and g loops. The $q_i^2 - m_i^2$ are propagators associated to the edges e_i, with q_i being the momentum flow through the edge, and $\delta^+(q_i^2 - m_i^2)$ means to take the residue of the original form $\omega := \frac{d^{Dg}x}{\prod_1^n (q_j^2 - m_j^2)}$ along the divisor $q_i^2 - m_i^2 = 0$ and then to integrate over the part of the real locus of that intersection of divisors where the energy $q_i^{(0)} > 0$. External momenta are placed near the threshold divisor associated to a pinch point of the intersection $\bigcap_1^r \{q_i^2 - m_i^2 = 0\}$. ("Pinch point" here refers to the singularity of the intersection of divisors $S_1 \cap \cdots \cap S_n$ given in local coordinates by equations (11.5) with $t_1 = 0$.)

Said another way, for fixed masses, the quadrics $q_i^2 - m_i^2$ are parametrized by external momenta associated to vertices of the graph. There will in general be a divisor D in the space of external momenta (threshold divisor) corresponding to configurations where the intersection of the quadrics $q_i^2 - m_i^2 = 0$, $1 \leq i \leq r$ is not transverse. We imagine a small loop in external momentum space winding around D, and $\mathrm{Var}(A_G)$ refers to the change in A_G.

The basic monodromy calculation was related to the classical Picard-Lefschetz formula in mathematics by Pham, [**21**]. Pham gives a complete description of the local topology of the union of propagator quadrics as the external momentum approaches a general threshold point. However, the physical formula (10.1) involves the real structure of the quadrics, and more work is necessary to link the topology to the Cutkosky formula. In the following I sketch the arguments involved in proving Cutkosky's formula for the case of *physical singularities* (definition 12.1).

The quadrics have the form $q^2 - m^2 = (e^\vee - c_e)^2 - m^2$ where $e \in E = E(G)$ is an edge, $c_e \in \mathbb{C}^D$ is a D-tuple of constants, and $e^\vee$ is a D-tuple of functions on $\mathbb{C}^{Dg}$:

$$\mathbb{C}^{Dg} = H_1(G, \mathbb{C}^D) \hookrightarrow (\mathbb{C}^D)^E \xrightarrow{e^\vee - c_e} \mathbb{C}^D \xrightarrow{()^2 - m^2} \mathbb{C}. \tag{10.2}$$

Write Q_e for the quadric in $\mathbb{C}^{Dg}$ defined by (10.2). Given edges $e_1, \ldots, e_r$, it is straightforward to check that the rD linear functions $(e_1^\vee, \ldots, e_r^\vee) : H_1(G, \mathbb{C}^D) \to \mathbb{C}^{rD}$ form part of a system of coordinates on $\mathbb{C}^{Dg}$ if and only if this map is surjective, and that this condition is independent of D. Taking $D = 1$, one sees that the linear functions do not form part of a system of coordinates if and only if the graph obtained from G by cutting $e_1, \ldots, e_r$ is not connected. On the other hand, if the linear functions form part of a system of coordinates and if none of the masses $m_i = 0$, it is clear that the intersection of the propagator quadrics cannot have a pinch point (irrespective of the c_{e_i}). We conclude

PROPOSITION 10.1. *With notation as above, assuming all masses $m_i \neq 0$, a necessary condition for the intersection of the quadrics*

$$Q_{e_1} \cap \cdots \cap Q_{e_r}$$

to have a pinch point for suitable values of the c_{e_i} is that the graph obtained from G by cutting $e_1, \ldots, e_r$ be disconnected.

For simplicity, in what follows I consider only the case $r = n$, i.e. the case when all the edges are placed on shell. The resulting cut graph is just the collection $V(G)$ of vertices and is certainly disconnected (assuming G has at least 2 vertices).

EXAMPLE 10.2 (Banana graphs). Suppose G has 2 vertices connected by n edges $e_1, \ldots, e_n$. The only possibility to disconnect G is to cut all the edges. Since cutting $e_1, \ldots, e_{n-1}$ leaves G connected, it follows as above that $e_i^\vee : \mathbb{C}^{D(n-1)} \to \mathbb{C}^D$, $1 \leq i \leq n-1$ are independent coordinates, and up to orientation $e_n^\vee = \sum_1^{n-1} e_i^\vee - a$ for some $a \in \mathbb{C}^D$. We can replace the $e_i^\vee$ by $e_i^\vee - c_i$ and arrange that the jacobian matrix for the n quadrics looks like

$$2\begin{pmatrix} e_1^\vee & 0 & 0 & \cdots \\ 0 & e_2^\vee & 0 & \cdots \\ \vdots & \vdots & \vdots & \vdots \\ 0 & \cdots & 0 & e_{n-1}^\vee \\ \sum_1^{n-1} e_i^\vee + a & \sum_1^{n-1} e_i^\vee + a & \sum_1^{n-1} e_i^\vee + a & \sum_1^{n-1} e_i^\vee + a \end{pmatrix} \tag{10.3}$$

(Note each entry is a D-vector.) We want this matrix to have rank $n - 1$ at some point where the n propagator quadrics vanish. A small exercise in linear algebra (for details, see [**2**], section 12) yields

$$|a| = \sum_1^n \mu(i) m_i; \quad \mu(i) = \pm 1. \tag{10.4}$$

(Here $|a| := \sqrt{a^2}$.)

REMARK 10.1. The referee points out that when the space-time dimension $D = 1$, we are looking at an intersection of n quadrics in $\mathbb{C}^{n-1}$ which is in general empty. One should assume here that $D \geq 2$.

11. Cutkosky rules: Pham's vanishing cycles

In this section I summarize briefly the calculus of vanishing cycles; first the classical theory and then the theory of F. Pham. Classically, $f : \mathcal{X} \to \Delta$ is a proper family of varieties over a disk $\Delta \subset \mathbb{C}$. We assume $\mathcal{X}$ is non-singular, and f is smooth away from $0 \in \Delta$. Further, $\mathcal{X}_0 = f^{-1}(0)$ has a single ordinary double point s_0 and is non-singular elsewhere. With these hypotheses, we can find analytic coordinates $x_1, \ldots, x_n$ near x_0 and a coordinate t on Δ such that locally the family is defined by

$$x_1^2 + \ldots + x_n^2 = t. \tag{11.1}$$

For $0 < \varepsilon << 1$, the homology of a ball B around s_0 intersected with the fibre $\mathcal{X}_\varepsilon$ is generated by the class of the real sphere $S_\varepsilon : \sum x_i^2 = \varepsilon,\ x_i \in \mathbb{R}$. The monodromy action on the smooth fibre $H^*(\mathcal{X}_\varepsilon, \mathbb{Q})$ is given by the Picard-Lefschetz formula

$$c \mapsto c + (-1)^{n(n+1)/2} \langle ex(c), S_\varepsilon \rangle i_* S_\varepsilon. \tag{11.2}$$

Here

$$ex : H_*(\mathcal{X}_\varepsilon) \to H_*(B \cap \mathcal{X}_\varepsilon, \partial(B \cap \mathcal{X}_\varepsilon)) \tag{11.3}$$

denotes excision, and $i_* : H_*(B \cap \mathcal{X}_\varepsilon) \to H_*(\mathcal{X}_\varepsilon)$ is the pushforward. The pairing $\langle ex(c), S_\varepsilon \rangle$ refers to the natural pairing

$$H_{n-1}(B \cap \mathcal{X}_\varepsilon, \partial(B \cap \mathcal{X}_\varepsilon)) \otimes H_{n-1}(B \cap \mathcal{X}_\varepsilon) \to \mathbb{Q}. \tag{11.4}$$

To explain Pham's vanishing cycles, we change notation. Let $f : M \to P$ be a smooth, proper map of algebraic varieties. Let $S = \bigcup_{i=1}^n S_i \subset M$ be a normal crossings divisor. Let $s_0 \in \bigcap_{i=1}^n S_i \subset S \subset M$, and let $p_0 = f(s_0)$. We will work locally near s_0 on M, so we may fix analytic coordinates $t_1, \ldots, t_k$ around p_0 on P and $x_1, \ldots, x_\ell$ around s_0 on M in such a way that $t_1, \ldots, t_k, x_1, \ldots, x_\ell$ form a full set of coordinates on M near s_0. We assume that locally

$$\begin{aligned} &S_i : x_i = 0; \quad i = 1, \ldots, n-1, \\ &S_n : t_1 - (x_1 + \cdots + x_{n-1} + x_n^2 + \cdots + x_\ell^2) = 0. \end{aligned} \tag{11.5}$$

For such a configuration of divisors, the point s_0 (origin) is called a *pinch point.* Notice that on the smallest stratum $\bigcap_1^n S_i$ of S, viewed as a variety fibred over P, we have a classical Picard-Lefschetz vanishing cycle local equation, which is a family of spheres $x_n^2 + \cdots + x_\ell^2 = t_1$ degenerating as $t \to 0$. For $t_1 = \varepsilon > 0$, the *vanishing sphere* $\text{vsphere}_\varepsilon$ is the real sphere

$$\text{vsphere}_\varepsilon : \varepsilon = x_n^2 + \cdots + x_\ell^2; \quad x_i \in \mathbb{R},\ n \leq i \leq \ell. \tag{11.6}$$

Classically, $\text{vsphere}_\varepsilon$ is referred to as the vanishing cycle, but we follow Pham here and distinguish 3 different topological chains, the vanishing sphere, the vanishing cell, and the vanishing cycle.

DEFINITION 11.1. With notation as above, the vanishing cell,

$$\text{vcell}_\varepsilon : x_i \geq 0,\ 1 \leq i \leq n-1; \quad \varepsilon - (x_n^2 + \cdots + x_\ell^2) \geq 0. \tag{11.7}$$

The vanishing cycle $\text{vcycle}_\varepsilon$ is the iterated tube

$$\tau_{1,\varepsilon_1}^* \tau_{2,\varepsilon_2}^* \cdots \tau_{n-1,\varepsilon_{n-1}}^* (\text{vsphere}_\varepsilon).$$

Here $\varepsilon >> \varepsilon_{n-1} >> \cdots >> \varepsilon_1 > 0$. The notation τ_{i,ε_i}^* refers to pulling back to the circle bundle of radius ε_i inside the (metrized) normal bundle for $S_1 \cap \cdots \cap S_i \subset$

$S_1 \cap \cdots \cap S_{i-1}$. This circle bundle is viewed as embedded in $S_1 \cap \cdots \cap S_{i-1}$ and not meeting $S_1 \cap \cdots \cap S_i$. The inequalities $\varepsilon_i << \varepsilon_{i+1}$ insure that $\text{vcycle}_\varepsilon$ is a closed chain on $M - \bigcup_{i=1}^n S_i$.

Here are two pictures in the case $n = 2$.

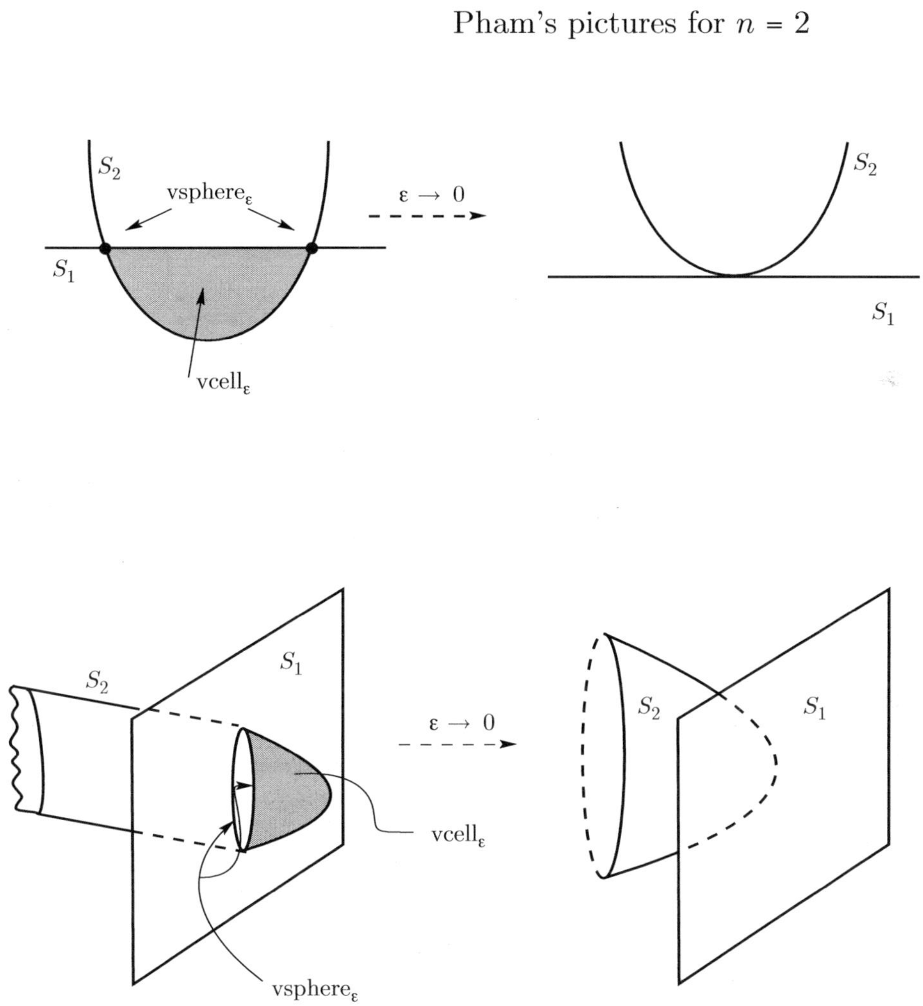

$\text{vcycle}_\varepsilon = S^1 \times S^1$-bundle over $\text{vsphere}_\varepsilon$ contained in B-$(S_1 \cup S_2)$.

The local structure of homology around s_0 is computed by Pham to be

THEOREM 11.2. *Let B be a small ball around s_0 in M. Let $p \in P$ be near p_0 and assume $t_1(p) = \varepsilon > 0$. Write $d = \dim_{\mathbb{C}} M_p$.*

(i) We have for reduced homology

$$\widetilde{H}_j\Big(\bigcap_1^n S_{i,p}\cap B,\mathbb{Z}\Big)=(0),\ j\neq d-n:=\dim_{\mathbb{C}}\Big(\bigcap_1^n S_{i,p}\Big).$$

$$\widetilde{H}_{d-n}\Big(\bigcap_1^n S_{i,p}\cap B,\mathbb{Z}\Big)=\mathbb{Z}\cdot \mathrm{vsphere}_\varepsilon.$$

(ii) In relative homology

$$H_j(B_p,S_p\cap B)=(0),\ j\neq d;\quad H_d(B_p,S_p\cap B)=\mathbb{Z}\cdot \mathrm{vcell}_\varepsilon.$$

(iii) For the homology of the complement

$$H_j(B_p-S_p\cap B)=(0),\ j\neq d;\quad H_d(B_p-S_p\cap B)=\mathbb{Z}\cdot \mathrm{vcycle}_\varepsilon.$$

Local Poincaré duality yields a pairing

$$\langle\cdot,\cdot\rangle : H_d(B_p-S_p\cap B)\otimes H_d(B_p,(S_p\cap B)\cup\partial B_p)\to H_{2d}(B_p,\partial B_p)\cong\mathbb{Z}. \tag{11.8}$$

Let γ be a simple closed path on P based at p, supported in a neighborhood of p_0, and looping once around the divisor $t_1=0$. Using Thom's isotopy lemma [**16**] and assuming that f is a submersion on strata except at the point s_0, Pham shows that the variation of monodromy

$$var:=\gamma_*-Id: H_*(M_p-S_p)\to H_*(M_p-S_p) \tag{11.9}$$

factors through excision and a local variation map var_{loc}

$$H_d(M_p-S_p)\xrightarrow{excision} H_d(B_p-S_p\cap B_p,\partial B_p)\xrightarrow{var_{loc}} H_d(B_p-S_p)\xrightarrow{i_*} H_d(M_p-S_p). \tag{11.10}$$

(The variation map is zero in homological degrees $\neq d$.)

The Picard-Lefschetz theorem in this setup is

THEOREM 11.3. *We have*

$$var_{loc}(\mathrm{excision}(c))=\mathrm{excision}(c)+(-1)^{(n+1)(n+2)/2}\langle\mathrm{excision}(c),\mathrm{vcell}_\varepsilon\rangle\mathrm{vcycle}_\varepsilon \tag{11.11}$$

PROOF. See [**21**]. □

EXAMPLE 11.4. In this example we compute the Picard-Lefschetz transformation arising in physics. One has a set of quadrics $Q_i: f_i=0,\ 1\le i\le n$ indexed by the edges of a graph. Our normal crossings divisor in this case is $\bigcup Q_i$. We consider the situation locally around a pinch point s_0, and we write in Pham's coordinates, $Q_i: x_i=0,\ 1\le i\le n-1$ and $Q_n: t_1-(x_1+\cdots+x_{n-1}+\sum_n^\ell x_i^2)=0$. Notice that in these coordinates, the bad fibre over p_0 (where $t_1=0$) of $\bigcap_1^n Q_{i,p_0}$ has an isolated $\mathbb{R}$-point at the origin. We will see (proposition 12.4) in the case of *physical singularities* that this is also the case for the space-time coordinates. Hence it is plausible to assume that at least for physical singularities Pham's local coordinates are defined over $\mathbb{R}$. We are interested in computing the monodromy on $M-\bigcup Q_i$ of the chain given by taking space-time coordinates in $\mathbb{R}$. As it stands this makes no sense, because our quadrics are defined using the Minkowski metric so the Q_i meet the real locus. The standard ploy to avoid this problem is to replace the defining equations $f_j = 0$ by $f_j + ia_j = 0$ with $1 >> a_j > 0$. Using theorem 11.3, we need to compute

$$\langle \text{excision}(\mathbb{R}^{Dg}), \text{vcell}_\varepsilon \rangle,$$

where D is the dimension of space-time and g is the number of loops of the graph.

In Pham's local coordinates, the deformed quadrics are defined by

$$x_j = -ia_j,\ 1 \le j \le n-1 \text{ and } \varepsilon + ia_n - (\sum_1^{n-1} x_j + \sum_n^{Dg} x_j^2) = 0.$$

We let x_j run over the path

$$x_j = (1-\rho_j)ia_j + \rho_j(\varepsilon + ia_n)/(n-1),\ 1 \le j \le n-1.$$

Here $0 \le \rho_j \le 1$. For points $x^0 = (x_1^0, \ldots, x_{n-1}^0)$ on those paths, the quadratic equation becomes

$$\sum_n^{Dg} x_j^2 = \varepsilon + ia_n - \sum_1^{n-1} x_j^0 =: r(x^0)e^{i\theta(x^0)}.$$

The locus

$$\{(e^{i\theta(x^0)/2}u_n, \ldots e^{i\theta(x^0)/2}u_{Dg}) \mid \sum_{k=n}^{Dg} u_k^2 \le r(x^0)\} \tag{11.12}$$

is a solid sphere, and the union of those solid spheres as the x_j run over those paths is the chain vcell_ε for the deformed quadrics.

According to theorem 11.3, the multiplicity of the vanishing cycle in the variation of the cycle $\mathbb{R}^{Dg}$ (or more precisely of $\mathbb{P}^{Dg}(\mathbb{R})$) around the given pinch point is the intersection of this real locus with vcell_ε.

Note that x_j is real only at the point $\rho_j = a_j/(a_j + a_n/(n-1)) \in (0,1)$. At the point with these coordinates, the quadratic equation reads $\sum_n^{Dg} x_j^2 = R + ia_n$ for some $R \in \mathbb{R}$. since $a_n \neq 0$, the only real point on the corresponding solid sphere (11.12) lies at the origin. Thus, $\text{vcell}_\varepsilon \cap \mathbb{R}^{Dg}$ is precisely one point. The intersection looks locally like the intersection $\mathbb{R}^N \cap e^{i\theta}\mathbb{R}^N \subset \mathbb{C}^N$, so it is transverse, and

$$\langle \text{excision}(\mathbb{P}^{Dg}(\mathbb{R})), \text{vcycle} \rangle = \pm 1.$$

We do not try to compute the sign.

This example is important because it proves that there really is non-trivial monodromy associated to the physical situation. Note the computation is very dependent on taking the $a_j > 0$.

12. Cutkosky's theorem

We want to prove (10.1) in the case of *physical singularities.* (Recall, we are also assuming for simplicity that we cut all edges so $r = n$ in (10.1).) To understand physical singularities, consider formula (5.20) for the integral A_G. Viewed as a function of external momenta, A_G will be non-singular unless the chain of integration $\mathbb{R}^{Dg} \times \sigma$ meets the polar locus $\sum_{e\in E} A_e f_e = 0$. When it does meet, it may still happen that the chain can be deformed away from the polar locus so A_G has no singularity. The logical place to look for branch points of A_G is thus at external momenta where $\sigma \times \mathbb{R}^{Dg}$ contains singular points of the universal quadric

$$Q : \sum_{e\in E} A_e f_e = 0. \tag{12.1}$$

DEFINITION 12.1. A point $p = (c_1, \ldots, c_n, p') \in Q(\mathbb{R}) \subset \mathbb{R}^n \times \mathbb{R}^{Dg}$ is a physical singularity if p is a singular point of Q and all $c_i > 0$. By extension, a point $p' \in \mathbb{R}^{Dg}$ is a physical singularity if there exists a physical singularity $(c, p') \in Q$.

LEMMA 12.2. *Let $p = (c_1, \ldots, c_n, p')$ be a physical singularity of the universal quadric Q. Then $p' \in \bigcap_1^n \{f_i = 0\}$ and p' is a singular point in that intersection of quadrics.*

PROOF. Let $f = \sum A_i f_i$. By assumption, p is singular so $f_i(p') = \frac{\partial}{\partial A_i} f(p) = 0$. Also, vanishing of the partials of f with respect to coordinates on $\mathbb{R}^{Dg}$ yields the relation

$$\sum_1^n c_i \mathrm{grad}_{p'}(f_i) = 0, \tag{12.2}$$

which implies that p' is singular on the intersection of quadrics. □

DEFINITION 12.3. A physical singularity p will be called non-degenerate if two conditions hold:

(i) The relation (12.2) is the only non-trivial relation among the gradients at p'.

(ii) The Hessian matrix at the singular point for the intersection of the quadrics is non-degenerate.

Condition (i) in the above definition means we can choose a partial set of local coordinates $x_1, \ldots, x_{n-1}$ at the singular point p' in such a way that the intersection of the quadrics is cut out locally near p' by $x_1 = \cdots = x_{n-1} = g = 0$ for some function g. Condition (ii) says that the matrix of second order partials of $g|\{x_1 = \cdots = x_{n-1} = 0\}$ is non-degenerate at p'.

In order to understand the $\mathbb{R}$-structure we borrow a standard picture for a Morse function f, [**17**].

The Hessian matrix of f at the points a, d is definite, while the Hessian matrix at b, c is indefinite. Note $f^{-1}f(a) = a$ and the fibre $f^{-1}(x)$ for x slightly above $f(a)$ is a sphere (a circle in the picture). This is the vanishing cycle which contracts to the point a as $x \to f(a)$. The picture at d is similar for y slightly below $f(d)$. On the other hand, $f^{-1}f(b)$ and $f^{-1}f(c)$ are figure-eights, and there are no vanishing cycles. In our case, if we know the Hessian for the function g at p' is either positive or negative definite, then Pham's vanishing cycle will coincide with the real sphere given by Morse theory.

PROPOSITION 12.4. *Let $p = (c_1, \ldots, c_n, p')$ be a non-degenerate physical singularity. Assume $m_i^2 > 0$ for all masses. Then the Hessian associated to $p' \in \bigcap_1^n \{f_i = 0\}$ is negative definite. In particular, Pham's vanishing sphere is real, and Cutkosky's theorem* (10.1) *holds for the variation of A_G.*

PROOF. Order the edges $e_1, \ldots, e_n$ such that $e_1^\vee, \ldots, e_g^\vee$ give coordinates on $H_1(G, \mathbb{R}^D)$. (Note each $e_i^\vee = (e_i^{\vee,1}, \ldots, e_i^{\vee,D})$ is a D-tuple of coordinate functions.) We change notation slightly and write

$$p = (c_1, \ldots, c_n, e_1^\vee(p'), \ldots, e_g^\vee(p')) \in \mathbb{R}^n \times \mathbb{R}^{Dg}.$$

The quadrics have the form $f_i = (e_i^\vee - s_i)^2 - m_i^2$ for some $s_i \in \mathbb{R}^D$. By a Euclidean change of variables, we can take $s_i = 0$ for $1 \leq i \leq g$. Because p is singular on the universal quadric Q, the sum $\sum c_i f_i$ is pure quadratic when expanded about

Morse Function Picture

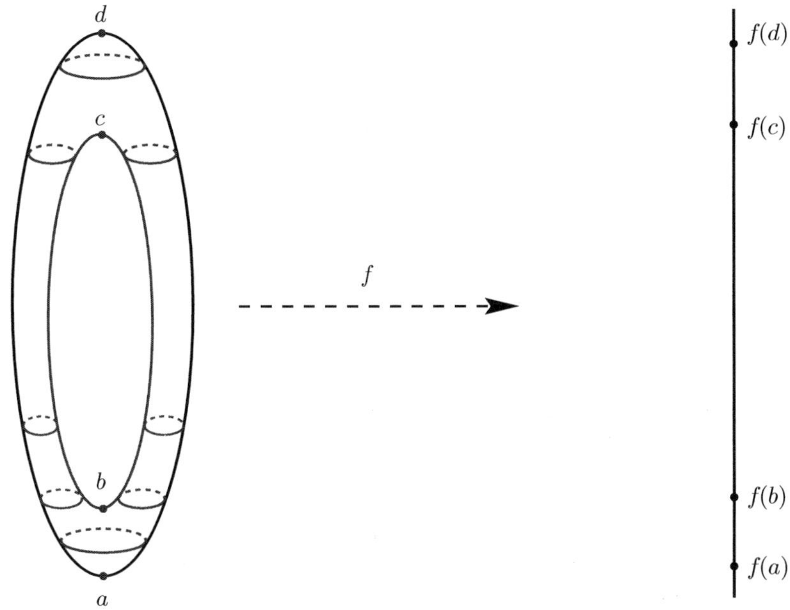

p', i.e. constant and linear terms vanish. Notice that the quadratic part of f_i is $(e_i^\vee)^2$ which is a Minkowski square. The quadratic form associated to the Hessian is then simply $\sum c_i(e_i^\vee)^2$. Of course, these are Minkowski squares, so it is not possible to say anything about the sign of the form on $\mathbb{R}^{Dg}$. However, what we want is the sign of the form restricted to the intersection of the tangent spaces at p' of the f_i. Among these are the tangent spaces at p' to $f_i = (e_i^\vee)^2 - m_i^2$ for $1 \leq i \leq g$. The tangent space at p' for such a f_i is $(e_i^\vee)^{-1}(e_i^\vee(p')^\perp) \subset \mathbb{R}^{Dg}$. (Here $e_i^\vee(p')^\perp \subset \mathbb{R}^D$. Note $(e_i^\vee(p'))^2 = m_i^2 > 0,\ 1 \leq i \leq g$.) Since the Minkowski metric has sign $(1, -1, \ldots, -1)$ on $\mathbb{R}^D$, we conclude that the sign of the metric on the intersection of these spaces is negative definite. In particular, $\sum_1^n c_i(e_i^\vee)^2$ is certainly negative semi-definite on the intersection of the tangent spaces. (Indeed, we can write $e_j^\vee = \sum_{i=1}^g \mu_j^i e_i^\vee$ with $\mu_j^i \in \mathbb{R}$. This decomposition is orthogonal, so $(e_j^\vee)^2 = \sum_{i=1}^g (\mu_j^i)^2(e_i^\vee)^2$. Since the $(e_i^\vee)^2$ on the right are negative definite on the intersection of the tangent spaces, $(e_j^\vee)^2$ is negative definite as well.) The only way $\sum_1^n c_i(e_i^\vee)^2$ could fail to be negative definite would be if there was a non-zero tangent vector t such that $e_i^\vee(t) = 0,\ 1 \leq i \leq n$. But this is not possible since the $e_i^\vee$, $1 \leq i \leq g$, give coordinates. □

References

[1] O. Amini, S. Bloch, J. I. Burgos Gil, and J. Fresán, *Feynman amplitudes and limits of heights*, Izv. Ross. Akad. Nauk Ser. Mat. **80** (2016), no. 5, 5–40, DOI 10.4213/im8492; English transl., Izv. Math. **80** (2016), no. 5, 813–848. MR3588803

[2] Spencer Bloch and Dirk Kreimer, *Feynman amplitudes and Landau singularities for one-loop graphs*, Commun. Number Theory Phys. **4** (2010), no. 4, 709–753, DOI 10.4310/CNTP.2010.v4.n4.a4. MR2793425

[3] Francis Brown and Oliver Schnetz, *A K3 in ϕ^4*, Duke Math. J. **161** (2012), no. 10, 1817–1862, DOI 10.1215/00127094-1644201. MR2954618

[4] James A. Carlson, Eduardo H. Cattani, and Aroldo G. Kaplan, *Mixed Hodge structures and compactifications of Siegel's space (preliminary report)*, Journées de Géometrie Algébrique d'Angers, Juillet 1979/Algebraic Geometry, Angers, 1979, Sijthoff & Noordhoff, Alphen aan den Rijn—Germantown, Md., 1980, pp. 77–105. MR605337

[5] Eduardo H. Cattani, *Mixed Hodge structures, compactifications and monodromy weight filtration*, Topics in transcendental algebraic geometry (Princeton, N.J., 1981/1982), Ann. of Math. Stud., vol. 106, Princeton Univ. Press, Princeton, NJ, 1984, pp. 75–100. MR756847

[6] R. E. Cutkosky, *Singularities and discontinuities of Feynman amplitudes*, J. Mathematical Phys. **1** (1960), 429–433, DOI 10.1063/1.1703676. MR118351

[7] Pierre Deligne, *Théorie de Hodge. III* (French), Inst. Hautes Études Sci. Publ. Math. **44** (1974), 5–77. MR498552

[8] R. J. Eden, P. V. Landshoff, D. I. Olive, and J. C. Polkinghorne, *The analytic S-matrix*, Cambridge University Press, London-New York-Ibadan, 1966. MR0204051

[9] Robin Hartshorne, *Deformation theory*, Graduate Texts in Mathematics, vol. 257, Springer, New York, 2010. MR2583634

[10] C. Itzykson, J.-B. Zuber, Quantum Field Theory, Dover Publ. (2005).

[11] K. Kato, C. Nakayama, S. Usui, Classifying Spaces of Degenerating Mixed Hodge Structures, J. Alg. Geom. 17 (2008), 401-479.

[12] Kazuya Kato, Chikara Nakayama, and Sampei Usui, *Classifying spaces of degenerating mixed Hodge structures, II: spaces of* SL(2)*-orbits*, Kyoto J. Math. **51** (2011), no. 1, 149–261, DOI 10.1215/0023608X-2010-023. MR2784750

[13] Kazuya Kato, Chikara Nakayama, and Sampei Usui, *Classifying spaces of degenerating mixed Hodge structures, III: Spaces of nilpotent orbits*, J. Algebraic Geom. **22** (2013), no. 4, 671–772, DOI 10.1090/S1056-3911-2013-00629-3. MR3084721

[14] Kazuya Kato and Sampei Usui, *Classifying spaces of degenerating polarized Hodge structures*, Annals of Mathematics Studies, vol. 169, Princeton University Press, Princeton, NJ, 2009. MR2465224

[15] L. D. Landau, *On analytic properties of vertex parts in quantum field theory*, Nuclear Phys. **13** (1959), 181–192. MR0141405

[16] J. Mather, Notes on Topological Stability, Harvard University Notes (available on the web), July, 1970.

[17] J. Milnor, *Morse theory*, Based on lecture notes by M. Spivak and R. Wells. Annals of Mathematics Studies, No. 51, Princeton University Press, Princeton, N.J., 1963. MR0163331

[18] Eric Patterson, *On the singular structure of graph hypersurfaces*, Commun. Number Theory Phys. **4** (2010), no. 4, 659–708, DOI 10.4310/CNTP.2010.v4.n4.a3. MR2793424

[19] Gregory J. Pearlstein, *Degenerations of mixed Hodge structure*, Duke Math. J. **110** (2001), no. 2, 217–251, DOI 10.1215/S0012-7094-01-11022-3. MR1865240

[20] Chris A. M. Peters and Joseph H. M. Steenbrink, *Mixed Hodge structures*, Ergebnisse der Mathematik und ihrer Grenzgebiete. 3. Folge. A Series of Modern Surveys in Mathematics [Results in Mathematics and Related Areas. 3rd Series. A Series of Modern Surveys in Mathematics], vol. 52, Springer-Verlag, Berlin, 2008. MR2393625

[21] F. Pham, *Introduction à l'étude topologique des singularités de Landau* (French), Mémorial des Sciences Mathématiques, Fasc. 164, Gauthier-Villars Éditeur, Paris, 1967. MR0229263

[22] Wilfried Schmid, *Variation of Hodge structure: the singularities of the period mapping*, Invent. Math. **22** (1973), 211–319, DOI 10.1007/BF01389674. MR382272

[23] Piotr Tourkine, *Tropical amplitudes*, Ann. Henri Poincaré **18** (2017), no. 6, 2199–2249, DOI 10.1007/s00023-017-0560-7. MR3649455

[24] Pierre Vanhove, *The physics and the mixed Hodge structure of Feynman integrals*, String-Math 2013, Proc. Sympos. Pure Math., vol. 88, Amer. Math. Soc., Providence, RI, 2014, pp. 161–194, DOI 10.1090/pspum/088/01455. MR3330287

5765 S Blackstone Ave., Chicago, Illinois 60637
Email address: `spencer_bloch@yahoo.com`

Clay Mathematics Proceedings
Volume **21**, 2020
https://doi.org/10.1090/cmip/21/477

Feynman integrals and periods in configuration spaces

Özgür Ceyhan and Matilde Marcolli

ABSTRACT. We describe two different constructions related to Feynman amplitudes in configuration spaces. One of them corresponds to the physical Feynman amplitude, while the second problem is, in a suitable sense, a complexification. We show that the two problems, while similar in form, lead to very different mathematical techniques and results. In the first case, we use analytic methods related to the expansion of the Green function of the Laplacian in Gegenbauer polynomials, while in the second case we use algebro-geometric techniques related to wonderful compactifications of configuration spaces. In the first case, we identify a contribution to the Feynman integral, which has a natural description in representation theoretic terms, which is expressible in terms of Mordell–Tornheim and Apostol–Vu series, even when the remaining terms of the integral might not be expressible in terms of multiple zeta values. In the second case, we can always express the resulting integral as an integral of an algebraic differential form over a variety whose motive is mixed Tate, even though the argument does not directly ensure that the algebraic differential form will be defined over the rationals.

1. Introduction

The past decade has seen the development of a new field of inquiry, aimed at understanding the geometric and algebraic structure underlying regularization and renormalization of Feynman integrals as well as the number theoretic properties of the residues of Feynman graphs obtained in perturbative quantum field theory. In particular, an unexpected connection between quantum field theory and Grothendieck's theory of motives of algebraic varieties was uncovered. The main idea behind this relation is that (regularized) integrals associated to Feynman graphs can be interpreted as periods of some algebraic varieties, and the nature of the motive of the variety can predict what class of numbers can arise as periods. An especially interesting class of numbers that occur as periods are the multiple zeta values, which occur as periods of an especially interesting class of motives, the mixed Tate motives. Thus, within the general framework of the relation between quantum field theory and motives, specific questions have often revolved around whether the specific motives are mixed Tate, and the periods expressible as combinations of multiple zeta values. This area of research has grown rapidly in the past

2010 *Mathematics Subject Classification.* Primary 81Q30, 81T18, 35J08, 42C05, 11M32, 14C15.

The first author is partially supported by EU grant PCIG11-GA-2012-322154. The second author is partially supported by NSF grants DMS-1007207, DMS-1201512, PHY-1205440.

few years, hence the list of relevant references is extensive. Without any claim of completeness, we refer the reader to some of the relevant literature, starting with the paper of Bloch, Esnault and Kreimer [**22**], including several other contributions to the subject: [**7**], [**8**], [**9**], [**10**], [**24**], [**27**], [**30**], [**31**], [**32**], [**33**], [**40**], [**44**], [**67**].

In the perturbative expansion in Feynman graphs of a Euclidean scalar massless quantum field theory, one can compute the contributions of individual Feynman graphs either in configuration or in momentum space. A good part of the literature on the subject of motives and periods in quantum field theory has focused on the momentum space picture, and especially the Feynman parametric form, where the contribution of a Feynman graph is interpreted as the (possibly divergent) integral of an algebraic differential form on the complement of a hypersurface (the graph hypersurface) in a projective space. In this setting, the relevant motivic question is the nature of the motive of the graph hypersurface. General results of Belkale and Brosnan [**16**] show that the graph hypersurfaces additively span a localized version of the Grothendieck ring of varieties, hence they can be arbitrarily far from the mixed Tate case, and more concrete recent results of Doryn, Brown and Doryn, and Schnetz, provided explicit non-mixed-Tate examples, [**30**], [**31**], [**44**].

In this paper, we will focus on the dual picture in configuration space. We will consider both the physical case of the Feynman amplitude and some mathematical generalizations that show similar but different behavior. Clearly, momentum and configuration space approaches are dual pictures, related by a Fourier transform of the propagators, as we will show more precisely in Section 3.4 below. However, Fourier transform is not an algebraic operation, hence there is no a priori reason why it should preserve the nature of motives and periods, other than a general philosophical belief in the consistency of the universe. In other words, Fourier transform preserves the L^2 but not the L^1 norm, hence it does not preserve period integrals. These also take place in different dimensions: due to momentum conservation, integration takes place in a space of dimension $b_1(\Gamma)$ in the momentum space formulation, and in dimension $\#V_\Gamma$ in the configuration space case. All this simply means that one cannot just appeal directly to Fourier duality to explain the equivalence of the two pictures, in terms of the arithmetic information encoded in the Feynman integrals. One has to develop a mathematical argument that shows what happens in both cases and how they are related.

1.1. Integrals in configuration spaces. In the present paper we will consider two different cases of integrals in configuration spaces. The first case, which we refer to as *the real case*, corresponds to the physically relevant Feynman integral, while the second case, which we call *the complexified case* is a simple mathematical generalization. We find it instructive to compare these two cases, because although they at first look very similar, they lead to different mathematical techniques and different results.

In both cases, for consistency, we use the following general setting and notation. Let Γ be a Feynman graph of the given Euclidean scalar massless quantum field theory. We assume chosen an orientation of Γ. Let $D \in \mathbb{N}$ be the spacetime dimension of the theory. Let X be a D-dimensional algebraic variety, which contains a dense big cell $\mathbb{A}^D \subset X$. Let $\mathrm{Conf}_\Gamma(X) = X^{V_\Gamma} \smallsetminus \cup_{e \in E_\Gamma} \Delta_e$ denote the configuration space of the graph Γ, where $\Delta_e = \{x = (x_v) \,|\, x_{s(e)} = x_{t(e)}\}$ are the diagonals. The geometry of these configuration spaces and of their wonderful compactifications is described in [**36**], [**61**], [**62**].

Note: here and elsewhere in the paper, for a space X and a finite set S we use the standard notation $X^S = \mathrm{Maps}(S, X) \cong X^{\#S}$, where $\#S$ is the cardinality, and the identification is through a choice of a bijection $S \to \{1, \dots, \#S\}$. Thus, in the above, any such bijection identifies $X^{V_\Gamma} \cong X^N$ with $N = \#V_\Gamma$.

1.1.1. *The real case.* In this case, we associate to the graph Γ the differential form of middle-dimensional degree DN, with $N = \#V_\Gamma$,

$$\omega_\Gamma = \prod_{e \in E_\Gamma} \frac{1}{\|x_{s(e)} - x_{t(e)}\|^{2\lambda}} \bigwedge_{v \in V_\Gamma} dx_v,$$

where $D = 2\lambda + 2$, and where $\|\cdot\|$ is the Euclidean norm on $\mathbb{A}^{DN}(\mathbb{C})$.

This a $\mathcal{C}^\infty$-differential form with singularities along diagonals $x_{s(e)} = x_{t(e)}$. The domain of integration in this case is the middle dimensional locus of real points

$$\sigma_\Gamma = X(\mathbb{R})^{V_\Gamma}.$$

For later purposes, let us emphasize the following two facts:

(1) ω_Γ is not an algebraic form;
(2) ω_Γ is not a closed form on $\mathrm{Conf}_\Gamma(X)$.

1.1.2. *The complexified case.* In this other case we consider the variety $Z = X \times X$ with projection $p : Z \to X$, $p : z = (x, y) \mapsto x$. We associate to the graph Γ a differential form of middle-dimensional degree $2DN$, with $N = \#V_\Gamma$,

$$\omega_\Gamma^{(Z)} = \prod_{e \in E_\Gamma} \frac{1}{\|x_{s(e)} - x_{t(e)}\|^{2D-2}} \bigwedge_{v \in V_\Gamma} dx_v \wedge d\bar{x}_v$$

where $\|x_{s(e)} - x_{t(e)}\| = \|p(z)_{s(e)} - p(z)_{t(e)}\|$. As chain of integration, in this case, we consider the middle dimensional locus

$$\sigma^{(Z,y)} = X^{V_\Gamma}(\mathbb{C}) = X^{V_\Gamma} \times \{y = (y_v)\} \subset Z^{V_\Gamma} = X^{V_\Gamma} \times X^{V_\Gamma}$$

for a fixed $y = (y_v \mid v \in V_\Gamma)$. In this case, we have the following situation:

(1) $\omega_\Gamma^{(Z)}$ is not an algebraic form;
(2) $\omega_\Gamma^{(Z)}$ is a closed form on $\mathrm{Conf}_\Gamma(X) \times X^{V_\Gamma}$.

1.2. Algebraic versus smooth differential forms. In the typical physical situation, the relevant integral is over a locus of real coordinates, corresponding to the physical momenta, or positions. It is in general convenient, in order to apply geometric methods, to extend the differential form from the real locus to an ambient space with complex coordinates. In general, there are inequivalent ways of doing this, which agree on the physically relevant locus of real coordinates. Since the extension to complex coordinates is an artifact and the values on the real locus are all that matters to the computation of the integral, the result is independent of the chosen extension. However, different extensions have different properties and lead to different methods of dealing with the question of the arithmetic nature of the Feynman integral.

1.2.1. *Algebraic formulations.* Typical algebraic formulations of Feynman integrals extend the integration form from real to complex variables using a quadratic form (algebraic) instead of a Euclidean norm. For example, in momentum space amplitudes, for $n = \#E_\Gamma$, one writes the amplitude at

$$U(\Gamma) = \int \frac{\delta(\sum_{i=1}^n \epsilon_{v,i} k_i + \sum_{j=1}^N \epsilon_{v,j} p_j)}{q_1 \cdots q_n} \, d^D k_1 \cdots d^D k_n,$$

where $\epsilon_{v,i}$ and $\epsilon_{v,j}$ are incidence matrices for internal and external edges, that express the linear constraints of momentum conservation at vertices, and the denominator is written as a product of quadratic form (in the case with non-zero mass m)

$$q_e(k_e) = \sum_{j=1}^{D} k_{e,j}^2 + m^2.$$

In configuration space, with mass $m = 0$, the analogous approach would correspond to an integral, for $N = \#V_\Gamma$,

$$U(\Gamma) = \int \frac{1}{Q_1 \cdots Q_n} d^D x_{v_1} \cdots d^D x_{v_N},$$

where the quadratic forms in the denominator are given by

$$Q_e(x_{s(e)}, x_{t(e)}) = \sum_{j=1}^{D} (x_{s(e),j} - x_{t(e),j})^2.$$

This point of view for configuration space integrals was discussed in [**36**].

The advantage of this algebraic approach to extending the form from real to complex coordinates is that one directly deals with an algebraic differential form. Thus, the interpretation of the Feynman integral as a period (modulo the convergence issue), is completely transparent, as it is manifestly written as an integral of an algebraic differential form on a locus defined by algebraic equations inside an algebraic variety.

The disadvantage of this method is that the singularities of the algebraic differential form occur along a hypersurface defined by the vanishing of the product of quadratic forms $Q_1 \cdots Q_n$, and typically the motive of this hypersurface is difficult to control.

1.2.2. *Non-algebraic formulation.* The non-algebraic formulation, which is our main focus in the present paper, extends the form from real to complex coordinates using a Euclidean norm. The resulting form

$$\omega_\Gamma = \prod_{e \in E_\Gamma} \frac{1}{\|x_{s(e)} - x_{t(e)}\|^{2\lambda}} \bigwedge_{v \in V_\Gamma} dx_v$$

is manifestly not an algebraic differential form. It is a smooth $\mathcal{C}^\infty$-form on the complement of the diagonals Δ_e, where it is singular.

The advantage of this approach is that now the locus of singularities of the form is not a hypersurface, but a union of diagonals, which, as we will see more in detail, is much easier to control in geometric and motivic terms.

The obvious disadvantage lies in the fact that, since the form is not algebraic, one cannot directly describe the integral as a period, and one should either use cohomological methods for replacing the form with an algebraic one in the same cohomology class (closed form case) or proceed via some explicit computation that identifies the result as a period by direct inspection (non-closed form case).

1.3. Structure of the paper. Section 2 gives a quick overview of the questions and methods developed to investigate the relation between Feynman integrals and motives from different viewpoints.

Section 3 introduces algebraic varieties $X^{\mathbf{V}_\Gamma}$ and $Z^{\mathbf{V}_\Gamma}$ that describe configuration spaces associated to a Feynman graph Γ, and smooth differential forms on these varieties, with singularities along diagonals, that extend the Feynman rules and Feynman propagators in configuration spaces to varieties containing a dense affine space that represents the "physical space" on which the Feynman diagram integration lives.

We consider three possible variants of the geometry, and of the corresponding differential form, respectively given by the forms (3.2), (3.9) and (3.12), because this will allow us to present different methods of regularization and integration and show, in different ways, how the relation between Feynman integrals and periods of mixed Tate motives arises in the setting of configuration spaces.

These include the "real case" and "complexified case" discussed above, as well as another variant that is related to some interesting integral kernels we will discuss in Section 8.

The geometric setting builds upon our previous work [**36**], and we will be referring to that source for several of the algebro-geometric arguments we need to use here. In terms of the Feynman amplitudes themselves, the main difference between the approach followed in this paper and the one of [**36**], as we mentioned above, is in the non-algebraic nature of the differential form. In our previous work [**36**] we extended the Feynman propagator to the configuration space $X^{\mathbf{V}_\Gamma}$ as an algebraic differential form, which then had singularities not only along the diagonals, but also along a quadric Z_Γ (the configuration space analog of the graph hypersurfaces describing singularities of Feynman amplitudes in momentum space). In this paper, we extend the Feynman propagator to a $\mathcal{C}^\infty$ (non-algebraic) differential form on $X^{\mathbf{V}_\Gamma}$, which then has singularities only along the diagonals.

In the real case, we will deal with the resulting integrals through analytic methods, and identify a term that is expressible through multiple zeta values. In the complexified case, we will appeal to cohomological arguments to replace the form with an algebraic one. In this second case, we will also use the fact that, having singularities only along diagonals allows us to keep a good control over the motive, which does not leave the mixed Tate class.

Section 4 deals with the "real case" (the physical Feynman amplitude). We consider the first form (3.2) as the Feynman amplitude and its restriction to the affine space $\mathbb{A}^{D\mathbf{V}_\Gamma}(\mathbb{R})$ in the real locus in the graph configuration space $X^{\mathbf{V}_\Gamma}$, with $X = \mathbb{P}^D(\mathbb{C})$. We first describe a setting generalizing a recent construction, by Spencer Bloch, of cycles in the configuration spaces $X^{\mathbf{V}_\Gamma}$. Our setting assigns a middle-dimensional relative cycle, with boundary on the union of diagonals, to any acyclic orientation on the graph. We then consider the development of the propagator into Gegenbauer orthogonal polynomials. This is a technique for the calculation of Feynman diagrams in configuration spaces (known as x-space technique) that is well established in the physics literature since at least the early '80s, [**38**]. We split these x-space integrals into contributions parameterized by the chains described above. The decompositions of the chains of integration, induces a corresponding decomposition of the Feynman integral into an angular and a radial part. In the case of dimension $D = 4$, and for polygon graphs these simplify into integrals over simplexes of polylogarithm functions, which give rise to zeta values. For more general graphs, each star of a vertex contributes a certain combination of integrals of triple integrals of spherical harmonics. These can be expressed in terms of isoscalar coefficients. We compute explicitly, in the case $D = 4$ the top term of this expansion, and show that, when pairs of half edges are joined together to form an edge of the graph, one obtains a combination of nested sums that can be expressed in terms of Mordell–Tornheim and Apostol–Vu series. This result means that, even when the overall Feynman amplitude would contain non-mixed Tate periods, one can isolate a contribution, which is defined in representation theoretic terms, which remains a mixed Tate period.

Starting with §5, we consider what we called the "complexified case". Here we deal with a different integration form, given by (3.9), which is a generalization of the Feynman amplitude previously considered. This is defined after performing a doubling of the dimension of the relevant configuration spaces $Z^{\mathbf{V}_\Gamma} \simeq (X \times X)^{\mathbf{V}_\Gamma}$. The advantage of passing to this formulation is that one is then dealing with a closed form, unlike the case of (3.2) considered in the previous section, hence cohomological arguments become available. We first describe the simple modifications to the geometry of configuration spaces, with respect to the results of our previous work [**36**], which are needed in order to deal with this doubling of dimension. We introduce the wonderful compactification $F(X,\Gamma)$ of the configuration space $Z^{\mathbf{V}_\Gamma}$, which works pretty much as in the case of the wonderful compactification $\overline{\mathrm{Conf}_\Gamma(X)}$ of $X^{\mathbf{V}_\Gamma}$ described in [**36**], [**61**], [**62**], with suitable modifications. The purpose of passing to the wonderful compactification is to realize the complement of the diagonals in the configuration space explicitly as the complement of a union of divisors intersecting transversely, so that we can describe de Rham cohomology classes in terms of representatives that are algebraic forms with logarithmic poles along the divisors. We also discuss, in this section, the properties of the motive of the wonderful compactification and we verify that, if the underlying variety X is a Tate motive, the wonderful compactification also is, and so are the unions and intersections of the divisors obtained in the construction. We also describe iterated Poincaré residues of these forms along the intersections of divisors, according to the general theory of iterated Poincaré residues of forms with logarithmic poles, [**2**], [**3**], [**42**], [**52**].

In §6, we consider a first regularization method for the Feynman integral based on the amplitude (3.9), which is essentially a cutoff regularization and is expressed in terms of principal-value currents and the theory of Coleff–Herrera residue currents, [**18**], [**39**]. We show that, when we regularize the Feynman integral as a principal value current, the ambiguity is expressed by residue currents supported on the intersections of the divisors of the wonderful compactification, which are related to the iterated Poincaré residues described in §5. In particular, when evaluated on algebraic test differential forms on these intersections, the currents describing the ambiguities take values that are periods of mixed Tate motives.

In §7, we introduce a more directly algebro-geometric method of regularization of both the Feynman amplitude form (3.9) and of the chain of integration, which is based on the *deformation to the normal cone* [**46**], which we use to separate the chain of integration from the locus of divergences of the form. We check again that the motive of the deformation space constructed using the deformation to the normal cone remains mixed Tate, and we show once again that the regularized Feynman integral obtained in this way gives rise to a period.

Section 8 is an Appendix, where we present additional material that can be read independently of the rest of the paper (or can be skipped by readers more directly interested in the main results on periods). Its main purpose is to explain the relation between the Feynman amplitudes (3.2), (3.9) and (3.12) and real and complex

Green functions and the Bochner–Martinelli integral kernel operator. In particular, we show that the form of the Feynman amplitude leads naturally to a Bochner–Martinelli type theorem, adapted to the geometry of graph configuration spaces, which combines the usual Bochner–Martinelli theorem on complex manifolds [**53**] and the notion of harmonic functions and Laplacians on graphs.

2. Motives and Periods in Feynman Integrals

This introductory section was added at the request of the volume editors as a short overview of the occurrence of motives and periods in quantum field theory.

The empirical observation of the pervasive presence of multiple zeta values in Feynman integral of massless scalar field theories illustrated in the extensive analysis carried out by Broadhurst and Kreimer [**28**] suggested a possible deeper relation between periods and Feynman integrals. Indeed, it was known by [**51**] that multiple zeta values can be realized as periods of motives (moduli spaces of curves of genus zero) that are mixed Tate, and it was conjectured and later proved by Brown [**29**] that indeed all periods of mixed Tate motives over $\mathbb{Z}$ are $\mathbb{Q}[(2\pi i)^{-1}]$-linear combinations of multiple zeta values.

At the level of formal categorical structures, the work of Connes and the second author [**40**] (see also Chapter 1 of [**41**]), showed that the divergences of a scalar field theory can be organized into a Tannakian category of irregular-singular differential systems, whose Tannakian Galois group has the same kind of pro-unipotent structure as the motivic Galois groups of mixed Tate motives. While this approach is also suggestive of a direct relation between quantum field theory and motives, it does not provide a way to associate a specific motive to a specific Feynman integral.

In order to analyze motives and periods arising from individual Feynman integrals, one can consider different ways of writing the (possibly divergent) integrals associated to individual Feynman graphs. For a (massive) Euclidean scalar field theory in dimension D, the unrenormalized Feynman integral of a Feynman graph is given, up to a divergent Γ-factor, by the expression

$$C \int \frac{\prod_{v\in V(\Gamma)} \delta(\sum_{e\in E_{int}(\Gamma)} \epsilon_{v,e} k_e + \sum_{j\in E_{ext}(\Gamma)} \epsilon_{v,j} p_j)}{\prod_{e\in E_{int}(\Gamma)} q_e(k_e)} \prod_{e\in E_{int}(\Gamma)} \frac{d^D k_e}{(2\pi)^D}$$

where $q_e(k_e)$ is a quadratic form in the momentum variables k_e and the mass parameters m_e associated to the internal edges $e \in E_{int}(\Gamma)$,

$$q_e(k_e) = \sum_{i=1}^{D} k_{e,i}^2 + m_e^2.$$

The constant is given by

$$C = \prod_v \lambda_v (2\pi)^{-D}$$

with λ_v the coupling constant of the $\phi^{\iota(v)}$ term in the Lagrangian, with $\iota(v)$ the valence of the vertex v, and where p_j are the momenta associated to the external edges $e_j \in E_{ext}(\Gamma)$ of the graph Γ. The matrix $(\epsilon_{v,e})_{v\in V(\Gamma), e\in E_{int}(\Gamma)}$ is the incidence matrix with entries ± 1 or 0 according to whether $v = s(e)$, or $v = t(e)$ or $v \notin \partial(e)$, and similarly for $\epsilon_{v,j}$, with e_j running over the external edges. The explicit derivation of the perturbative expansion with this form of the Feynman integrals from the Lagrangian of the quantum field theory and the formal infinite dimensional integral is recalled briefly in the first chapter of [**65**], or in the first chapter of [**41**], see also standard quantum field theory textbooks like [**20**], [**57**].

Under the assumption that the parameters p_j and λ_v are rational numbers, the integral above (if convergent) could be seen as a period of the complement of a union of quadrics. It is important to notice that, in general, in the *massive* case with $m = (m_e) \neq 0$, one does not expect to find mixed Tate motives even for very simple graphs like the sunset graph and other so called banana graphs (two vertices with a set of parallel edges between them). Indeed, it was shown in [**23**], [**23**], [**85**] that non-mixed-Tate motives and periods occur in the case of the sunset and the 3-banana graph. This massive case should be contrasted with the massless case, where the banana graphs provide a very simple example where one can compute explicitly the motive (at least at the level of classes in the Grothendieck ring) and see easily that it is mixed Tate, [**6**].

Indeed the case of the massless Feynman integrals is significantly different from the massive case, in terms of their motivic properties and one encounters mixed-Tate motives for all graphs up to a significant size. The integral computation, in the massless case, can be reformulated in Feynman's parametric form (see [**20**], [**71**], and also Chapter 3 of [**65**]). When rewritten in terms of Feynman parameters, the Feynman integrals for a massless field theory take the form

$$C \int_{\sigma_n} \frac{P_\Gamma(t,p)^{-n+D\ell/2}}{\Psi_\Gamma(t)^{-n+D(\ell+1)/2}} \omega_n$$

where $C \in \mathbb{Q}$ is a constant, p are the external momenta, $n = \#E_{int}(\Gamma)$ and $\ell = b_1(\Gamma)$, with locus of integration σ_n the n-simplex, and volume form ω_n. The integrand is a rational function with polynomials Ψ_Γ and P_Γ consisting of the Kirchhoff (or first Symanzik) polynomial

$$\Psi_\Gamma(t) = \sum_{T \subset \Gamma} \prod_{e \notin T} t_e$$

and the second Symanzik polynomial

$$P_\Gamma(t,p) = \sum_{C \subset \Gamma} s_C \prod_{s \in C} t_e$$

with the sum over cut-sets C of Γ (collections of $b_1(\Gamma) + 1$ edges consisting of the complement of a spanning tree plus one edge), with s_C a quadratic function of the external momenta. In the case where all the parameters are rational and $-n + D\ell/2 \geq 0$, the integral (if convergent) would be identified with a period of the complement of the graph hypersurface $X_\Gamma = \{\Psi_\Gamma = 0\}$.

Kontsevich conjectured that the motives of the graph hypersurfaces X_Γ would always be mixed Tate, which would explain the occurrence of multiple zeta values. This conjecture was disproved by Belkale and Brosnan [**16**], who showed that the graph hypersurfaces generate the localization of the Grothendieck ring at the multiplicative subset S generated by $\mathbb{L}^n - \mathbb{L}$, where $\mathbb{L} = [\mathbb{A}^1]$ is the Lefschetz motive, as a module over the localization $S^{-1}\mathbb{Z}[\mathbb{L}]$. This implies that, for sufficiently large graph, not only the classes of the graph hypersurfaces will leave the mixed Tate world but they will become arbitrarily complicated as motives. This can be contrasted with the remarkably simple behavior of the Grothendieck classes of the graph hypersurfaces in the quotient of the Grothendieck ring obtained by killing the Lefschetz motive, as described in [**11**].

Even though the general result of [**16**] shows that the classes of the graph hypersurfaces are not always mixed-Tate, explicit examples are difficult to come by and only occur for graphs that are considered to be very large from the quantum field theory perspective. Indeed, it was first proved by Stembridge [**81**] that all graphs up to 12 edges contribute mixed Tate motives, and the first explicit non-mixed-Tate examples, found by Doryn [**44**] and Schnetz [**75**] occur only at 14 and 16 edges.

Most of the results on the relation between quantum fields and motives have been obtained so far using the Feynman parametric form of the integrals and studying the graph hypersurfaces, and the problem of divergences has been approached in terms of blowups and deformations of the graph hypersurface, see [**22**], [**24**], [**66**]. Other methods involved mapping the integrals to the complement of determinant hypersurfaces and their compactifications, [**7**], [**67**]. In all of these cases, one is dealing with the formulation of Feynman integrals in momentum variables. In this paper instead, following the approach developed in [**36**], [**37**], we consider instead Feynman integrals written in terms of position variables in configuration spaces.

3. Feynman amplitudes in configuration spaces and complexifications

In the following we let X be a D-dimensional smooth projective variety, which contains a dense subset isomorphic to an affine space $\mathbb{A}^D$, whose set of real points $\mathbb{A}^D(\mathbb{R})$ we identify with Euclidean D-dimensional spacetime. For instance, we can take $X = \mathbb{P}^D$. We assume that D is even and we write $D = 2\lambda + 2$.

3.1. Feynman graphs.

A *graph* Γ is a one-dimensional finite CW-complex. We denote the set of vertices of Γ by V_Γ, the set of edges by E_Γ, and the boundary map by $\partial_\Gamma : E_\Gamma \to (V_\Gamma)^2$.

When an orientation is chosen on Γ, we write $\partial_\Gamma(e) = \{s(e), t(e)\}$, the source and target vertices of the oriented edge e.

A *looping* edge is an edge for which the two boundary vertices coincide and *multiple edges* are edges between the same pair of vertices. We assume that our graphs have no looping edges.

3.1.1. *Induced subgraphs and quotient graphs.* A subgraph $\gamma \subseteq \Gamma$ is called an *induced subgraph* if its set of edges E_γ is equal to $\partial_\Gamma^{-1}((V_\gamma)^2)$, that is, γ is determined by a choice of a set of vertices $V_\gamma \subseteq V_\Gamma$ with all the edges of Γ on the same set of vertices V_γ. We will denote by $\mathbf{SG}(\Gamma)$ the set of all induced subgraphs of Γ and by

$$\mathbf{SG}_k(\Gamma) = \{\gamma \in \mathbf{SG}(\Gamma) \mid |V_\gamma| = k\}, \tag{3.1}$$

the subset $\mathbf{SG}_k(\Gamma) \subseteq \mathbf{SG}(\Gamma)$ of all induced subgraphs with k vertices.

For $\gamma \in \mathbf{SG}(\Gamma)$, we denote by $\Gamma//\gamma$ the graph obtained from Γ by shrinking each connected component of γ to a separate vertex. The quotient graph $\Gamma//\gamma$ does not have looping edges since we consider only induced subgraphs.

3.2. Feynman amplitude.

When computing Feynman integrals in configuration space, one considers singular differential forms on X^{V_Γ}, integrated on the real locus of this space.

DEFINITION 3.1. The Euclidean massless Feynman amplitude in configuration space is given by the differential form

$$\omega_\Gamma = \prod_{e \in E_\Gamma} \frac{1}{\|x_{s(e)} - x_{t(e)}\|^{2\lambda}} \bigwedge_{v \in V_\Gamma} dx_v. \tag{3.2}$$

The form (3.2) defines a $\mathcal{C}^\infty$-differential form on the configuration space

$$\mathrm{Conf}_\Gamma(X) = X^{V_\Gamma} \smallsetminus \cup_{e\in E_\Gamma} \Delta_e, \tag{3.3}$$

with singularities along the diagonals

$$\Delta_e = \{(x_v \,|\, v \in V_\Gamma) \mid x_{s(e)} = x_{t(e)}\}. \tag{3.4}$$

REMARK 3.2. The diagonals (3.4) corresponding to edges between the same pair of vertices are the same, consistently with the fact that the notion of degeneration that defines the diagonals is based on “collisions of points” and not on contraction of the edges connecting them. This suggests that we may want to exclude graphs with multiple edges. However, multiple edges play a role in the definition of Feynman amplitudes (see Definition 3.1 above and §4 of [**36**]) hence we allow this possibility. On the other hand, the definition of configuration space is void in the presence of looping edges, since the diagonal Δ_e associated to a looping edge is the whole space X^{V_Γ}, and its complement is empty. Thus, our assumption that graphs have no looping edges is aimed at excluding this geometrically trivial case. Looping edges can also occur in Feynman graphs, for instance in tadpole graphs, but the geometry of the configuration spaces $\mathrm{Conf}_\Gamma(X)$ is just not suitable for treating these cases.

REMARK 3.3. As we already discussed in the Introduction, our choice of the (non-algebraic) Euclidean norm $\|x_{s(e)} - x_{t(e)}\|^2$ in the Feynman propagator differs from the customary choice of propagator (see for instance [**36**]) where the algebraic quadratic form $(x_{s(e)} - x_{t(e)})^2$ is used instead, but these two expressions agree on the locus $X^{V_\Gamma}(\mathbb{R})$ of real points, which is the chain of integration of the Feynman amplitude. The latter choice gives a manifestly algebraic differential form, but at the cost of introducing a hypersurface $\mathcal{Z}_\Gamma$ in X^{V_Γ} where the singularities of the form occur, which makes it difficult to control explicitly the nature of the motive. Our choice here only gives a $\mathcal{C}^\infty$ differential form, but the domain of definition is now simply $\mathrm{Conf}_\Gamma(X)$, whose geometry is much easier to understand than that of $\mathrm{Conf}_\Gamma(X) \smallsetminus \mathcal{Z}_\Gamma$, see [**36**].

Formally (before considering the issue of divergences), the Feynman integral is obtained by integrating the form (3.2) on the locus of real points of the configuration space.

DEFINITION 3.4. The chain of integration for the Feynman amplitude is taken to be the set of real points of this configuration space,

$$\sigma_\Gamma = X(\mathbb{R})^{V_\Gamma}. \tag{3.5}$$

The form (3.2) defines a top form on σ_Γ. There are two sources of divergences in integrating the form (3.2) on σ_Γ: the intersection between the chain of integration and the locus of divergences of ω_Γ,

$$\sigma_\Gamma \cap \cup_e \Delta_e = \cup_e \Delta_e(\mathbb{R})$$

and the behavior at infinity, on $\Delta_\infty := X \smallsetminus \mathbb{A}^D$. In physics terminology, these correspond, respectively, to the *ultraviolet* and *infrared* divergences.

We will address these issues in §4 below.

3.3. Complexification and other variations upon a theme. In addition to the form (3.2) considered above, we will also consider other variants, which will allow us to discuss different possible methods to address the question of periods and in particular the occurrence of multiple zeta values in generalizations of Feynman integrals in configuration spaces. As mentioned in the Introduction, comparing two apparently very similar cases of the Feynman amplitude (3.2) and of the complexified version (3.9) will allow us to illustrate how these lead to the use of very different mathematical methods.

3.3.1. *Complexified case.* In this setting, instead of the configuration space X^{V_Γ} and its locus of real points $\sigma = X^{V_\Gamma}(\mathbb{R})$, we will work with a slightly different space, obtained as follows.

As above, let X be a smooth projective variety, and Z denote the product $X \times X$. Let $p : Z \to X$, $p : z = (x, y) \mapsto x$ be the projection. This trivial doubling of the space X and projection are introduced purely as a simple artifact to ensure that (as in the real case) the differential form and locus of integration we will be considering in this complexified setting will still be middle-dimensional. In essence, the "complexification" consists of the fact that we will be replacing real affine spaces $\mathbb{A}^D(\mathbb{R})$ in the chain of integration with complex $\mathbb{A}^D(\mathbb{C})$.

Given a graph Γ, the *configuration space* $F(X, \Gamma)$ of Γ in Z is the complement

$$Z^{V_\Gamma} \setminus \bigcup_{e \in E_\Gamma} \Delta_e^{(Z)} \cong (X \times X)^{V_\Gamma} \setminus \bigcup_{e \in E_\Gamma} \Delta_e^{(Z)}, \tag{3.6}$$

in the cartesian product $Z^{V_\Gamma} = \{(z_v \mid v \in V_\Gamma)\}$ of the *diagonals associated to the edges* of Γ,

$$\Delta_e^{(Z)} \cong \{(z_v \mid v \in V_\Gamma) \in Z^{V_\Gamma} \mid p(z_{s(e)}) = p(z_{t(e)})\}. \tag{3.7}$$

The relation between the configuration space $F(X, \Gamma)$ and the previously considered $\mathrm{Conf}_\Gamma(X)$ of (3.3) is described by the following.

LEMMA 3.5. *The configuration space $F(X, \Gamma)$ is isomorphic to*

$$F(X, \Gamma) \simeq \mathrm{Conf}_\Gamma(X) \times X^{V_\Gamma}, \tag{3.8}$$

and the diagonals (3.7) *and related to those of* (3.4) *by* $\Delta_e^{(Z)} \cong \Delta_e \times X^{V_\Gamma}$.

PROOF. The map realizing the identification (3.8) is induced by the identification $Z = X \times X$ through (3.6), so that

$$F(X, \Gamma) = Z^{V_\Gamma} \setminus \bigcup_{e \in E_\Gamma} \Delta_e^{(Z)}$$

$$= (X \times X)^{V_\Gamma} \setminus \bigcup_{e \in E_\Gamma} \Delta_e^{(Z)} = (X^{V_\Gamma} \smallsetminus \bigcup_{e \in E_\Gamma} \Delta_e) \times X^{V_\Gamma},$$

since $\Delta_e^{(Z)} = \Delta_e \times X^{V_\Gamma}$. □

3.3.2. *Amplitudes in the complexified case.* We assume that the smooth projective variety Z contains a dense open set isomorphic to affine space $\mathbb{A}^{2D}$, with coordinates $z = (x_1, \ldots, x_D, y_1, \ldots, y_D)$.

In principle, our setting could also be generalized by just requiring that the variety X (hence $Z = X \times X$) contain a dense open subset Ω that is isomorphic to an open set in an affine space $\mathbb{A}^D$, that is, by considering the case of a rational X. However, in this more general case one will have to consider Green functions for the Laplacian on the domain Ω, for a suitable elliptic boundary value problem, and those will in general be less explicitly accessible than the Green functions on the whole affine space $\mathbb{A}^D$.

DEFINITION 3.6. Given a Feynman Γ with no looping edges, we define the corresponding Feynman amplitude (weight) as

$$\omega_\Gamma^{(Z)} = \prod_{e\in E_\Gamma} \frac{1}{\|x_{s(e)} - x_{t(e)}\|^{2D-2}} \bigwedge_{v\in V_\Gamma} dx_v \wedge d\bar{x}_v, \tag{3.9}$$

where $\|x_{s(e)} - x_{t(e)}\| = \|p(z)_{s(e)} - p(z)_{t(e)}\|$ and where the differential forms dx_v and $d\bar{x}_v$ denote, respectively, the holomorphic volume form $dx_{v,1} \wedge \cdots \wedge dx_{v,D}$ and its conjugate. The chain of integration is given, in this case, by the range of the projection

$$\sigma^{(Z,y)} = X^{V_\Gamma} \times \{y = (y_v)\} \subset Z^{V_\Gamma} = X^{V_\Gamma} \times X^{V_\Gamma}, \tag{3.10}$$

for a fixed choice of a point $y = (y_v \,|\, v \in V_\Gamma)$.

The form (3.9) is a closed $\mathcal{C}^\infty$ differential form on $F(X,\Gamma)$ of degree $2\dim_{\mathbb{C}} X^{V_\Gamma} = \dim_{\mathbb{C}} Z^{V_\Gamma} = 2DN$, with $N = \#V_\Gamma$, hence it defines a cohomology class in the group $H^{2DN}(F(X,\Gamma))$, and it gives a top form on the locus $\sigma^{(Z,y)}$.

In this case, the divergences of $\int_{\sigma^{(Z,y)}} \omega_\Gamma^{(Z)}$ are coming from the union of the diagonals $\cup_{e\in E_\Gamma}\Delta_e^{(Z)}$ and from the divisor at infinity

$$\Delta_{\infty,\Gamma}^{(Z)} := \sigma^{(Z,y)} \setminus \mathbb{A}^{DN}(\mathbb{C}) \times \{y\}. \tag{3.11}$$

with $N = \#V_\Gamma$. We will address the behavior of the integrand near the loci $\cup_{e\in \mathbf{E}_\Gamma}\Delta_e^{(Z)}$ and $\Delta_{\infty,\Gamma}^{(Z)}$ and the appropriate regularization of the Feynman amplitude and the chain of integration in §5 below. When convenient, we will choose coordinates so as to identify the affine space $\mathbb{A}^{DN}(\mathbb{C}) \times \{y\} \subset \sigma^{(Z,y)}$ with a real affine space $\mathbb{A}^{2DN}(\mathbb{R})$, with $N = \#V_\Gamma$.

3.3.3. *Complexified amplitudes and complex Green forms.* A variant of the amplitude $\omega_\Gamma^{(Z)}$ of (3.9), which we will discuss briefly in §8, is related to the complex Green forms. In this setting, instead of (3.9) one considers the closely related form

$$\hat{\omega}_\Gamma = \prod_{e\in E_\Gamma} \frac{1}{\|x_{s(e)} - x_{t(e)}\|^{2D-2}} \bigwedge_{v\in V_\Gamma} \sum_{k=1}^{D} (-1)^{k-1} dx_{v,[k]} \wedge d\bar{x}_{v,[k]}, \tag{3.12}$$

where the forms $dx_{v,[k]}$ and $d\bar{x}_{v,[k]}$ denote

$$dx_{v,[k]} = dx_{v,1} \wedge \cdots \wedge \widehat{dx_{v,k}} \wedge \cdots \wedge dx_{v,D},$$

$$d\bar{x}_{v,[k]} = d\bar{x}_{v,1} \wedge \cdots \wedge \widehat{d\bar{x}_{v,k}} \wedge \cdots \wedge d\bar{x}_{v,D},$$

respectively, with the factor $dx_{v,k}$ and $d\bar{x}_{v,k}$ removed.

Notice how, unlike the form considered in (3.9), which is defined on the affine $\mathbb{A}^{2DN} \subset Z^{V_\Gamma}$, with $N = \#V_\Gamma$, the form (3.12) has the same degree of homogeneity $2D-2$ in the numerator and denominator, when the graph Γ has no multiple edges, hence it is invariant under rescaling of the coordinates by a common non-zero scalar factor.

3.3.4. *Distributional interpretation.* In the cases discussed above, the amplitudes defined by (3.2) and (3.9) can be given a distributional interpretation, as a pairing of a distribution

$$\prod_{e\in E_\Gamma} \frac{1}{\|x_{s(e)}-x_{t(e)}\|^\alpha}, \tag{3.13}$$

where α is either $2\lambda = D-2$ or $2D-2$, or $2D$, and test forms given in the various cases, respectively, by

(1) forms $\phi(x_v)\bigwedge_{v\in V_\Gamma} dx_v$, where ϕ is a test function in $\mathcal{C}^\infty(\mathbb{A}^{DN}(\mathbb{R}))$, with $N=\#V_\Gamma$;
(2) forms $\phi(z_v)\bigwedge_{v\in V_\Gamma} dx_v\wedge d\bar{x}_v$, with $\phi(z_v)=\phi(p(z_v))=\phi(x_v)$ a test function in $\mathcal{C}^\infty(X^{V_\Gamma}(\mathbb{C}))$.

This distributional interpretation will be used later in the paper, when we introduce Principal Value Currents, see Definition 6.6.

3.3.5. *Relation between complexified amplitudes.* We have introduced here above two simple mathematical generalization of the Feynman amplitude (3.2), respectively given by (3.9) and (3.12). These are related by a simple operation.

Let τ_Γ denote the form

$$\tau_\Gamma = \bigwedge_{v\in\mathbf{V}_\Gamma}\left(\sum_{k=1}^{D}(-1)^{D+k}dx_{v,k}\wedge d\bar{x}_{v,k}\right). \tag{3.14}$$

LEMMA 3.7. *The forms $\omega_\Gamma^{(Z)}$ and $\hat{\omega}_\Gamma$ of (3.9) and (3.12) are related by*

$$\omega_\Gamma^{(Z)} = \tau_\Gamma\wedge\hat{\omega}_\Gamma. \tag{3.15}$$

PROOF. We have

$$\bigwedge_{w\in\mathbf{V}_\Gamma}\left(\sum_{j=1}^{D}(-1)^{D-1}(-1)^{j-1}dx_{w,j}\wedge d\bar{x}_{w,j}\right)\wedge\bigwedge_{v\in\mathbf{V}_\Gamma}\left(\sum_{k=1}^{D}(-1)^{k-1}dx_{v,[k]}\wedge d\bar{x}_{v,[k]}\right)$$
$$=\bigwedge_{v\in\mathbf{V}_\Gamma} dx_v\wedge d\bar{x}_v.$$

□

3.4. Amplitudes and Fourier transform. We expand here, briefly, on the comment made in the Introduction regarding amplitudes in momentum and configuration space and Fourier duality between the propagators. This was discussed in greater detail in [**37**].

In real momentum space, with momentum variables $k_e\in\mathbb{A}^D(\mathbb{R})$, for each edge $e\in E_\Gamma$, the form of the Feynman amplitude is a product of edge-propagators

$$\frac{1}{(m^2+\|k_e\|^2)}.$$

In configuration space, as we have seen above, one considers in (3.2) a product of propagators of the form

$$G_{0,\mathbb{R}}(x_{s(e)}-x_{t(e)}) := \frac{1}{\|x_{s(e)}-x_{t(e)}\|^{2\lambda}}, \quad \text{with} \quad D=2\lambda+2, \tag{3.16}$$

in the massless case. The massive case (which we will not consider in this paper) has a somewhat more complicated expression

$$G_{m,\mathbb{R}}(x_{s(e)} - x_{t(e)}) := \frac{m^\lambda}{(2\pi)^{(\lambda+1)}} \|x_{s(e)} - x_{t(e)}\|^{-\lambda} \mathcal{K}_\lambda(m\|x_{s(e)} - x_{t(e)}\|),$$

with $\mathcal{K}_\nu(z)$ the modified Bessel function.

The expression $G_{0,\mathbb{R}}(x_{s(e)} - x_{t(e)})$ of (3.16) is the Green function of Laplacian on $\mathbb{A}^D(\mathbb{R})$, with $D = 2\lambda + 2$:

$$G_{\mathbb{R}}(x, y) = \frac{1}{\|x - y\|^{2\lambda}}$$

Similarly the expression $G_{m,\mathbb{R}}(x_{s(e)} - x_{t(e)})$ above is fundamental solution of Helmholtz equation $(\Delta + m^2)G = \delta$.

For test functions $\varphi \in \mathcal{S}(\mathbb{R}^D)$, Fourier transform of convolution with the Green function gives

$$\widehat{(G_{0,\mathbb{R}} \star \varphi)}(k) = \frac{4\pi^{D/2}}{\Gamma(\lambda)} \frac{1}{\|k\|^2} \widehat{\varphi}(k)$$
$$\widehat{(G_{m,\mathbb{R},} \star \varphi)}(k) = \frac{1}{(m^2 + \|k\|^2)} \widehat{\varphi}(k).$$

This shows that, in the real case, the amplitude (3.2) we are considering is indeed the dual picture, under Fourier transform, of the momentum space Feynman integral.

When we pass from the Feynman amplitude (3.2) to the complexified amplitude (3.9), we replace the real Laplacian on $\mathbb{A}^D(\mathbb{R})$ with the complex Laplacian on $\mathbb{A}^D(\mathbb{C})$,

$$\Delta = \sum_{k=1}^{D} \frac{\partial^2}{\partial x_k \partial \bar{x}_k}.$$

The Green function of the complex Laplacian is given by

$$G_{\mathbb{C}}(x, y) = \frac{-(D-2)!}{(2\pi i)^D \|x - y\|^{2D-2}}, \tag{3.17}$$

which is the expression that replaces (3.16) when passing from (3.2) to (3.9). This shows why (3.9) can be regarded as a natural generalization of (3.2).

4. Feynman amplitudes: integration over the real locus

In this section we consider the physical Feynman amplitudes ω_Γ, defined as in (3.2), and the domain σ_Γ defined in (3.5). We give an explicit formulation of the integral in terms of an expansion of the real Green functions $\|x_{s(e)} - x_{t(e)}\|^{-2\lambda}$ in Gegenbauer polynomials, based on a technique well known to physicists (the x-space method, see [**38**]). We consider the integrand restricted to the real locus $X(\mathbb{R})^{\mathbf{V}_\Gamma}$, and express it in polar coordinates, separating out an angular integral and a radial integral. We identify a natural subdivision of the domain of integration into chains that are indexed by acyclic orientations of the graph. In the special case of dimension $D = 4$, we express the integrand in terms closely related to multiple polylogarithm functions.

Spencer Bloch recently introduced a construction of cycles in the relative homology $H_*(X^{\mathbf{V}_\Gamma}, \cup_e \Delta_e)$ of the graph configuration spaces, that explicitly yield multiple zeta values as periods [**21**]. We use here a variant of his construction, which will have a natural interpretation in terms of the x-space method for the computation of the Feynman amplitudes in configuration spaces.

4.0.1. *Directed acyclic graph structures.* We recall some basic facts about directed acyclic orientations on graphs.

DEFINITION 4.1. Let Γ be a finite graph without looping edges. Let $\Omega(\Gamma)$ denote the set of edge orientations on Γ such that the resulting directed graph is a directed acyclic graph.

It is well known that all finite graphs without looping edges admit such orientations. In fact, the number of possible orientations that give it the structure of a directed acyclic graph are given by $(-1)^{\mathbf{V}_\Gamma} P_\Gamma(-1)$, where $P_\Gamma(t)$ is the chromatic polynomial of the graph Γ, see [**79**].

The following facts about directed acyclic graphs are also well known (see for instance §2.6 of [**58**]), and we recall them here for later use.

- Each orientation $\mathbf{o} \in \Omega(\Gamma)$ determines a partial ordering on the vertices of the graph Γ, by setting $w \geq_{\mathbf{o}} v$ whenever there is an oriented path of edges from v to w in the directed graph $(\Gamma, \mathbf{o})$.
- In every directed acyclic graph there is at least a vertex with no incoming edges and at least a vertex with no outgoing edges.

4.0.2. *Relative cycles from directed acyclic structures.* Given a graph Γ we consider the space $X^{\mathbf{V}_\Gamma}$. On the dense subset $\mathbb{A}^D(\mathbb{R}) = X(\mathbb{R}) \setminus \Delta_\infty(\mathbb{R})$ of the chain of integration σ_Γ, we use polar coordinates with $x_v = r_v \omega_v$, with $r_v \in \mathbb{R}_+$ and $\omega_v \in S^{D-1}$.

DEFINITION 4.2. Let $\mathbf{o} \in \Omega(\Gamma)$ be an acyclic orientation. Define the chain associated to the acyclic orientation $\mathbf{o}$ of Γ as

$$\Sigma_{\mathbf{o}} := \{(x_v) \in X^{\mathbf{V}_\Gamma}(\mathbb{R}) \,:\, r_w \geq r_v \text{ whenever } w \geq_{\mathbf{o}} v\}. \tag{4.1}$$

It is a middle dimensional relative homology class in $X^{\mathbf{V}_\Gamma}$.

Points on a diagonal $\Delta_e = \{x_{s(e)} = x_{t(e)}\}$ have $r_{s(e)} = r_{t(e)}$, hence the locus $\cup_e \Delta_e$ of singularities of the Feynman amplitudes is contained in the boundary $\partial\Sigma_{\mathbf{o}}$.

The following simple observation will be useful in the Feynman integral calculation we describe later in this section.

LEMMA 4.3. *Let $\mathbf{o} \in \Omega(\Gamma)$ be an acyclic orientation and $\Sigma_{\mathbf{o}}$ the chain defined in* (4.1). *Then $\Sigma_{\mathbf{o}} \setminus \cup_v \{r_v = 0\}$ is a bundle with fiber $(S^{D-1})^{\mathbf{V}_\Gamma}$ over a base*

$$\overline{\Sigma}_{\mathbf{o}} = \{(r_v) \in (\mathbb{R}_+^*)^{\mathbf{V}_\Gamma} \,:\, r_w \geq r_v \textit{ whenever } w \geq_{\mathbf{o}} v\}. \tag{4.2}$$

PROOF. This is immediate from the polar coordinate form $x_v = r_v \omega_v$, with $r_v \in \mathbb{R}_+^*$ and $\omega_v \in S^{D-1}$. □

4.1. Gegenbauer polynomials and angular integrals. One of the techniques developed by physicists to compute Feynman amplitudes, by passing from momentum to configuration space, relies on the expansion in Gegenbauer polynomials, see for instance [**38**] and the recent [**72**].

The Gegenbauer polynomials (or ultraspherical polynomials) are defined through the generating function

$$\frac{1}{(1-2tx+t^2)^\lambda} = \sum_{n=0}^{\infty} C_n^{(\lambda)}(x) t^n, \tag{4.3}$$

for $|t| < 1$. For $\lambda > -1/2$, they satisfy

$$\int_{-1}^{1} C_n^{(\lambda)}(x) C_m^{(\lambda)}(x)\,(1-x^2)^{\lambda-1/2} dx = \delta_{n,m} \frac{\pi 2^{1-2\lambda}\Gamma(n+2\lambda)}{n!(n+\lambda)\Gamma(\lambda)^2}. \tag{4.4}$$

We use what is known in the physics literature as the x-space method (see [**38**]) to reformulate the integration involved in the Feynman amplitude calculation in a way that involves the relative chains of Definition 4.2.

THEOREM 4.4. *In even dimension $D = 2\lambda + 2$, the integral $\int_{\sigma_\Gamma} \omega_\Gamma$ of the form (3.2) on the chain σ_Γ can be rewritten in the form*

$$\sum_{\mathbf{o}} \in \Omega(\Gamma) m_{\mathbf{o}} \int_{\Sigma_{\mathbf{o}}} \prod_{e \in \mathbf{E}_\Gamma} r_{t_{\mathbf{o}}(e)}^{-2\lambda} \left(\sum_n (\frac{r_{s_{\mathbf{o}}(e)}}{r_{t_{\mathbf{o}}(e)}})^n C_n^{(\lambda)}(\omega_{s_{\mathbf{o}}(e)} \cdot \omega_{t_{\mathbf{o}}(e)}) \right) dV, \tag{4.5}$$

for some positive integers $m_{\mathbf{o}}$, and with volume element

$$dV = \prod_v d^D x_v = \prod_v r_v^{D-1} dr_v\, d\omega_v.$$

PROOF. We write the integral in polar coordinates, with

$$d\omega = \sin^{D-2}(\phi_1)\sin^{D-3}(\phi_2)\cdots\sin(\phi_{D-2}) d\phi_1 \cdots d\phi_{D-1}$$

the volume element on the sphere S^{D-1} and $d^D x_v = r_v^{D-1} dr_v\, d\omega_v$.

In dimension $D = 2\lambda + 2$, by (4.3), the Newton potential has an expansion in Gegenbauer polynomials, so that

$$\begin{aligned} \frac{1}{\|x_{s(e)} - x_{t(e)}\|^{2\lambda}} &= \frac{1}{\rho_e^{2\lambda}(1 + (\frac{r_e}{\rho_e})^2 - 2\frac{r_e}{\rho_e}\omega_{s(e)} \cdot \omega_{t(e)})^\lambda} \\ &= \rho_e^{-2\lambda} \sum_{n=0}^{\infty} (\frac{r_e}{\rho_e})^n C_n^{(\lambda)}(\omega_{s(e)} \cdot \omega_{t(e)}), \end{aligned} \tag{4.6}$$

where $\rho_e = \max\{\|x_{s(e)}\|, \|x_{t(e)}\|\}$ and $r_e = \min\{\|x_{s(e)}\|, \|x_{t(e)}\|\}$ and with $\omega_v \in S^{D-1}$.

We can subdivide the integration into open sectors where, for each edge, either $r_{s(e)} < r_{t(e)}$ or the converse holds, so that each term $\rho_e^{-2\lambda}$ is $r_{t(e)}^{-2\lambda}$ (or $r_{s(e)}^{-2\lambda}$) and each term $(r_e/\rho_e)^n$ is $(r_{s(e)}/r_{t(e)})^n$ (or its reciprocal). In other words, let $\mathbf{b}$ denote an assignment of either $r_{s(e)} < r_{t(e)}$ or $r_{s(e)} > r_{t(e)}$ at each edge, which we write simply as $\mathbf{b}(r_{s(e)}, r_{t(e)})$, and let

$$\bar{\mathcal{R}}_{\mathbf{b}} = \{(r_v) \in (\mathbb{R}_+^*)^{\mathbf{V}_\Gamma} \,|\, \mathbf{b}(r_{s(e)}, r_{t(e)}) \text{ for } e \in \mathbf{E}_\Gamma\}$$

and $\mathcal{R}_{\mathbf{b}} = \bar{\mathcal{R}}_{\mathbf{b}} \times (S^{D-1})^{\mathbf{V}_\Gamma}$. Then we identify the domain of integration with $\cup_{\mathbf{b}} \mathcal{R}_{\mathbf{b}}$, up to a set of measure zero. The set $\mathcal{R}_{\mathbf{b}}$ is empty unless the assignment $\mathbf{b}$ defines a strict partial ordering of the vertices of Γ, in which case $\mathbf{b}$ determines an acyclic orientation $\mathbf{o} = \mathbf{o}(\mathbf{b})$ of Γ, as described in Definition 4.1. In this case, then, the chain of integration corresponding to the sector $\mathcal{R}_{\mathbf{b}}$ is the chain $\Sigma_{\mathbf{o}}$ of Definition 4.2, with $\rho_e = r_{t_{\mathbf{o}(e)}}$ and $r_e = r_{s_{\mathbf{o}(e)}}$. Thus, the domain of integration can be identified with $\cup_{\mathbf{o} \in \Omega(\Gamma)} m_{\mathbf{o}} \Sigma_{\mathbf{o}}$, where $m_{\mathbf{o}}$ is a multiplicity, taking into account the fact that different strict partial orderings may define the same acyclic orientation (for example, the cyclic graph with four vertices has 24 different possible vertex sequences and 14 acyclic orientations). □

The integral (4.5) can be approached by first considering a family of angular integrals

$$\mathcal{A}_{(n_e)_{e\in\mathbf{E}_\Gamma}} = \int_{(S^{D-1})^{\mathbf{V}_\Gamma}} \prod_e C^{(\lambda)}_{n_e}(\omega_{s(e)}\cdot\omega_{t(e)}) \prod_v d\omega_v, \tag{4.7}$$

labelled by all choices of integers n_e for $e \in \mathbf{E}_\Gamma$. The evaluation of these angular integrals will lead to an expression $\mathcal{A}_{n_e}$ in the n_e, so that one obtains a radial integral

$$\sum_{\mathbf{o}\in\Omega(\Gamma)} m_{\mathbf{o}} \int_{\bar{\Sigma}_{\mathbf{o}}} \prod_{e\in\mathbf{E}_\Gamma} \mathcal{F}(r_{s_{\mathbf{o}(e)}}, r_{t_{\mathbf{o}(e)}}) \prod_v r_v^{D-1} dr_v \tag{4.8}$$

where

$$\mathcal{F}(r_{s_{\mathbf{o}(e)}}, r_{t_{\mathbf{o}(e)}}) = r_{t_{\mathbf{o}}(e)}^{-2\lambda} \sum_{n_e} \mathcal{A}_{n_e} (\frac{r_{s_{\mathbf{o}(e)}}}{r_{t_{\mathbf{o}}(e)}})^{n_e}. \tag{4.9}$$

4.2. Polygons and polylogarithms. We first discuss the very simple example of a polygon graph, where one sees polylogarithms and zeta values arising in the expression (4.9) and its integration on the domains $\bar{\Sigma}_{\mathbf{o}}$. In the following subsections we will analyze the more general structure of these integrals for more complicated graphs.

4.2.1. *The angular integral for polygons in arbitrary dimension.* The angular integral for polygon graphs has the following explicit expression.

PROPOSITION 4.5. *Let Γ be a polygon with k edges. Then the angular integral (4.7) depends on a single variable $n \in \mathbb{N}$ and is given by*

$$\mathcal{A}_n = \left(\frac{\lambda 2\pi^{\lambda+1}}{\Gamma(\lambda+1)(n+\lambda)}\right)^k \cdot \dim \mathcal{H}_n(S^{2\lambda+1}), \tag{4.10}$$

and $\mathcal{H}_n(S^{2\lambda+1})$ is the space of harmonic functions of degree n on the sphere $S^{2\lambda+1}$.

PROOF. The angular integral, in this case, is simply given by

$$\int_{(S^{D-1})^{\mathbf{V}_\Gamma}} C^{(\lambda)}_{n_1}(\omega_{v_k}\cdot\omega_{v_1}) C^{(\lambda)}_{n_2}(\omega_{v_1}\cdot\omega_{v_2}) \cdots C^{(\lambda)}_{n_k}(\omega_{v_{k-1}}\cdot\omega_{v_k}) \prod_{v\in\mathbf{V}_\Gamma} d\omega_v,$$

which is independent of the orientation. We then use the fact that the Gegenbauer polynomials satisfy ([**14**] Vol.2, Lemma 4, §11.4)

$$\int_{S^{D-1}} C^{(\lambda)}_m(\omega_1\cdot\omega) C^{(\lambda)}_n(\omega\cdot\omega_2) d\omega = \delta_{n,m} \frac{\lambda\, Vol(S^{D-1})}{n+\lambda} C^{(\lambda)}_n(\omega_1\cdot\omega_2), \tag{4.11}$$

with $Vol(S^{D-1}) = 2\pi^{\lambda+1}/\Gamma(\lambda+1)$, for $D = 2\lambda+2$. Thus, we obtain

$$\mathcal{A}_n = \left(\frac{\lambda 2\pi^{\lambda+1}}{\Gamma(\lambda+1)}\right)^{k-1} \frac{1}{(n+\lambda)^{k-1}} \int_{S^{D-1}} C^{(\lambda)}_n(\omega\cdot\omega)\, d\omega, \tag{4.12}$$

where $n = n_1 = \cdots = n_k$ and where the remaining integral is just

$$\int_{S^{D-1}} C^{(\lambda)}_n(\omega\cdot\omega) d\omega = C^{(\lambda)}_n(1)\, Vol(S^{D-1}).$$

The value of $C_n^{(\lambda)}(1)$ can be seen using the fact that the Gegenbauer polynomials are related to the *zonal spherical harmonics* (see §4 of **[80]** and also **[49]**, **[70]**, **[87]**) $Z^{(n)}_{\omega_1}(\omega_2)$ by

$$C_n^{(\lambda)}(\omega_1 \cdot \omega_2) = c_{D,n}\, Z^{(n)}_{\omega_1}(\omega_2), \tag{4.13}$$

for $D = 2\lambda + 2$, with $\omega_1, \omega_2 \in S^{D-1}$, where the coefficient $c_{D,n}$ is given by

$$c_{D,n} = \frac{Vol(S^{D-1})\,(D-2)}{2n + D - 2}. \tag{4.14}$$

In turn, the zonal spherical harmonics are expressed in terms of an orthonormal basis $\{Y_j\}$ of the Hilbert space $\mathcal{H}_n(S^{D-1})$ of spherical harmonics on S^{D-1} of degree n, as

$$Z^{(n)}_{\omega_1}(\omega_2) = \sum_{j=1}^{\dim \mathcal{H}_n(S^{D-1})} Y_j(\omega_1)\overline{Y_j(\omega_2)}. \tag{4.15}$$

The dimension of the space $\mathcal{H}_n(S^{D-1})$ of spherical harmonics is given by

$$\dim \mathcal{H}_n(S^{D-1}) = \binom{D-1+n}{n} - \binom{D-3+n}{n-2}.$$

Using (4.15), we then have

$$\int_{S^{D-1}} C_n^{(\lambda)}(\omega\cdot\omega)d\omega = c_{D,n} \int_{S^{D-1}} Z^{(n)}_{\omega}(\omega)d\omega$$

$$= c_{D,n} \sum_{j=1}^{\dim \mathcal{H}_n(S^{D-1})} \int_{S^{D-1}} |Y_j(\omega)|^2 d\omega = c_{D,n} \dim \mathcal{H}_n(S^{D-1}),$$

which gives $C_n^{(\lambda)}(1) = \frac{2\lambda \dim \mathcal{H}_n(S^{D-1})}{2(n+\lambda)}$, for $D = 2\lambda + 2$. □

4.2.2. *Polygon amplitudes in dimension four.* We now specialize to the case where $D = 4$ (hence $\lambda = 1$) and we show how one obtains integrals of polylogarithm functions

$$\mathrm{Li}_s(z) = \sum_{n=1}^{\infty} \frac{z^n}{n^s}.$$

PROPOSITION 4.6. *Let Γ be a polygon with k edges and let $D = 4$. Then the Feynman amplitude is given by the integral*

$$(2\pi^2)^k \sum_{\mathbf{o}} m_{\mathbf{o}} \int_{\bar\Sigma_{\mathbf{o}}} \mathrm{Li}_{k-2}\Big(\prod_i \frac{r^2_{w_i}}{r^2_{v_i}}\Big) \prod_v r_v\, dr_v, \tag{4.16}$$

where the vertices v_i and w_i are the sources and tails of the oriented paths determined by $\mathbf{o}$.

PROOF. We write the terms in the integrand (4.5) as

$$\prod_{e\in \mathbf{E}_\Gamma} \rho_e^{-2\lambda} \left(\sum_n (\frac{r_e}{\rho_e})^n\, C_n^{(\lambda)}(\omega_{s(e)}\cdot \omega_{t(e)}) \right) = \tag{4.17}$$

$$(\rho_1\cdots\rho_k)^{-2\lambda} \sum_{n_1,\ldots,n_k} (\frac{r_1}{\rho_1})^{n_1}\cdots(\frac{r_k}{\rho_N})^{n_k}\, C_{n_1}^{(\lambda)}(\omega_{s(e_1)}\cdot\omega_{t(e_1)})\cdots C_{n_k}^{(\lambda)}(\omega_{s(e_k)}\cdot\omega_{t(e_k)})$$

and we perform the angular integral as in (4.10). In the case $D = 4$ we have

$$\dim \mathcal{H}_n(S^3) = \binom{n+3}{n} - \binom{n+1}{n-2} = (n+1)^2.$$

Thus, the angular integral of Proposition 4.5 becomes

$$\mathcal{A}_n = \frac{(2\pi^2)^k}{(n+1)^{k-2}}. \tag{4.18}$$

We then write the radial integrand as in (4.8). An acyclic orientation $\mathbf{o} \in \Omega(\Gamma)$, subdivides the polygon Γ into oriented paths γ_i such that $s_{\gamma_i} = s_{\gamma_{i-1}}$ and $t_{\gamma_i} = t_{\gamma_{i+1}}$ or $s_{\gamma_i} = s_{\gamma_{i+1}}$ and $t_{\gamma_i} = t_{\gamma_{i-1}}$. Correspondingly, the set of vertices is subdivided into $\mathbf{V}_\Gamma = \{v_i\} \cup \{w_i\} \cup \{v \notin \{v_i, w_i\}\}$ with v_i the sources and w_i the tails of the oriented paths and the remaining vertices partitioned into internal vertices of each oriented path. We then have

$$\begin{aligned}\mathcal{F}(r_{s(e)}, r_{t(e)}) &= (2\pi^2)^k (\prod_e r_{t(e)}^{-2}) \sum_{n\geq 0} \frac{1}{(n+1)^{k-2}} (\prod_i \frac{r_{v_i}^2}{r_{w_i}^2})^n \\ &= (2\pi^2)^k (\prod_e r_{t(e)}^{-2}) (\prod_i \frac{r_{w_i}^2}{r_{v_i}^2}) \sum_{n\geq 1} \frac{1}{n^{k-2}} (\prod_i \frac{r_{v_i}^2}{r_{w_i}^2})^n.\end{aligned}$$

Since each w_i is counted twice as target of an edge and the internal vertices of the oriented paths are counted only once, we obtain

$$\prod_e \mathcal{F}(r_{s(e)}, r_{t(e)}) \prod_v r_v^3 dv = (2\pi^2)^k \cdot \mathrm{Li}_{k-2}(\prod_i \frac{r_{w_i}^2}{r_{v_i}^2}) \prod_v r_v dr_v. \tag{4.19}$$

□

4.2.3. *Zeta values.* After a cutoff regularization, these integrals produce combinations of zeta values with coefficients that are rational combinations of powers of $2\pi i$. To see how this happens, we look explicitly at the contribution of an acyclic orientation of the polygon consisting of just two oriented paths γ_1 and γ_2 with source v and target w, respectively with k_1 and k_2 internal vertices. The other summands can be handled similarly. By changing variables to $t = r_v^2/r_w^2$, $t_i = r_{v_i}^2/r_w^2$ for v_i the internal edges of γ_1 and $s_i = r_{v_i}^2/r_w^2$ for v_i the internal edges of γ_2, we obtain

$$\bigwedge_{v\in\mathbf{V}_\gamma} r_v\, dr_v = \pm 2^{1-k} r_w^{2k+1} dr_w\, dt \bigwedge_i dt_i \wedge ds_i.$$

After factoring out a divergence along $\Delta_\infty = X \smallsetminus \mathbb{A}^4$, coming from the integration of the r_w term, which gives a pole along the divisor Δ_∞, one obtains an integral of the form

$$2\pi^{2k} \int_{\bar\Sigma_1\cap\bar\Sigma_2} \mathrm{Li}_{k-2}(t)\, dt \prod_i dt_i ds_i,$$

where $\bar\Sigma_{\mathbf{o}} = \bar\Sigma_1 \cap \bar\Sigma_2$ with $\bar\Sigma_1 = \{(t, t_i, s_i)\,|\, t \leq t_1 \leq \cdots \leq t_{k_1-1} \leq 1\}$ and $\bar\Sigma_2 = \{(t, t_i, s_i)\,|\, t \leq s_1 \leq \cdots \leq s_{k_2-1} \leq 1\}$. One can use the relation [**45**]

$$\int x^m \,\mathrm{Li}_n(x)\, dx = \frac{1}{m+1} x^{m+1}\, \mathrm{Li}_n(x) - \frac{1}{m+1}\int x^m\, \mathrm{Li}_{n-1}(x) dx, \tag{4.20}$$

to reduce the integral to a combination of zeta values. Further results on zeta values and multiple zeta values, beyond the relations in [**45**], can be found in [**1**], [**26**].

We will return to discuss cutoff regularizations in §5.8 and §6, in the case of the complexified amplitudes.

4.3. Stars of vertices and isoscalars. To see how more complicated expressions can arise in the integrands, which eventually lead to the presence of multiple zeta values, it is convenient to regard graphs as being built out of stars of vertices pasted together by suitably matching the half edges, where by the *star of a vertex* we mean a single vertex v of valence k with k half-edges e_j attached to it. One can then built the Feynman integral by first identifying the contribution of the star of a vertex, which is obtained by integrating in the variables r_v and ω_v of the central vertex v, and results in a function of variables r_{v_j} and ω_{v_j}, for each of the edges e_j. One then obtains the integral for the graph, which gives a number (possibly after a regularization), by matching the half edges and identifying the corresponding variables and integrating over them.

We first introduce the analog of the angular integral (4.7) for the case of a graph with half-edges. While in the case of a usual graph, where all half-edges are paired to form edges, the angular integral (4.7) is a number, in the case with open (unpaired) half-edges it is a function of the variables of the half edges, which we denote by $\mathcal{A}_{\underline{n}}(\underline{\omega})$, where $\underline{n} = (n_1, \dots, n_\ell)$ and $\underline{\omega} = (\omega_1, \dots, \omega_\ell)$ are vectors of integers $n_j \in \mathbb{N}$, for each half-edge e_j, and of variables $\omega_j \in S^{D-1}$. We will sometime denote these variables by ω_{v_j}, where v_j simply denotes the end of the half-edge e_j. We will equivalently use the notation $\mathcal{A}_{\underline{n}}(\underline{\omega})$ or $\mathcal{A}_{(n_j)}(\omega_{v_j})$. In the case of the star of a vertex, the angular integral is of the form

$$\mathcal{A}_{\underline{n}}(\underline{\omega}) = \int_{S^{D-1}} \prod_j C^{(\lambda)}_{n_j}(\omega_j \cdot \omega)\, d\omega, \text{ with } \underline{n} = (n_j)_{e_j \in \mathbf{E}_\Gamma}, \ \underline{\omega} = (\omega_j)_{e_j \in \mathbf{E}_\Gamma}. \tag{4.21}$$

LEMMA 4.7. *Let Γ be the star of a valence k vertex v. Then the angular integral (4.21) is given by the function*

$$\begin{gathered} \mathcal{A}_{(n_j)}(\omega_{v_j}) = c_{D,n_1} \cdots c_{D,n_k}\, \tilde{\mathcal{A}}_{(n_j)}(\omega_{v_j}), \\ \tilde{\mathcal{A}}_{(n_j)}(\omega_{v_j}) = \sum_{\ell_1,\dots,\ell_k} \overline{Y^{(n_1)}_{\ell_1}(\omega_1) \cdots Y^{(n_k)}_{\ell_k}(\omega_k)} \int_{S^{D-1}} Y^{(n_1)}_{\ell_1}(\omega) \cdots Y^{(n_k)}_{\ell_k}(\omega)\, d\omega, \end{gathered} \tag{4.22}$$

where $\{Y^{(n)}_\ell\}_{\ell=1,\dots,d_n}$ is an orthonormal basis of the space $\mathcal{H}_n(S^{D-1})$ of spherical harmonics of degree n, and $d_n = \dim \mathcal{H}_n(S^{D-1})$, with the coefficients $c_{D,n}$ as in (4.14).

PROOF. Consider first the angular integral as written in (4.7). Using (4.11), we can write each edge contribution $C^{(\lambda)}_{n_e}(\omega_{s(e)} \cdot \omega_{t(e)})$ in (4.7) in terms of an integral

$$\frac{n_e + \lambda}{\lambda \operatorname{Vol}(S^{D-1})} \int_{S^{D-1}} C^{(\lambda)}_{n_e}(\omega_{s(e)} \cdot \omega) C^{(\lambda)}_{n_e}(\omega \cdot \omega_{t(e)})\, d\omega.$$

Thus, we can think of the edge contribution $C^{(\lambda)}_{n_e}(\omega_{s(e)} \cdot \omega_{t(e)})$ as obtained by a contraction (integration on a common variable) of contributions $C^{(\lambda)}_{n_e}(\omega_{s(e)} \cdot \omega)$ and $C^{(\lambda)}_{n_e}(\omega \cdot \omega_{t(e)})$ associated to two half-edges, with a factor depending on n_e, λ, D. This gives us an angular integral associated to a star of half-edges, which is of the form (4.21). Using the relation (4.13), (4.15) between the Gegenbauer polynomials and the spherical harmonics, we then rewrite the angular integral (4.21) in the form (4.22). □

Thus, the evaluation of the angular integrals (4.22) for stars of vertices relates to the well known problem of evaluating coupling coefficients for spherical harmonics,

$$\langle Y^{(n_1)}_{\ell_1}, \dots, Y^{(n_k)}_{\ell_k} \rangle_D := \int_{S^{D-1}} Y^{(n_1)}_{\ell_1}(\omega) \cdots Y^{(n_k)}_{\ell_k}(\omega)\, d\omega. \tag{4.23}$$

In the following we will be using the standard labeling of the basis $\{Y^{(n)}_\ell\}$ where, for fixed n, the indices ℓ run over a set of $(D-2)$-tuples

$$(m_{D-2}, m_{D-1}, \dots, m_2, m_1) \quad \text{with} \quad n \geq m_{D-2} \geq \cdots \geq m_2 \geq |m_1|.$$

The spherical harmonics $Y^{(n)}_\ell$ have the symmetry

$$\overline{Y^{(n)}_\ell} = (-1)^{m_1} Y^{(n)}_{\bar{\ell}}, \tag{4.24}$$

where, for $\ell = (m_{D-2}, m_{D-1}, \dots, m_2, m_1)$, one has

$$\bar{\ell} := (m_{D-2}, m_{D-1}, \dots, m_2, -m_1).$$

In the simplest case of a tri-valent vertex, these coefficients are also referred to as the Gaunt coefficients, and have been extensively studied, see for instance [**5**], [**59**], [**77**]. The Gaunt coefficients arising from the integration of three harmonic functions determine the coefficients of the expansion formula

$$Y^{(n_1)}_{\ell_1} Y^{(n_2)}_{\ell_2} = \sum_{n,\ell} \mathcal{K}_{D,n_i,n,\ell_i,\ell}\, Y^{(n)}_\ell \tag{4.25}$$

that expresses the product of two harmonic functions in terms of a linear combination of other harmonic functions, with the cases where some of the factors are conjugated taken care of by the symmetry (4.24). In the more general case (4.23) one can therefore repeatedly apply (4.25), hence we focus here on the example of the star of a tri-valent vertex.

The Gaunt coefficients $\langle Y^{(n_1)}_{\ell_1}, Y^{(n_2)}_{\ell_2}, Y^{(n_3)}_{\ell_3} \rangle_D$ can be computed via Racah's factorization lemma ([**5**], [**59**]) in terms of *isoscalar factors* and the Gaunt coefficients for $D-1$, according to

$$\langle Y^{(n_1)}_{\ell_1}, Y^{(n_2)}_{\ell_2}, Y^{(n_3)}_{\ell_3} \rangle_D = \begin{pmatrix} n_1 & n_2 & n_3 \\ n'_1 & n'_2 & n'_3 \end{pmatrix}_{D:D-1} \langle Y^{(n'_1)}_{\ell'_1}, Y^{(n'_2)}_{\ell'_2}, Y^{(n'_3)}_{\ell'_3} \rangle_{D-1}, \tag{4.26}$$

where $\ell_i = (n'_i, \ell'_i)$ with $n'_i = m_{D-2,i}$ and $\ell'_i = (m_{D-3,i}, \dots, m_{1,i})$. An explicit expression of the isoscalar factors

$$\begin{pmatrix} n_1 & n_2 & n_3 \\ n'_1 & n'_2 & n'_3 \end{pmatrix}_{D:D-1} \tag{4.27}$$

is given in [**5**], [**59**]. We will discuss this more in detail in §4.4 and §4.5 below.

4.3.1. *Gaunt versus Clebsch–Gordon.* There is some inconsistency of terminology in the literature on spherical harmonics and coupling coefficients between what is referred to as "Gaunt coefficients" and what is termed "Clebsch–Gordon coefficients". The two notions are very closely related, but the respective explicit expressions considered in the literature vary by a factor and the terminology adopted is not always consistent. For example, the definition of Gaunt coefficients adopted in [**77**], as the integral of three spherical harmonics, differs from the notion of Gaunt coefficient used in [**89**] (see equation (10) of [**89**]) by a factor as in equation (21) of [**77**]. In turn, the Gaunt coefficients considered in [**89**] differ from the Clebsch–Gordon coefficients computed for instance in [**86**] as indicated in equation (46) of [**89**]. Given the frequent inconsistencies of terminology in the literature, we urge the reader to check for the specific conventions followed by specific authors and not simply consider Gaunt and Clebsch–Gordon as interchangeable synonyms.

4.4. Gluing two stars along an edge. We now consider the effect of patching together two trivalent stars by gluing two half edges with matching orientations.

LEMMA 4.8. *Let $\mathcal{A}_{n,n_1,n_2}(\omega,\omega_1,\omega_2)$ and $\mathcal{A}_{n',n_3,n_4}(\omega',\omega_3,\omega_4)$ be the angular integrals associated to two trivalent stars, as in Lemma 4.7. Then the angular integral of the graph obtained by joining the two stars at an edge is*

$$\mathcal{A}_{(n_i)_{i=1,\dots,4}}((\omega_i)_{i=1,\dots,4},n)=\sum_{\ell_i}\prod_{i=1}^{4}c_{D,n_i}\,\overline{Y^{(n_i)}_{\ell_i}(\omega_i)}\,\mathcal{K}_{n_i,\ell_i}(n), \tag{4.28}$$

with coefficients $\mathcal{K}_{\underline{n},\underline{\ell}}(n)$ given by

$$\mathcal{K}_{n_i,\ell_i}(n)=c^2_{D,n}\sum_{\ell=1}^{d_n}\langle Y^{(n)}_\ell,Y^{(n_1)}_{\ell_1},Y^{(n_2)}_{\ell_2}\rangle_D\cdot\langle Y^{(n)}_\ell,Y^{(n_3)}_{\ell_3},Y^{(n_4)}_{\ell_4}\rangle_D. \tag{4.29}$$

PROOF. As shown in Lemma 4.7 above, using (4.11) one reduces the contribution $C^{(\lambda)}_n(\omega_v\cdot\omega_{v'})$ of an edge e with $s(e)=v$ and $t(e)=v'$ to an integration along an angular variable ω of a product $C^{(\lambda)}_n(\omega_v\cdot\omega)C^{(\lambda)}_n(\omega\cdot\omega_{v'})$ (with a proportionality factor that depends on D,n,λ). One should think here of the variable ω as associated to a half-edge attached to the vertex v and contracted with the corresponding variable associated to a half-edge attached to the vertex v' when the half edges are joined to form the edge e. That is the meaning of (4.11). By rewriting the Gegenbauer polynomials in terms of spherical harmonics one can identify, as shown in Lemma 4.7, the contribution to the angular integral of a star of half-edges associated to a vertex, in the form written in (4.22). In particular, consider a trivalent star of half-edges attached to a vertex. According to Lemma 4.7, the expression associated to this trivalent star is of the form

$$\overline{Y^{(n)}_\ell(\omega)Y^{(n_1)}_{\ell_1}(\omega_1)Y^{(n_2)}_{\ell_2}(\omega_2)}\int_{S^{D-1}}Y^{(n)}_\ell(\omega_v)Y^{(n_1)}_{\ell_1}(\omega_v)Y^{(n_2)}_{\ell_2}(\omega_v)d\omega_v,$$

where $\omega_v\in S^{D-1}$ is the angular variable associated to the vertex v, and ω,ω_1,ω_2 are the angular variables associated to the three half-edges. This can be visualized graphically as in the figure, with the Gaunt coefficient

$$\langle Y^{(n)}_\ell,Y^{(n_1)}_{\ell_1},Y^{(n_2)}_{\ell_2}\rangle_D=\int_{S^{D-1}}Y^{(n)}_\ell(\omega_v)Y^{(n_1)}_{\ell_1}(\omega_v)Y^{(n_2)}_{\ell_2}(\omega_v)d\omega_v$$

associated to the vertex v of the star and the spherical harmonics $\overline{Y^{(n)}_\ell(\omega)}$, $\overline{Y^{(n_1)}_{\ell_1}(\omega_1)}$ and $\overline{Y^{(n_2)}_{\ell_2}(\omega_2)}$ associated to each of the half-edges, respectively.

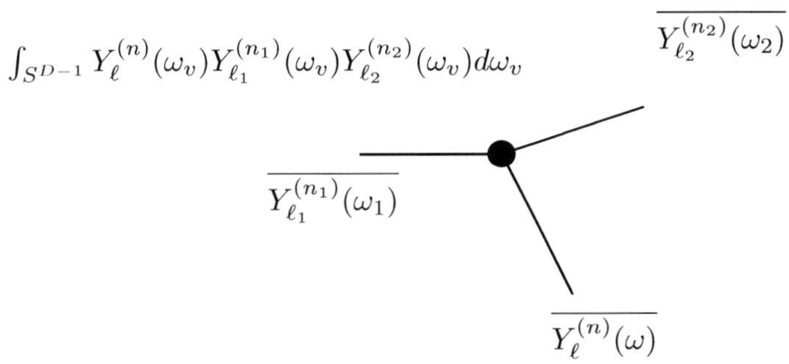

Consider then a second trivalent star, for which the expression of Lemma 4.7 will be of the form

$$\overline{Y_{\ell'}^{(n')}(\omega') Y_{\ell_3}^{(n_3)}(\omega_3) Y_{\ell_4}^{(n_4)}(\omega_4)} \int_{S^{D-1}} Y_{\ell'}^{(n')}(\omega_{v'}) Y_{\ell_3}^{(n_3)}(\omega_{v'}) Y_{\ell_4}^{(n_4)}(\omega_{v'}) d\omega_{v'}.$$

We again represent this graphically with the Gaunt coefficient

$$\langle Y_{\ell'}^{(n')}, Y_{\ell_3}^{(n_3)}, Y_{\ell_4}^{(n_4)} \rangle = \int_{S^{D-1}} Y_{\ell'}^{(n')}(\omega_{v'}) Y_{\ell_3}^{(n_3)}(\omega_{v'}) Y_{\ell_4}^{(n_4)}(\omega_{v'}) d\omega_{v'}$$

associated to the vertex v' of the star and with the spherical harmonics $\overline{Y_{\ell'}^{(n')}(\omega')}$, $\overline{Y_{\ell_3}^{(n_3)}(\omega_3)}$ and $\overline{Y_{\ell_4}^{(n_4)}(\omega_4)}$ placed on the half-edges, as in the figure.

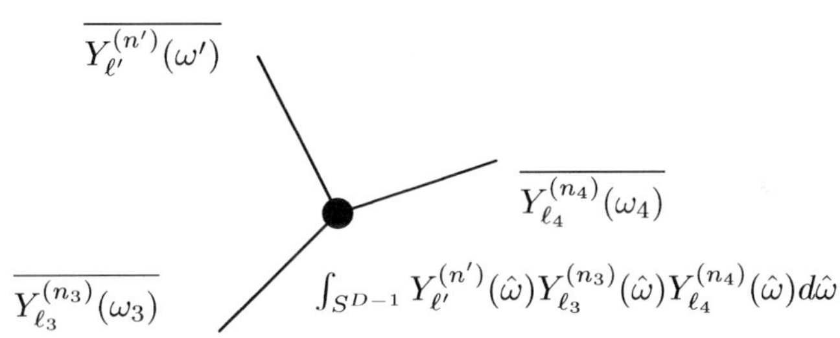

Gluing together the half-edges labeled, respectively, by the spherical harmonics $\overline{Y_\ell^{(n)}(\omega)}$ and $\overline{Y_{\ell'}^{(n')}(\omega')}$ corresponds to the integration of (4.11), which corresponds to

$$\int_{S^{D-1}} \overline{Y_\ell^{(n)}(\omega) Y_{\ell'}^{(n')}(\omega')}\, d\omega.$$

This is represented graphically in the following picture.

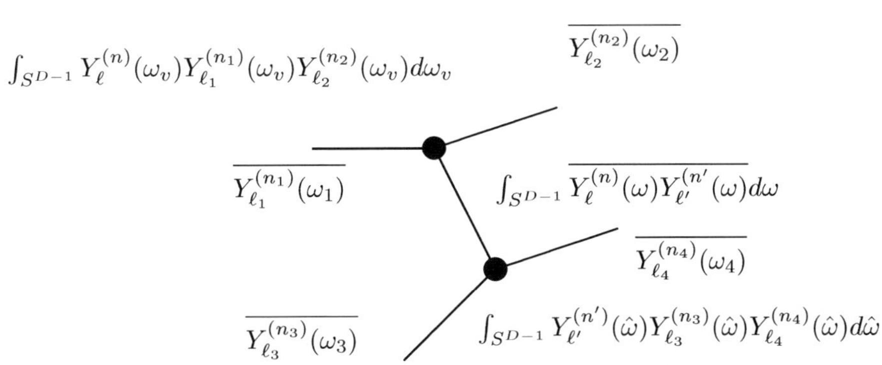

Thus, the angular integral $\mathcal{A}_{(n_i)_{i=1,\dots,4}}((\omega_i)_{i=1,\dots,4},n)$ is obtained by integrating along the variables of the matched half edges,

$$\mathcal{A}_{(n_i)_{i=1,\dots,4}}((\omega_i)_{i=1,\dots,4},n)=c_{D,n}c_{D,n'}(\prod_{i=1}^{4}c_{D,n_i})\cdot\tilde{\mathcal{A}}_{(n_i)_{i=1,\dots,4}}((\omega_i)_{i=1,\dots,4},n),$$

where $\tilde{\mathcal{A}}_{(n_i)_{i=1,\dots,4}}((\omega_i)_{i=1,\dots,4},n)$ is given by

$$\int_{S^{D-1}}d\omega\sum_{\ell,\ell',\ell_i}\overline{Y_\ell^{(n)}(\omega)Y_{\ell'}^{(n')}(\omega)}\prod_i\overline{Y_{\ell_i}^{(n_i)}(\omega_i)}\begin{pmatrix} n & n_1 & n_2\\ \ell & \ell_1 & \ell_2\end{pmatrix}_D\begin{pmatrix} n' & n_3 & n_4\\ \ell' & \ell_3 & \ell_4\end{pmatrix}_D$$

where we used the shorthand notation

$$\begin{pmatrix} n & n_1 & n_2\\ \ell & \ell_1 & \ell_2\end{pmatrix}_D:=\langle Y_\ell^{(n)},Y_{\ell_1}^{(n_1)},Y_{\ell_2}^{(n_2)}\rangle_D. \tag{4.30}$$

Using the orthogonality relations for the spherical harmonics, this gives

$$(\prod_{i=1}^{4}c_{D,n_i})\sum_{\ell_1,\ell_2,\ell_3,\ell_4}\overline{Y_{\ell_1}^{(n_1)}(\omega_1)Y_{\ell_2}^{(n_2)}(\omega_2)Y_{\ell_3}^{(n_3)}(\omega_3)Y_{\ell_4}^{(n_4)}(\omega_4)}\ \mathcal{K}_{\underline{n},\underline{\ell}}(n),$$

with the coefficients as in (4.29). □

REMARK 4.9. Note that, since according to (4.11) each edge e carries an integral of a product of Gegenbauer polynomials $C_n^{(\lambda)}(\omega_v\cdot\omega)C_n^{(\lambda)}(\omega\cdot\omega_{v'})$, when viewed as a contraction of two half-edges this corresponds to viewing stars of vertices as carrying a Gaunt coefficient at the vertex (of order equal to the valence of the vertex) and spherical harmonics attached to the half edges, as in (4.22). Thus, for the case of gluing two trivalent stars along one edge we only need triple Gaunt coefficients and not quadruple ones, which would instead occur in the presence of valence four vertices.

The coefficients $\mathcal{K}_{\underline{n},\underline{\ell}}(n)$ are usually very involved to compute explicitly (see (3.3), (3.6) and (4.7) of [**5**]). However, some terms simplify greatly in the case $D=4$, and that will allow us to show the occurrence of functions closely related to multiple polylogarithm functions in §4.5 below. For later use, we give here the explicit computation, in dimension $D=4$, of the coefficients $\mathcal{K}_{\underline{n},\underline{\ell}}(n)$, in the particular case with $\underline{\ell}=0$.

PROPOSITION 4.10. *In the case where $D=4$, the coefficient $\mathcal{K}_{\underline{n},\underline{\ell}}(n)$ with $\ell_i=0$ has the form*

$$\mathcal{K}_{\underline{n},\underline{0}}^{(D=4)}(n)=(\prod_{i=1}^{4}\frac{1}{(n_i+1)^{1/2}})\frac{4\pi^4}{(n+1)^3}, \tag{4.31}$$

in the range where $n+n_1+n_2$ and $n+n_3+n_4$ are even and the inequalities $|n_j-n_k|\le n_i\le n_j+n_k$ hold for (n_i,n_j,n_k) equal to (n,n_1,n_2) or (n,n_3,n_4) and transpositions, and are equal to zero outside of this range.

PROOF. We use the fact that ([**5**], (4.9) and [**59**], (22) and (23)) the coefficients $\langle Y_0^{(n_1)},Y_0^{(n_2)},Y_0^{(n_3)}\rangle_D$ are zero outside the range where

$$\sum_i n_i \quad\text{is even and}\quad |n_j-n_k|\le n_i\le n_j+n_k, \tag{4.32}$$

while within this range they are given by the expression
(4.33)

$$\epsilon_D \frac{1}{\Gamma(D/2)} \left(\frac{(J+D-3)!}{(D-3)!\,\Gamma(J+D/2)} \prod_i \frac{(n_i+\frac{D}{2}-1)\,\Gamma(J-n_i+\frac{D}{2}-1)}{d^{(D)}_{n_i}\,(J-n_i)!} \right)^{1/2},$$

where ϵ_D is a sign, $J=\frac{1}{2}\sum_i n_i$, and $d^{(D)}_{n_i}=\dim \mathcal{H}_{n_i}(S^{D-1})$. In the particular case where $D=4$, the expression (4.33) reduces to

$$\begin{pmatrix} n_1 & n_2 & n_3 \\ 0 & 0 & 0 \end{pmatrix}_4 = \epsilon_4 \prod_i \frac{(n_i+1)^{1/2}}{(d^{(4)}_{n_i})^{1/2}} = \epsilon_4 \prod_i (n_i+1)^{-1/2},$$

using again the fact that $\dim \mathcal{H}_n(S^3)=(n+1)^2$. Thus, we obtain

$$\mathcal{K}^{(D=4)}_{n_i,\ell_i=0}(n) = c^2_{4,n} \begin{pmatrix} n & n_1 & n_2 \\ 0 & 0 & 0 \end{pmatrix}_4 \begin{pmatrix} n & n_3 & n_4 \\ 0 & 0 & 0 \end{pmatrix}_4 =$$

$$= \left(\frac{2\,Vol(S^3)}{2(n+1)} \right)^2 \frac{1}{(n+1)} \prod_{i=1}^4 \frac{1}{(n_i+1)^{1/2}} = \prod_{i=1}^4 (n_i+1)^{-1/2} \frac{4\pi^4}{(n+1)^3}.$$

□

4.5. Gluing stars of vertices. We now consider the full integrand, including the radial variables and again look at the effect of gluing together two half edges of two trivalent stars. We will see that one can explicitly identify the leading term in the resulting expression in the integrand with a function closely related to multiple polylogarithms.

LEMMA 4.11. *Consider the star of a trivalent vertex, and let $D=4$. After a change of variables $t_i=r_{v_i}/r$, with $r=r_v$ for v the central vertex of the star, the integrand (4.8), for an orientation $\mathbf{o}$, can be written as an expression $\mathcal{I}_{\mathbf{o}}(r,t_1,t_2,t_3,\omega_1,\omega_2,\omega_3)\,dr dt_1 dt_2 dt_3$ of the form*

$$r^9 \prod_{i=1}^3 t_i^{\alpha_i} \sum_{n_1,n_2,n_3} \mathcal{A}_{(n_1,n_2,n_3)}(\omega_1,\omega_2,\omega_3) t_1^{\epsilon_1 n_1} t_2^{\epsilon_2 n_2} t_3^{\epsilon_3 n_3}\, dr \prod_{i=1}^3 dt_i, \tag{4.34}$$

where $\alpha_i=1$ and $\epsilon_i=1$ if the half-edge e_i is outgoing in the orientation $\mathbf{o}$ and $\alpha_i=3$ and $\epsilon_i=-1$ if it is incoming, and where $\mathcal{A}_{\underline{n}}(\underline{\omega})=\mathcal{A}_{(n_j)}(\omega_j)$ is the angular integral of Lemma 4.7 with $D=4$.

PROOF. The integrand of (4.8), for the case of a trivalent star, is of the form

$$\prod_{i=1}^3 \mathcal{F}(r_{s(e_i)}, r_{t(e_i)}) r^3 dr \prod_{i=1}^3 r^3_{v_i} dr_{v_i}, \tag{4.35}$$

with

$$\prod_{i=1}^3 \mathcal{F}(r_{s(e_i)}, r_{t(e_i)}) = (\prod_{i=1}^3 r^{-2}_{t(e_i)}) \sum_{\underline{n}} \mathcal{A}_{\underline{n}}(\underline{\omega}) \left(\frac{r_{s(e_1)}}{r_{t(e_1)}}\right)^{n_1} \left(\frac{r_{s(e_2)}}{r_{t(e_2)}}\right)^{n_2} \left(\frac{r_{s(e_3)}}{r_{t(e_3)}}\right)^{n_3}.$$

When combined with the volume form as in (4.35), this can be rewritten as

$$r^{\alpha_0} r_1^{\alpha_1} r_2^{\alpha_2} r_3^{\alpha_3} \sum_{\underline{n}} \mathcal{A}_{\underline{n}}(\underline{\omega}) \left(\frac{r_1}{r}\right)^{\epsilon_1 n_1} \left(\frac{r_2}{r}\right)^{\epsilon_2 n_2} \left(\frac{r_3}{r}\right)^{\epsilon_3 n_3} dr\, dr_1 dr_2 dr_3, \tag{4.36}$$

where the exponents α_i are given by the table

	$\mathbf{o}_0$	$\mathbf{o}_1$	$\mathbf{o}_2$	$\mathbf{o}_3$
α_0	-3	3	-1	1
α_1	1	3	1	1
α_2	1	3	3	1
α_3	1	3	3	3

where the orientation $\mathbf{o}_0$ has all the half-edges of the star pointing outward, $\mathbf{o}_1$ all pointing inward, $\mathbf{o}_2$ has e_1 outward and e_2 e_3 inward and $\mathbf{o}_3$ has e_1 and e_2 outward and e_3 inward. All the other cases are obtained by relabeling of indices. After we change variables to $t_i = r_{v_i}/r$, we obtain $dr \wedge \wedge_{i=1}^3 dr_i = r^3 dr \wedge \wedge_{i=1}^3 dt_i$ and (4.36) becomes

$$\mathcal{I}_{\mathbf{o}}(r,(t_i),\underline{\omega}) = r^9 t_1^{\alpha_1} t_2^{\alpha_2} t_3^{\alpha_3} \sum_{\underline{n}} \mathcal{A}_{\underline{n}}(\underline{\omega})\ (t_1)^{\epsilon_1 n_1}\ (t_2)^{\epsilon_2 n_2}\ (t_3)^{\epsilon_3 n_3}\ dr\, dt_1 dt_2 dt_3. \tag{4.37}$$

□

We can now perform the gluing of two stars by matching an oriented half-edge of one trivalent star to an oriented half edge of the other, so that one obtains an oriented edge. We rewrite (4.37) as

$$\mathcal{I}_{\mathbf{o}}(r,(t_i),\underline{\omega}) = \tilde{\mathcal{I}}_{\mathbf{o}}((t_i),\underline{\omega})\, r^9\, dr \tag{4.38}$$

where we set

$$\tilde{\mathcal{I}}_{\mathbf{o}}((t_i),\underline{\omega}) = t_1^{\alpha_1} t_2^{\alpha_2} t_3^{\alpha_3} \sum_{\underline{n}} \mathcal{A}_{\underline{n}}(\underline{\omega})\ (t_1)^{\epsilon_1 n_1}\ (t_2)^{\epsilon_2 n_2}\ (t_3)^{\epsilon_3 n_3}\ dt_1 dt_2 dt_3. \tag{4.39}$$

LEMMA 4.12. *The gluing of two stars by matching an oriented half-edge corresponds to an integration*

$$\mathcal{I}_{\mathbf{o}}(t_i,\omega_i) := \int_{\bar{\Sigma}} \int_{S^{D-1}} \tilde{\mathcal{I}}_{\mathbf{o}}(t,t_1,t_2,\omega,\omega_1,\omega_2) \tilde{\mathcal{I}}_{\mathbf{o}}(t,t_3,t_4,\omega,\omega_3,\omega_4)\, dt d\omega, \tag{4.40}$$

where the domain of integration $\bar{\Sigma} = \bar{\Sigma}(t_1,t_2,t_3,t_4)$ for the variable t is given by

$$\bar{\Sigma} = \cap_{i,j:t(e_i)=s(e),s(e_j)=t(e)} \{t \,|\, t_i \le t \le t_j\},$$

and a cutoff regularized divergence

$$\int_0^\Lambda \mathcal{I}_{\mathbf{o}}(t_i,\omega_i)\, r^9\, dr \sim_{\Lambda\to\infty} \frac{\Lambda^{10}}{10}\, \mathcal{I}_{\mathbf{o}}(t_i,\omega_i). \tag{4.41}$$

PROOF. Gluing of two stars by matching an oriented half-edge means integrating

$$\int_0^\infty \int_{\bar{\Sigma}} \int_{S^{D-1}} \mathcal{I}_{\mathbf{o}}(r,t,t_1,t_2,\omega,\omega_1,\omega_2) \mathcal{I}_{\mathbf{o}}(r,t,t_3,t_4,\omega,\omega_3,\omega_4)\, dr\, dt\, d\omega. \tag{4.42}$$

This gives rise to an overall divergent factor (4.41) arising from the integration of (4.42) in the variable r, which can be taken care of by a cutoff regularization. The integration in the remaining variables gives the integrand $\mathcal{I}(t_1,t_2,t_3,t_4,\omega_1,\omega_2,\omega_3,\omega_4)$, as the result of gluing two trivalent stars by matching oriented half-edges to form an oriented edge e, which is given by (4.40). □

In the following we write $\underline{t} = (t_1,t_2,t_3,t_4)$ and similarly for $\underline{\omega}$, $\underline{n}$ and $\underline{\ell}$.

By combining (4.28) with (4.22), we can rephrase (4.40) in terms of isoscalars. This gives a decomposition of $\mathcal{I}_{\mathbf{o}}(t_i, \omega_i)$ into a sum of terms of the form

$$\mathcal{I}_{\mathbf{o}}(t_i, \omega_i) = \sum_{\underline{n},\underline{\ell}} \mathcal{I}_{\mathbf{o},\underline{n},\underline{\ell}}(t_i, \omega_i).$$

We denote by $\mathcal{I}_{\mathbf{o},0}(\underline{t}, \underline{\omega})$ the leading term

$$\mathcal{I}_{\mathbf{o},0}(t_i, \omega_i) = \sum_{\underline{n}} \mathcal{I}_{\mathbf{o},\underline{n},\underline{0}}(t_i, \omega_i), \tag{4.43}$$

involving only the isoscalars with all $\ell_i = 0$. We have the following result computing the terms $\mathcal{I}_{\mathbf{o},0}(t_i, \omega_i)$.

LEMMA 4.13. *In the case $D = 4$, the integrands $\mathcal{I}_{\mathbf{o},0}(\underline{t}, \underline{\omega})$ are given by*

$$\mathcal{I}_{\mathbf{o},0}(\underline{t}, \underline{\omega}) = \sum_{\underline{n}} (\prod_{i=1}^{4} c_{D,n_i} \overline{Y_0^{(n_i)}(\omega_i) \frac{t_i^{\alpha_i + \epsilon_i n_i}\, dt_i}{(n_i+1)^{1/2}}}) \int_{\bar{\Sigma}} t^4\, dt \sum_n \frac{4\pi^2}{(n+1)^3} t^{\epsilon n}, \tag{4.44}$$

where the sum over the indices n and $\underline{n}$ is restricted by the constraints $n + n_1 + n_2$ and $n + n_3 + n_4$ are even and the inequalities $|n_j - n_k| \leq n_i \leq n_j + n_k$ hold for (n_i, n_j, n_k) equal to (n, n_1, n_2) or (n, n_3, n_4) and transpositions.

PROOF. Using (4.40) and (4.28), (4.29), (4.34), we obtain for $\mathcal{I}_{\mathbf{o}}(\underline{t}, \underline{\omega})$ the expression

$$\sum_{\underline{n}} \sum_{\underline{\ell}} (\prod_{i=1}^{4} c_{D,n_i} \overline{Y_{\ell_i}^{(n_i)}(\omega_i) t_i^{\alpha_i + \epsilon_i n_i}\, dt_i}) \int_{\bar{\Sigma}} t^4\, dt \sum_n \mathcal{K}_{\underline{n},\underline{\ell}}(n)\, t^{\epsilon n}.$$

The expression (4.44) then follows directly from the form (4.31) of the coefficients $\mathcal{K}_{\underline{n},\underline{0}}^{(D=4)}(n)$. The factor t^4 in the integral comes from the exponents $\alpha = 1$ and $\alpha = 3$ of the two half edges, which have matching orientations. $\square$

Notice that, without the constraints on the summation range of the indices n, n_i, we would obtain again an integral of the general form (4.20), involving polylogarithm functions $\mathrm{Li}_s(t^\epsilon)$, with $s = 3 = k - 2$ as in the case of polygons analyzed above. However, because not all values of n, n_i are allowed and one needs to impose the constraints of the form (4.32), one obtains more interesting expressions. We first introduce some notation.

4.5.1. *Summation domains and even condition.* In the following we let $\mathcal{R}$ denote a domain of summation for integers $(n_1, \dots, n_k)$. We consider in particular the cases

$$\begin{aligned} \mathcal{R} = \mathcal{R}_P^{(k)} &:= \{(n_1, \dots, n_k) \,|\, n_i > 0,\ \ i = 1, \dots, k\} \\ \mathcal{R} = \mathcal{R}_{MP}^{(k)} &:= \{(n_1, \dots, n_k) \,|\, n_k > \cdots > n_2 > n_1 > 0\} \\ \mathcal{R} = \mathcal{R}_T^{(3)} &:= \{(n_1, n_2, n_3) \,|\, n_2 > n_1,\ \ n_2 - n_1 < n_3 < n_2 + n_1\}. \end{aligned} \tag{4.45}$$

We denote by $\mathrm{Li}^{\mathcal{R}}_{s_1,\dots,s_k}(z_1, \dots, z_k)$ the associated series

$$\mathrm{Li}^{\mathcal{R}}_{s_1,\dots,s_k}(z_1, \dots, z_k) = \sum_{(n_1,\dots,n_k)\in\mathcal{R}} \frac{z_1^{n_1} \cdots z_k^{n_k}}{n_1^{s_1} \cdots n_k^{s_k}}. \tag{4.46}$$

In the first two cases of (4.45), this is, respectively, a product of polylogarithms $\mathrm{Li}^{\mathcal{R}_P}_{s_1,\dots,s_k}(z_1,\dots,z_k)=\prod_j \mathrm{Li}_{s_j}(z_j)$ and a multiple polylogarithm

$$\mathrm{Li}^{\mathcal{R}_{MP}}_{s_1,\dots,s_k}(z_1,\dots,z_k)=\mathrm{Li}_{s_1,\dots,s_k}(z_1,\dots,z_k).$$

We will discuss the third case more in detail below. We then define

(4.47)
$$\begin{aligned}\mathrm{Li}^{\mathcal{R},\mathrm{even}}_{s_1,\dots,s_k}(z_1,\dots,z_k) &:= \tfrac{1}{2}\left(\mathrm{Li}^{\mathcal{R}}_{s_1,\dots,s_k}(z_1,\dots,z_k)+\mathrm{Li}^{\mathcal{R}}_{s_1,\dots,s_k}(-z_1,\dots,-z_k)\right)\\ \mathrm{Li}^{\mathcal{R},\mathrm{odd}}_{s_1,\dots,s_k}(z_1,\dots,z_k) &:= \tfrac{1}{2}\left(\mathrm{Li}^{\mathcal{R}}_{s_1,\dots,s_k}(z_1,\dots,z_k)-\mathrm{Li}^{\mathcal{R}}_{s_1,\dots,s_k}(-z_1,\dots,-z_k)\right).\end{aligned}$$

The odd $\mathrm{Li}^{\mathcal{R},\mathrm{odd}}_{s_1,\dots,s_k}(z_1,\dots,z_k)$ is a direct generalization of the Legendre χ function, while the even $\mathrm{Li}^{\mathcal{R},\mathrm{even}}_{s_1,\dots,s_k}(z_1,\dots,z_k)$ corresponds to summing only over those indices in $\mathcal{R}$ whose sum is even,

$$\mathrm{Li}^{\mathcal{R},\mathrm{even}}_{s_1,\dots,s_k}(z_1,\dots,z_k)=\sum_{(n_1,\dots,n_k)\in\mathcal{R},\ \sum_i n_i\in 2\mathbb{N}}\frac{z_1^{n_1}\cdots z_k^{n_k}}{n_1^{s_1}\cdots n_k^{s_k}}. \tag{4.48}$$

More generally, one can also consider summations of the form

$$\mathrm{Li}^{\mathcal{R},\mathcal{E}_1,\dots,\mathcal{E}_k}_{s_1,\dots,s_k}(z_1,\dots,z_k)=\sum_{(n_1,\dots,n_k)\in\mathcal{R},\ n_i\in\mathcal{E}_i}\frac{z_1^{n_1}\cdots z_k^{n_k}}{n_1^{s_1}\cdots n_k^{s_k}}, \tag{4.49}$$

where, for each $i=1,\dots,k$, $\mathcal{E}_i=2\mathbb{N}$ or $\mathcal{E}_i=\mathbb{N}\smallsetminus 2\mathbb{N}$, that is, some of the summation indices are even and some odd.

4.5.2. *Matching half-edges.* We now illustrate in one sufficiently simple and explicit case, what the leading $\ell=0$ term looks like when all the half-edges of stars are joined together. We look at the case of two stars of trivalent vertices with the half edges pairwise joined, that is, the 3-banana graph (two vertices and three parallel edges between them).

PROPOSITION 4.14. *In the case of $D=4$, consider the graph with two vertices and three parallel edges between them. The $\underline{\ell}=0$ amplitude $\mathcal{I}_{\mathbf{o},0}$ is given by*

$$\mathcal{I}_{\mathbf{o},0}=\int_0^1 t^9\left(2^6\,\mathrm{Li}^{\mathcal{R}_{MP},odd,even}_{6,3}(t,t)+2\,\mathrm{Li}^{\mathcal{R}_T,even}_{3,3,3}(t,t,t)\right)dt. \tag{4.50}$$

PROOF. There is a unique acyclic orientation of this graph, with the three edges oriented in the same direction. Thus, there is a single variable $t\in[0,1]=\bar{\Sigma}$ in the integrand of $\mathcal{I}_{\mathbf{o},0}$, and the latter has the form

$$\sum_{(n_1,n_2,n_3)\in\mathcal{D}}\mathcal{K}_{n_1,n_2,n_3}t^{n_1+n_2+n_3},$$

where the coefficients $\mathcal{K}_{n_1,n_2,n_3}$ are given by

$$\mathcal{K}_{n_1,n_2,n_3}=\frac{c^2_{4,n_1}c^2_{4,n_2}c^4_{4,n_3}}{(n_1+1)(n_2+1)(n_3+1)}=\frac{(4\pi^2)^3}{(n_1+1)^3(n_2+1)^3(n_3+1)^3},$$

according to Proposition 4.10, and the fact that all the half-edges of the two trivalent stars are matched. The summation domain $\mathcal{D}$ is given by

$$\mathcal{D}=\{(n_1,n_2,n_3)\,|\,n_i\geq 0\ |n_j-n_k|\leq n_i\leq n_j+n_k,\ \sum_i n_i \text{ even}\}.$$

We subdivide this into separate domains $\mathcal{D} = \mathcal{D}_1 \cup \mathcal{D}_2 \cup \mathcal{D}_3 \cup \mathcal{D}_4 \cup \mathcal{D}_5$, where

$$\begin{aligned}
\mathcal{D}_1 &= \{n_1 = 0,\ n_2 \geq 0,\ n_3 = n_2\}\\
\mathcal{D}_2 &= \{n_2 = 0,\ n_1 > 0,\ n_3 = n_1\}\\
\mathcal{D}_3 &= \{n_1 > 0,\ n_2 = n_1,\ 0 \leq n_3 \leq 2n_1,\ n_3 \text{ even }\}\\
\mathcal{D}_4 &= \{0 < n_2 < n_1,\ n_1 - n_2 \leq n_3 \leq n_1 + n_2,\ \textstyle\sum_i n_i \text{ even }\}\\
\mathcal{D}_5 &= \{0 < n_1 < n_2,\ n_2 - n_1 \leq n_3 \leq n_1 + n_2,\ \textstyle\sum_i n_i \text{ even }\}.
\end{aligned}$$

We have

$$\sum_{(n_1,n_2,n_3)\in\mathcal{D}_1} \frac{t^{n_1+n_2+n_3}}{(n_1+1)^3(n_2+1)^3(n_3+1)^3} = t^{-2}\sum_{n\geq 1}\frac{t^{2n}}{n^6} = t^{-2}\operatorname{Li}_6(t^2),$$

$$\sum_{(n_1,n_2,n_3)\in\mathcal{D}_2} \frac{t^{n_1+n_2+n_3}}{(n_1+1)^3(n_2+1)^3(n_3+1)^3} = t^{-2}\sum_{n\geq 2}\frac{t^{2n}}{n^6} = t^{-2}\operatorname{Li}_6(t^2) - 1$$

$$\sum_{(n_1,n_2,n_3)\in\mathcal{D}_3} \frac{t^{n_1+n_2+n_3}}{(n_1+1)^3(n_2+1)^3(n_3+1)^3} = \sum_{n>0, 0\leq\ell\leq n} \frac{t^{2(n+\ell)}}{(2\ell+1)^3(n+1)^6}$$

$$= -1 + 2^6 t^{-3} \sum_{n\geq 0,\, 0\leq \ell\leq n} \frac{t^{2n+2+2\ell+1}}{(2\ell+1)^3(2n+2)^6}$$

$$= -1 + 2^6 t^{-3} \sum_{\substack{0<m_1<m_2\\ m_1 \text{ odd },\, m_2 \text{ even}}} \frac{t^{m_1+m_2}}{m_1^6 m_2^3} = -1 + 2^6 t^{-3}\operatorname{Li}_{6,3}^{\mathcal{R}_{MP},\text{odd},\text{even}}(t,t)$$

$$\sum_{(n_1,n_2,n_3)\in\mathcal{D}_4} \frac{t^{n_1+n_2+n_3}}{(n_1+1)^3(n_2+1)^3(n_3+1)^3} = t^{-3}\operatorname{Li}_{3,3,3}^{\mathcal{R}_T,\text{even}}(t,t,t) + 1 - t^{-2}\operatorname{Li}_6(t^2)$$

$$\sum_{(n_1,n_2,n_3)\in\mathcal{D}_5} \frac{t^{n_1+n_2+n_3}}{(n_1+1)^3(n_2+1)^3(n_3+1)^3} = t^{-3}\operatorname{Li}_{3,3,3}^{\mathcal{R}_T,\text{even}}(t,t,t) + 1 - t^{-2}\operatorname{Li}_6(t^2),$$

where in the last two cases the term $t^{-2}\operatorname{Li}_6(t^2) - 1$ corresponds to the summation over $m_2 = 1$, $m_1 > 1$ and $m_3 = m_1$ (respectively, $m_1 = 1$, $m_2 > 1$, $m_3 = m_2$), with $m_i = n_1 + 1$. The integrand has a factor of t^4 for each edge, as in Lemma 4.13, which gives a power of t^{12} that combines with the t^{-3} factor in the result of the sum of the terms above to give the t^9 factor in (4.50). □

For more general graphs, where more vertices and more stars are involved, one gets summations involving several "triangular conditions" $|n_j - n_k| \leq n_i \leq n_j + n_k$ around each vertex, and the integrand can correspondingly be expressed in terms of series with a higher depth. Moreover, notice that we have focused here on the leading terms $\mathcal{K}^{D=4}_{\underline{n},\underline{\ell}=0}(n)$ only. When one includes all the other terms $\mathcal{K}_{\underline{n},\underline{\ell}}(n)$ with $\ell_i \neq 0$, the expressions become much more involved, as these coefficients are expressed in terms of the isoscalars (4.27) and of the standard $3j$-symbols for $SO(3)$, through the factorization (4.26) that expresses the triple $SO(4)$ Gaunt coefficients in terms of isoscalar factors and triple $SO(3)$ Gaunt coefficients. The isofactors are known explicitly [**5**], [**59**] so the computation can in principle be carried out in full, but it becomes much more cumbersome.

Next we show that the functions $\operatorname{Li}^{\mathcal{R}_T}_{s_1,s_2,s_3}(z_1,z_2,z_3)$ that appear in these Feynman amplitude computations can be related, via the Euler–Maclaurin summation formula, to some well known generalizations of multiple zeta values and multiple polylogarithms.

4.5.3. *Mordell–Tornheim and Apostol–Vu series.* We consider two generalizations of the multiple polylogarithm series, which arise in connection to the Mordell–Tornheim and the Apostol–Vu multiple series. The Mordell–Tornheim multiple series is given by [**69**], [**83**]

$$\zeta_{MT,k}(s_1,\dots,s_k;s_{k+1}) = \sum_{(n_1,\dots,n_k)\in\mathcal{R}_P^{(k)}} n_1^{-s_1}\cdots n_k^{-s_k}(n_1+\cdots+n_k)^{-s_{k+1}}, \tag{4.51}$$

with an associated multiple polylogarithm-type function

$$\mathrm{Li}^{MT}_{s_1,\dots,s_k;s_{k+1}}(z_1,\dots,z_k;z_{k+1}) = \sum_{(n_1,\dots,n_k)\in\mathcal{R}_P^{(k)}} \frac{z_1^{n_1}\cdots z_k^{n_k} z_{k+1}^{(n_1+\cdots+n_k)}}{n_1^{s_1}\cdots n_k^{s_k}(n_1+\cdots+n_k)^{s_{k+1}}}. \tag{4.52}$$

Similarly, the Apostol–Vu multiple series [**12**] is defined as

$$\zeta_{AV,k}(s_1,\dots,s_k;s_{k+1}) = \sum_{(n_1,\dots,n_k)\in\mathcal{R}_{MP}^{(k)}} n_1^{-s_1}\cdots n_k^{-s_k}(n_1+\cdots+n_k)^{-s_{k+1}}, \tag{4.53}$$

and we consider the associated multiple polylogarithm-type series

$$\mathrm{Li}^{AV}_{s_1,\dots,s_k;s_{k+1}}(z_1,\dots,z_k;z_{k+1}) = \sum_{(n_1,\dots,n_k)\in\mathcal{R}_{MP}^{(k)}} \frac{z_1^{n_1}\cdots z_k^{n_k} z_{k+1}^{(n_1+\cdots+n_k)}}{n_1^{s_1}\cdots n_k^{s_k}(n_1+\cdots+n_k)^{s_{k+1}}}. \tag{4.54}$$

4.5.4. *Euler–Maclaurin formula.* A way to understand better the behavior of the functions (4.48) with $\mathcal{R} = \mathcal{R}_T^{(3)}$ that appear in this result, is in terms of the Euler–Maclaurin summation formula.

LEMMA 4.15. *Let $f(t) = x^t t^{-s}$. Then*

$$f^{(k)}(t) = \sum_{j=0}^{k}(-1)^{k-j}\binom{k}{j}\binom{s+k-j-1}{k-j}(k-j)!t^{-(s+k-j)}x^t\log(x)^j. \tag{4.55}$$

PROOF. Inductively, we have

$$f^{(k)}(t) = \sum_{j=0}^{k}(-1)^{k-j}\binom{k}{j}s(s+1)\cdots(s+k-j-1)\,t^{-(s+k-j)}\,x^t\,\log(x)^j,$$

where $s(s+1)\cdots(s+k-j-1) = \binom{s+k-j-1}{k-j}(k-j)!$. □

The Euler–Maclaurin summation formula gives

$$\begin{aligned}\sum_{n=a}^{b} f(n) &= \int_a^b f(t)dt + \frac{1}{2}(f(b)+f(a)) \\ &+\sum_{k=2}^{N}\frac{b_k}{k!}(f^{(k-1)}(b)-f^{(k-1)}(a)) \\ &-\int_a^b \frac{B_N(t-[t])}{N!}f^{(N)}(t)\,dt,\end{aligned} \tag{4.56}$$

where b_k are the Bernoulli numbers and B_k the Bernoulli polynomials. We then have the following result.

PROPOSITION 4.16. *Consider the series* $\mathrm{Li}^{\mathcal{R}}_{s_1,s_2,s_3}(z_1,z_2,z_3)$ *defined as in* (4.46), *with* $\mathcal{R}=\mathcal{R}^{(3)}_T$. *When applying the Euler–Maclaurin formula to the innermost sum, the summation terms in* (4.56) *give rise to terms of the form*

$$\pm F_{j,k}(s_3,z_3)\ \mathrm{Li}^{AV}_{s_1,s_2;s_3+k-j}(z_1,z_2;z_3) \tag{4.57}$$

or

$$\pm F_{j,k}(s_3,z_3)\ \mathrm{Li}^{MT}_{s_1,s_3+k-j;s_2}(z_1,z_2;z_3), \tag{4.58}$$

where

$$F_{j,k}(s,z)=\frac{b_k}{k!}\binom{k}{j}\binom{s+k-j-1}{k-j}(k-j)!\ \log(z)^j \tag{4.59}$$

PROOF. For $\mathrm{Li}^{\mathcal{R}}_{s_1,s_2,s_3}(z_1,z_2,z_3)$, with $\mathcal{R}=\mathcal{R}^{(3)}_T$, the summation

$$\sum_{n_2-n_1<n_3<n_2+n_1}\frac{z_3^{n_3}}{n_3^{s_3}} \tag{4.60}$$

can be expressed, using Lemma 4.15, through the Euler–Maclaurin summation formula (4.56). Up to a sign, each summation term in the right-hand-side of (4.56) is the product of a function of z_3 of the form $F_{j,k}(s_3,z_3)$, as in (4.59), and a term of the form

$$\frac{z_3^{n_2+n_1}}{(n_2+n_1)^{s_3+k-j}}\quad\text{or}\quad\frac{z_3^{n_2-n_1}}{(n_2-n_1)^{s_3+k-j}}.$$

When inserted back into the summation on the remaining indices $n_2>n_1$, this gives summations of the form

$$\sum_{n_2>n_1>0}\frac{z_1^{n_1}z_2^{n_2}z_3^{n_1+n_2}}{n_1^{s_1}n_2^{s_2}(n_1+n_2)^{s_3+k-j}}, \tag{4.61}$$

in the first case, or in the second case, after a change of variables $m=n_2-n_1$, $n=n_1$ in the indices

$$\sum_{n>0,m>0}\frac{z_1^{n}z_3^{m}z_2^{n+m}}{n^{s_1}m^{s_3+k-j}(n+m)^{s_2}}, \tag{4.62}$$

which are respectively of the form (4.52) and (4.54). □

4.6. Conclusion. In this section we have seen that, in the "real case" of the physical Feynman amplitude (3.2), in dimension $D=4$, one can identify a specific contribution to the integral, which, after applying (4.26) that expresses the $SO(4)$ Gaunt coefficients in terms of isoscalars and $SO(3)$ Gaunt coefficients, is described naturally in representation theoretic terms as the $\ell=0$ (deepest) term in the $SO(3)$ isofactors. This term contributes an integral that is always expressible in terms of summations like Mordell-Tornheim and Apostol-Vu series, which are ultimately expressible in terms of multiple zeta values. Thus, even when, for sufficiently large graphs for which the Feynman integral would involve non-mixed Tate periods, this specific term remains within the class of mixed Tate periods. We do not have, at present, a direct motivic interpretation of this term.

Note, however, that here we only see this mixed Tate term at the level of the period integrals and not at the level of the motive. Thus, for instance, we can keep track of graph operations such as pasting two half edges through what happens to the integral, as illustrated for instance in Proposition 4.14, but not at the level of the motive itself. So, in particular, what we see in terms of the integrals does not directly imply any kind of "motivic decomposition" in the motive that would split off a mixed Tate summand. The question of whether anything of that sort may be the case would require a different investigation.

5. Wonderful compactifications and complexified amplitudes

We move now to discuss a different problem of regularization and renormalization of integrals in configuration spaces, based on our generalization of the Feynman amplitude (3.2) given by the complexified version (3.9) introduced in §3.3.2. As in Definition 3.6, the locus of integration is, in this case, the complex variety $X^{V_\Gamma} \times \{y = (y_v)\}$, for a fixed choice of a point $y = (y_v)$, inside the configuration space Z^{V_Γ} with $Z = X \times X$.

We described in detail in our previous work [**36**] the geometry of the wonderful compactifications of the configuration spaces $\mathrm{Conf}_\Gamma(X)$. We recall here the main definitions and statements, adapted from $\mathrm{Conf}_\Gamma(X)$ to $F(X, \Gamma)$. The arguments are essentially the same as in [**36**].

5.1. Arrangements of diagonals. A *simple arrangement* of subvarieties of a smooth quasi-projective ambient variety Y is a finite collection of nonsingular closed subvarieties $\mathcal{S} = \{S_i \subset Y, i \in I\}$ such that

- all nonempty intersections $\bigcap_{i\in J} S_i$ for $J \subset I$ are in the collection $\mathcal{S}$.
- for any pair $S_i, S_j \in \mathcal{S}$, the intersection $S_i \cap S_j$ is *clean*, that is, the tangent bundle of the intersection is the intersection of the restrictions of the tangent bundles.

5.1.1. *Diagonals of induced subgraphs and their arrangement.* For each induced and not necessarily connected subgraph $\gamma \subseteq \Gamma$, the corresponding (poly)diagonal is

$$\Delta_\gamma^{(Z)} = \{z = (z_v) \in Z^{V_\Gamma} \mid p(z_v) = p(z_{v'}) \text{ for } \{v, v'\} = \partial_\Gamma(e),\ e \in E_\gamma\}. \tag{5.1}$$

We have the following simple description.

LEMMA 5.1. *The diagonal $\Delta_\gamma^{(Z)}$ is isomorphic to $X^{V_{\Gamma//\gamma}} \times X^{V_\Gamma}$, and has dimension*

$$\dim \Delta_\gamma^{(Z)} = \dim X^{V_{\Gamma//\gamma}} \times X^{V_\Gamma} = (2N_\Gamma - N_\gamma + b_0(\gamma)) \dim(X) \tag{5.2}$$

with $N_\Gamma = \#V_\Gamma$, and with $b_0(\gamma)$ the number of connected components of γ.

We then observe that the diagonals form an arrangement of subvarieties. This is the analog of Lemma 5 of [**36**].

LEMMA 5.2. *For a given graph Γ, the collection*

$$\mathcal{S}_\Gamma = \{\Delta_\gamma^{(Z)} \mid \gamma \in \mathbf{SG}(\Gamma)\}, \tag{5.3}$$

with $\mathbf{SG}(\Gamma)$ the set of all induced subgraphs of Γ, is a simple arrangement of diagonal subvarieties in Z^{V_Γ}.

PROOF. Let γ_1 and γ_2 be a pair of induced subgraphs. If $\gamma_1 \cap \gamma_2 = \emptyset$, then $\gamma = \gamma_1 \cup \gamma_2$ is in $\mathbf{SG}(\Gamma)$, and the corresponding diagonal $\Delta_\gamma^{(Z)}$ is given by the intersection $\Delta_{\gamma_1}^{(Z)} \cap \Delta_{\gamma_2}^{(Z)}$. On the other hand, if $\gamma_1 \cap \gamma_2 \neq \emptyset$, we consider the connected components γ_j of the union γ. Then, the intersection $\Delta_{\gamma_1}^{(Z)} \cap \Delta_{\gamma_2}^{(Z)}$ can be written as $\cap_j \Delta_{i(\gamma_j)}^{(Z)}$ where $i(\gamma_j)$ is the smallest induced subgraph of Γ containing

γ_j. All diagonals are smooth and the ideal sheaf of intersection $\Delta_\gamma^{(Z)}$ is the direct sum of the ideal sheaves of the intersecting diagonals $\Delta_{\gamma_j}^{(Z)}$. By Lemma 5.1 of [**62**], their intersections are clean. □

5.1.2. *Building set of the arrangements of diagonals.* A subset $\mathcal{G} \subset \mathcal{S}$ is called a *building set* of the simple arrangement $\mathcal{S}$ if for any $S \in \mathcal{S}$, the minimal elements in $\{G \in \mathcal{G} : G \supseteq S\}$ intersect transversely and the intersection is S.

A $\mathcal{G}$-building set for the arrangement $\mathcal{S}_\Gamma$ can be identified by considering further combinatorial properties of graphs. A graph Γ is *2-vertex-connected* (or *biconnected*) if it cannot be disconnected by the removal of a single vertex along with the open star of edges around it. The graph consisting of a single edge is assumed to be biconnected. We then have the analog of Proposition 1 of [**36**].

PROPOSITION 5.3. *For a given graph Γ, the set*

$$\mathcal{G}_\Gamma = \{\Delta_\gamma^{(Z)} \mid \gamma \subseteq \Gamma \text{ induced, biconnected}\} \tag{5.4}$$

is a $\mathcal{G}$-building set for the arrangement $\mathcal{S}_\Gamma$ of (5.3).

PROOF. The intersection of a bi-connected subgraph of Γ with an induced subgraph is either empty or a union of bi-connected induced subgraphs attached at cut-vertices. We decompose induced subgraphs into bi-connected components. The diagonals corresponding to these bi-connected components are the minimal elements in the collection $\mathcal{S}_\Gamma$. For a pair of bi-connected induced subgraphs γ_1, γ_2 with $\gamma = \gamma_1 \cup \gamma_2$, we have the following equalities due to Lemma 5.1; $\dim \Delta_{\gamma_1}^{(Z)} + \dim \Delta_{\gamma_2}^{(Z)} - \dim \Delta_\gamma^{(Z)} = \dim(X^{V_{\Gamma//\gamma_1}} \times X^{V_\Gamma}) + \dim(X^{V_{\Gamma//\gamma_2}} \times X^{V_\Gamma}) - \dim(X^{V_{\Gamma//\gamma}} \times X^{V_\Gamma}) = 2\dim(X)N_\Gamma = \dim Z^{V_\Gamma}$, with $N_\Gamma = \#V_\Gamma$. These guarantee the transversality of the intersection $\Delta_{\gamma_1}^{(Z)} \cap \Delta_{\gamma_2}^{(Z)}$. □

5.2. The wonderful compactifications of arrangements. To be able to analyze the residues of Feynman integrals, we need a compactification $Z[\Gamma]$ of the configuration space $F(X,\Gamma)$ satisfying certain properties. In particular, $Z[\Gamma]$ must contain $F(X,\Gamma)$ as the top dimensional stratum, and the complement $Z[\Gamma]\backslash F(X,\Gamma)$ of this principal stratum must be a union of transversally intersecting divisors in $Z[\Gamma]$. The transversality is essential for the use of iterated Poincaré residues, which we will discuss in §7 below.

There is a smooth wonderful compactification $Z[\Gamma]$ of the configuration space $F(X,\Gamma)$ which is a generalization of the Fulton–MacPherson compactification [**48**]. The construction is completely analogous to the construction of the wonderful compactifications $\overline{\mathrm{Conf}_\Gamma}(X)$ considered in our previous paper [**36**]. Again, we illustrate here briefly what changes in passing from the case of $\mathrm{Conf}_\Gamma(X)$ to the case of $F(X,\Gamma)$.

5.2.1. *The iterated blowup description.* As in the case of $\mathrm{Conf}_\Gamma(X)$ (see §2.3 of [**36**]), the wonderful compactification $Z[\Gamma]$ is obtained by an iterated sequence of blowups. The following is the direct analog of Proposition 2 of [**36**].

Let $N = \#V_\Gamma$ and let $\mathcal{G}_{k,\Gamma} \subseteq \mathcal{G}_\Gamma$ be the subcollection

$$\mathcal{G}_{k,\Gamma} = \{\Delta_\gamma^{(Z)} \mid \gamma \in \mathbf{SG}_k(\Gamma) \text{ and biconnected}\} \quad \text{for} \quad k = 1, \ldots, N-1. \tag{5.5}$$

Let $Y_0 = Z^{V_\Gamma}$ and let Y_k be the blowup of Y_{k-1} along the (iterated) dominant transform of $\Delta_\gamma^{(Z)} \in \mathcal{G}_{N-k+1,\Gamma}$. If Γ is itself biconnected, then Y_1 is the blowup of Y_0 along the deepest diagonal $\Delta_\Gamma^{(Z)}$, and otherwise $Y_1 = Y_0$. Similarly, we have

$Y_k = Y_{k-1}$ if there are no biconnected induced subgraphs with exactly $N-k+1$ vertices. The resulting sequence of blowups

$$Y_{N-1} \to \cdots \to Y_2 \to Y_1 \to Z^{V_\Gamma} \tag{5.6}$$

does not depend on the order in which the blowups are performed along the (iterated) dominant transforms of the diagonals $\Delta_\gamma^{(Z)}$, for $\gamma \in \mathcal{G}_{N-k+1,\Gamma}$, for a fixed k. Thus, the intermediate varieties Y_k in the sequence (5.6) are all well defined. The variety Y_{N-1} obtained through this sequence of iterated blowups is called the wonderful compactification;

$$Z[\Gamma] := Y_{N-1}. \tag{5.7}$$

Note that $Z[\Gamma]$ is a smooth quasi-projective variety as can be seen through its iterated blow-up construction, see [**62**].

5.2.2. *Divisors and their intersections.* A flag $\mathcal{F}$ in the arrangement $\mathcal{S}_\Gamma$ consists of a sequence $\hat{\Delta}_{\gamma_1} \subseteq \hat{\Delta}_{\gamma_2} \subseteq \cdots \hat{\Delta}_{\gamma_r}$ of (poly)diagonals associated to disjoint unions of induced subgraphs γ_i. Recall from [**62**] and Definition 3.7.1 of [**68**] that a $\mathcal{G}_\Gamma$-*nest* $\mathcal{N}$ for the arrangement $\mathcal{S}_\Gamma$ is a collection $\{\gamma_1, \dots, \gamma_\ell\}$ of biconnected induced subgraphs with the property that there exists a flag $\mathcal{F}$ such that $\mathcal{N}$ is the set of all the $\mathcal{G}_\Gamma$-factors of the flag. The $\mathcal{G}_\Gamma$-factors of an element $\hat{\Delta}_{\gamma_i}$ of the flag are defined as the minimal elements of the building set $\mathcal{G}_\Gamma$ that contain $\hat{\Delta}_{\gamma_i}$.

The flags $\mathcal{F}$ of the arrangement $\mathcal{S}_\Gamma$ are in one-to-one correspondence with the forests of nested subgraphs $\mathcal{T}_\mathcal{F}$ (Proposition 3 of [**36**] and Proposition 3.7.3 of [**68**]). The latter are finite collections of rooted trees, where each component is a finite tree, where the vertices are labelled by connected induced subgraphs of Γ, in such a way that, if there is an edge from a vertex v_j to a vertex v_i (oriented away from the root), then the corresponding connected induced subgraphs satisfy $\gamma_i \subset \gamma_j$. Graphs assigned to vertices on different branches of a tree, or on different trees in the forest, are disjoint.

Given a subgraph $\gamma \subseteq \Gamma$, we denote by $\iota(\gamma)$ the smallest induced subgraph containing γ. By Proposition 3.7.3 of [**68**], given two biconnected induced subgraphs γ, γ', the set $\mathcal{N} = \{\Delta_\gamma, \Delta_{\gamma'}\}$ is a $\mathcal{G}_\Gamma$-nest if and only the graphs satisfy one of the following conditions:

(1) $\gamma \cap \gamma' = \emptyset$;
(2) $\iota(\gamma \cup \gamma')$ is not biconnected;
(3) $\gamma \subset \gamma'$ or $\gamma' \subset \gamma$.

REMARK 5.4. In [**62**] and in Proposition 3 of [**36**], the $\mathcal{G}_\Gamma$-nests are characterized in terms of a list of properties similar to the above, with the possibility of $\iota(\gamma \cup \gamma')$ not biconnected replaced by $\gamma \cap \gamma' = \{v\}$, a single vertex. However, the latter condition does not always correspond to a $\mathcal{G}_\Gamma$-nest, as observed by Martin and Sabbah, [**68**], [**73**]. This fact does not affect the results recalled below on the strata of the wonderful compactification.

We then have the following analog of Proposition 4 of [**36**] and Proposition 3.8.1 of [**68**].

PROPOSITION 5.5. *For a given biconnected induced subgraph $\gamma \subseteq \Gamma$, let $D_\gamma^{(Z)}$ be the divisor obtained as the iterated dominant transform of $\Delta_\gamma^{(Z)}$ in the iterated* *blowup construction* (5.6) *of* $Z[\Gamma]$. *Then*

$$Z[\Gamma] \setminus F(X,\Gamma) = \bigcup_{\Delta_\gamma^{(Z)} \in \mathcal{G}_\Gamma} D_\gamma^{(Z)}. \tag{5.8}$$

The divisors $D_\gamma^{(Z)}$ *have the property that*

$$D_{\gamma_1}^{(Z)} \cap \cdots \cap D_{\gamma_\ell}^{(Z)} \neq \emptyset \Leftrightarrow \{\gamma_1, \ldots, \gamma_\ell\} \text{ is a } \mathcal{G}_\Gamma\text{-nest.} \tag{5.9}$$

5.3. Motives of wonderful compactifications. As in the case of the wonderful compactifications $\overline{\mathrm{Conf}_\Gamma}(X)$ analyzed in [**36**], one can obtain the explicit formula for the motive of the compactifications $Z[\Gamma]$ directly from the formula for the motive of blow-ups and the iterated construction of §5.2.1.

We first introduce the following notation as in [**36**], [**61**]. Given a $\mathcal{G}_\Gamma$-nest $\mathcal{N}$, and a biconnected induced subgraph γ such that $\mathcal{N}' = \mathcal{N} \cup \{\gamma\}$ is still a $\mathcal{G}_\Gamma$-nest, we set

$$r_\gamma = r_{\gamma,\mathcal{N}} := \dim(\cap_{\gamma' \in \mathcal{N}: \gamma' \subset \gamma} \Delta_{\gamma'}) - \dim \Delta_\gamma, \tag{5.10}$$

$$M_\mathcal{N} := \{(\mu_\gamma)_{\Delta_\gamma \in \mathcal{G}_\Gamma} \,:\, 1 \leq \mu_\gamma \leq r_\gamma - 1,\ \mu_\gamma \in \mathbb{Z}\}, \tag{5.11}$$

$$\|\mu\| := \sum_{\Delta_\gamma \in \mathcal{G}_\Gamma} \mu_\gamma. \tag{5.12}$$

We write here $\mathfrak{m}(X)$ for the motive in the Voevodsky category. This corresponds to the notation M_{gm} of [**88**].

The following result is the analog of Proposition 8 of [**36**], see also [**61**] for the formulation in the case of Chow motives.

PROPOSITION 5.6. *Let* X *be a smooth projective variety. The Voevodsky motive* $\mathfrak{m}(Z[\Gamma])$ *of the wonderful compactification is given by*

$$\mathfrak{m}(Z[\Gamma]) = \mathfrak{m}(Z^{V_\Gamma}) \oplus \bigoplus_{\mathcal{N} \in \mathcal{G}_\Gamma\text{-nests}} \ \bigoplus_{\mu \in M_\mathcal{N}} \mathfrak{m}(X^{V_{\Gamma/\delta_\mathcal{N}(\Gamma)}} \times X^{V_\Gamma})(\|\mu\|)[2\|\mu\|] \tag{5.13}$$

where $\Gamma/\delta_\mathcal{N}(\Gamma)$ *is the quotient* $\Gamma/\delta_\mathcal{N}(\Gamma) := \Gamma//(\gamma_1 \cup \cdots \cup \gamma_r)$ *for the* $\mathcal{G}_\Gamma$*-nest* $\mathcal{N} = \{\gamma_1, \ldots, \gamma_r\}$.

PROOF. Let $\widetilde{Y} \to Y$ be the blow-up of a smooth scheme Y along a smooth closed subscheme $V \subset Y$. Then $\mathfrak{m}(\widetilde{Y})$ is canonically isomorphic to

$$\mathfrak{m}(Y) \oplus \bigoplus_{k=1}^{\mathrm{codim}_Y(V)-1} \mathfrak{m}(V)(k)[2k],$$

see Proposition 3.5.3 in [**88**]. The result then follows by applying this blow-up formula for Voevodsky's motives to the iterated blow-up construction given in Section 5.2.1. □

A simplified form of (5.13) (in the case of the Grothendieck class) was given in Theorem 4.3.3 of [**68**].

We obtain from Proposition 5.6 the following simple corollary (see §3.2 of [**36**]).

COROLLARY 5.7. *If the motive of the smooth projective variety X is mixed Tate, then the motive of $Z[\Gamma]$ is mixed Tate, for all graphs Γ. Moreover, the exceptional divisors $D_\gamma^{(Z)}$ associated to the biconnected induced subgraphs $\gamma \subseteq \Gamma$ and the intersections $D_{\gamma_1}^{(Z)} \cap \cdots \cap D_{\gamma_\ell}^{(Z)}$ associated to the $\mathcal{G}_\Gamma$-nests $\{\gamma_1, \ldots, \gamma_\ell\}$ are also mixed Tate.*

PROOF. This is an immediate consequence of the construction of $Z[\Gamma]$ since the motive of $Z[\Gamma]$ depends upon the motive of X only through products, Tate twists, sums, and shifts. All these operations preserve the subcategory of mixed Tate motives. The reason why the intersections $D_{\gamma_1}^{(Z)} \cap \cdots \cap D_{\gamma_\ell}^{(Z)}$ are also mixed Tate is because one has an explicit stratification, as described in [**36**] and [**62**], where one has a description of the intersections of diagonals in terms of configuration spaces of quotient graphs and by repeated use of the blowup formula for motives. □

REMARK 5.8. One can also see easily that, if the variety X is defined over $\mathbb{Z}$, then so is $Z[\Gamma]$ and so are the $D_\gamma^{(Z)}$ and their unions and intersections. Moreover, all these varieties then satisfy the unramified criterion of §3.5 and Proposition 3.10 of [**50**].

5.4. Complexified amplitude and wonderful compactifications. We now consider the form $\omega_\Gamma^{(Z)}$ defined as in (3.9) and discuss its behavior when pulled back from Z^{V_Γ} to the wonderful compactification $Z[\Gamma]$.

5.4.1. *Loci of divergence.* For massless scalar Euclidean field theories, the pole locus $\{\omega_\Gamma^{(Z)} = \infty\}$ in $Z^{\mathbf{V}_\Gamma}$ is

$$\mathcal{Z}_\Gamma := \left\{ \prod_{e \in E_\Gamma} \|p(z_{v_{s(e)}}) - p(z_{v_{t(e)}})\|^2 = 0 \right\}. \tag{5.14}$$

Since we have used the Euclidean norm in the definition of the amplitudes, rather than an algebraic quadratic form, the locus of divergences is simply a union of $\Delta_e^{(Z)} = \Delta_e \times X^{V_\Gamma}$, hence we can restate the above as follows.

LEMMA 5.9. *The divergent locus of the density $\omega_\Gamma^{(Z)}$ of (3.9) in Z^{V_Γ} is given by the union $\bigcup_{e \in E_\Gamma} \Delta_e^{(Z)}$.*

5.4.2. *Order of singularities in the blowups.* Let $\pi_\gamma^*(\omega_\Gamma^{(Z)})$ denote the pullbacks of the form $\omega_\Gamma^{(Z)}$ of (3.9) to the iterated blowups of $Z^{\mathbf{V}_\Gamma}$ along the (dominant transforms of) the diagonals $\Delta_\gamma^{(Z)}$, for $\gamma \subset \Gamma$ a biconnected induced subgraph.

PROPOSITION 5.10. *Let Γ be a connected graph. Then for every biconnected induced subgraph $\gamma \subset \Gamma$, the pullback $\pi_\gamma^*(\omega_\Gamma^{(Z)})$ of $\omega_\Gamma^{(Z)}$ to the blowup along the (dominant transform of) $\Delta_\gamma^{(Z)}$ has singularities of order*

$$\begin{aligned} \operatorname{ord}_\infty(\pi_\gamma^*(\omega_\Gamma), D_\gamma^{(Z)}) &= (2D-2)\#E_\gamma - 2D(\#V_\gamma - 1) + 2 \\ &= 2Db_1(\gamma) - 2\#E_\gamma + 2 \end{aligned} \tag{5.15}$$

along the exceptional divisors $D_\gamma^{(Z)}$ in the blowup. Here $b_1(\gamma)$ denotes the first Betti number of graph γ.

PROOF. Let $m = DN_\Gamma$, with $N_\Gamma = \#V_\Gamma$ and $L \subset \mathbb{A}^{2m}$ be the coordinate subspace given by the equations $\{x_1 = \cdots = x_k = 0\}$, and $\pi : \widetilde{\mathbb{A}}^{2m} \to \mathbb{A}^{2m}$ be the blowup along $L \subset \mathbb{A}^{2m}$. If one chooses the coordinates w_i in the blow up, so that $w_i = x_i$, for $i = k, \dots, 2m$, and $w_i w_k = x_i$ for $i < k$. The exceptional divisor is given by $w_k = 0$ in these coordinates. Then, one obtains

$$\pi^*(dx_1 \wedge d\bar{x}_1 \wedge \cdots \wedge dx_m \wedge d\bar{x}_m) = |w|^{2(k-1)} dw_1 \wedge d\bar{w}_1 \wedge \cdots \wedge dw_m \wedge d\bar{w}_m.$$

This form has a zero of order $2 \cdot (\mathrm{codim}(L) - 1)$ along the exceptional divisor of the blowup.

The codimension of the diagonal $\Delta_\gamma \subset X^{V_\Gamma}$ associated to a connected subgraph $\gamma \subset \Gamma$ is $D\ (N_\gamma - 1)$, with $N_\gamma = \# V_\gamma$ On the other hand, the form $\omega_\Gamma^{(Z)}$ has singularity along $\Delta_\gamma^{(Z)}$ of order $(2D-2)\# E_\gamma$. Hence,

$$\begin{aligned} \mathrm{ord}_\infty(\pi_\gamma^*(\omega_\Gamma^{(Z)}), D_\gamma^{(Z)}) &= (\text{order of } \infty) - (\text{order of zeros}) \\ &= (2D-2)n_\gamma - 2D(N_\gamma - 1) + 2. \end{aligned}$$

with $n_\gamma = \# E_\gamma$ and $N_\gamma = \# V_\gamma$. □

REMARK 5.11. Note that the orders of pole are different from the case of the form ω_Γ on $\overline{\mathrm{Conf}}_\Gamma(X)$, see §4.3 of [**36**]. In particular, note that we have order of singularity $(2D-2)n_\gamma$ along $\Delta_\gamma^{(Z)}$, with $n_\gamma = \# E_\gamma$, because the form $\omega_\Gamma^{(Z)}$ of (3.9) is built using the Green functions of the complex Laplacian, (3.17), which has exponent $2D-2$ instead of the exponent $D-2$ of the real case. In particular, in the real case of the physical Feynman amplitude one has the familiar log divergent condition for graphs, given by the relation $(D-2)n_\gamma = D(N_\gamma - 1)$, with $N_\gamma = \# V_\gamma$, or equivalently $n_\gamma = b_1(\gamma)D/2$. In the complexified case we are considering here, this relation is replaced by the condition $(D-1)n_\gamma = D(N_\gamma - 1)$, or equivalently $n_\gamma = b_1(\gamma)D$. This is compatible with the usual log divergent condition, since in the complexified case D is the complex dimension.

On the basis of the remark here above, we make the following definition, which replaces the usual definition of log divergent graphs of the real case.

DEFINITION 5.12. A graph Γ is primitive $\mathbb{C}$-log divergent if it satisfies the following conditions:

(1) $\# E_\Gamma = D b_1(\Gamma)$
(2) all subgraphs $\gamma \subseteq \Gamma$ satisfy $\# E_\gamma \geq D b_1(\gamma)$.

Lemma 5.9 and Proposition 5.10 then imply the following.

COROLLARY 5.13. *Let $\pi_\Gamma^*(\omega_\Gamma^{(Z)})$ denote the pullback of $\omega_\Gamma^{(Z)}$ to the wonderful compactification $Z[\Gamma]$. The divergence locus of $\pi_\Gamma^*(\omega_\Gamma^{(Z)})$ in $Z[\Gamma]$ is given by the union of divisors*

$$\mathcal{D}_\Gamma = \bigcup_{\Delta_\gamma^{(Z)} \in \mathcal{G}_\Gamma} D_\gamma^{(Z)}. \tag{5.16}$$

5.5. Chain of integration and divergence locus. When pulling back the form $\omega_\Gamma^{(Z)}$ along the projection $\pi_\Gamma : Z[\Gamma] \to Z^{V_\Gamma}$, one also replaces the chain of integration $\sigma_\Gamma^{(Z,y)} = X^{V_\Gamma} \times \{y\}$ of (3.10) with $\tilde{\sigma}_\Gamma^{(Z,y)} \subset Z[\Gamma]$ with $\pi_\Gamma(\tilde{\sigma}_\Gamma^{(Z,y)}) = \sigma_\Gamma^{(Z,y)}$, which gives

$$\tilde{\sigma}_\Gamma^{(Z,y)} = \overline{\mathrm{Conf}}_\Gamma(X) \times \{y\} \subset Z[\Gamma] = \overline{\mathrm{Conf}}_\Gamma(X) \times X^{V_\Gamma}. \tag{5.17}$$

LEMMA 5.14. *The chain of integration* $\tilde{\sigma}_\Gamma^{(Z,y)}$ *of* (5.17) *intersects the locus of divergence* (5.16) *in*

$$\bigcup_{\Delta_\gamma^{(Z)} \in \mathcal{G}_\Gamma} D_\gamma \times \{y\} \subset \overline{\mathrm{Conf}}_\Gamma(X) \times \{y\}. \tag{5.18}$$

PROOF. This follows directly from Corollary 5.13 and (5.17). □

Notice that, since $\mathcal{G}_\Gamma$-factors intersect transversely (see Proposition 2.8 of [**62**] and Proposition 4 of [**36**]), the intersection (5.18) of the chain of integration $\tilde{\sigma}_\Gamma^{(Z,y)}$ with the locus of divergence consists of a union of divisors D_γ inside X^{V_Γ} intersecting transversely, with $D_{\gamma_1} \cap \cdots \cap D_{\gamma_\ell} \neq \emptyset$ whenever $\{\gamma_1, \ldots, \gamma_\ell\}$ form a $\mathcal{G}_\Gamma$-nest (see [**36**] and [**62**]).

5.6. Smooth and algebraic differential forms. Consider the restriction of the amplitude $\pi_\Gamma^*(\omega_\Gamma^{(Z)})$ to the chain $\tilde{\sigma}_\Gamma^{(Z,y)}$. It is defined on the complement of the divergence locus, namely on

$$\tilde{\sigma}_\Gamma^{(Z,y)} \smallsetminus \left(\bigcup_{\Delta_\gamma^{(Z)} \in \mathcal{G}_\Gamma} D_\gamma \times \{y\} \right) \simeq \overline{\mathrm{Conf}}_\Gamma(X) \smallsetminus \left(\bigcup_{\Delta_\gamma^{(Z)} \in \mathcal{G}_\Gamma} D_\gamma \right), \tag{5.19}$$

which is a copy of $\mathrm{Conf}_\Gamma(X)$ inside $Z[\Gamma]$. The form $\pi_\Gamma^*(\omega_\Gamma^{(Z)})$ is a closed form of top dimension on this domain.

Our next goal is to replace the form $\pi_\Gamma^*(\omega_\Gamma^{(Z)})$ with a cohomologous form with logarithmic poles along the divisor $\mathcal{D}_\Gamma = \cup_{\Delta_\gamma^{(Z)} \in \mathcal{G}_\Gamma} \mathcal{D}_\gamma^{(Z)}$ of the compactification. As we see below, this could be done easily if one remains within the class of $\mathcal{C}^\infty$-forms, by directly applying the Griffiths-Schmid theorem of [**54**]. However, in view of identifying our regularized integrals with periods, we want something stronger, namely an *algebraic* form with logarithmic poles. We first need to recall some general well known facts about de Rham cohomology and forms with logarithmic poles.

5.6.1. *Forms with logarithmic poles.* Let $\mathcal{X}$ be a smooth projective variety of dimension m and let $\mathcal{D}$ be a union of smooth hypersurfaces intersecting transversely (strict normal crossings divisor). Let $\mathcal{U} = \mathcal{X} \smallsetminus \mathcal{D}$ and let $j : \mathcal{U} \hookrightarrow \mathcal{X}$ denote the inclusion. Let $\mathcal{M}^\star_{\mathcal{D},\mathcal{X}}$ denote the meromorphic de Rham complex, namely the complex of meromorphic forms on $\mathcal{X}$ with poles (of arbitrary order) along $\mathcal{D}$. Let $\Omega^\star_{\mathcal{X}}(\log \mathcal{D})$ denote the complex of differential forms on $\mathcal{X}$ with logarithmic poles along $\mathcal{D}$. We recall some general facts about cohomology classes, algebraic forms, and forms with logarithmic poles:

(1) **Grothendieck Comparison Theorem**, [**55**]: the natural morphism (de Rham morphism)

$$\mathcal{M}^\star_{\mathcal{D},\mathcal{X}} \to Rj_* \mathbb{C}_{\mathcal{U}}$$

is a quasi-isomorphism. This implies that the de Rham cohomology $H^\star_{dR}(\mathcal{U})$ is computed by the hypercohomology of the meromorphic de Rham complex. If $\mathcal{U}$ is affine, then the hypercohomology can be replaced by the ordinary cohomology of the complex of global sections.

(2) **Logarithmic Comparison Theorem**, [**43**]: for $\mathcal{D}$ a simple normal crossings divisor, the natural morphism

$$\Omega^\star_{\mathcal{X}}(\log \mathcal{D}) \to \mathcal{M}^\star_{\mathcal{D},\mathcal{X}}$$

is also a quasi-isomorphism. Combined with the previous statement, this implies that the de Rham cohomology $H^\star_{dR}(\mathcal{U})$ is computed by the hypercohomology of the logarithmic de Rham complex,

$$H^\star_{dR}(\mathcal{U}) \simeq \mathbb{H}^\star(\mathcal{X}, \Omega^\star_{\mathcal{X}}(\log \mathcal{D})). \tag{5.20}$$

However, even in the affine case, it is not always possible to replace the hypercohomology in (5.20) with ordinary cohomology and global sections, see Example 5.15.

(3) **Stein open sets** (Lemma 2.5 of [**34**]): the Logarithmic Comparison Theorem is equivalent to the statement that, for all Stein open sets $\mathcal{V} \subset \mathcal{X}$, there are isomorphisms $H^\star(\Gamma(\mathcal{V}, \Omega^\star_{\mathcal{X}}(\log \mathcal{D}))) \simeq H^\star_{dR}(\mathcal{V} \smallsetminus \mathcal{D})$. Namely, the hypercohomology in (5.20) can be replaced with ordinary cohomology of global sections when restricted to Stein open sets.

(4) **Hodge filtration**, [**43**]: there is a mixed Hodge structure on $\mathcal{U}$ with Hodge filtration given by

$$F^p H^k_{dR}(\mathcal{U}) = \mathrm{Im}(\mathbb{H}^k(\mathcal{X}, \Omega^{\geq p}_{\mathcal{X}}(\log \mathcal{D})) \to \mathbb{H}^k(\mathcal{X}, \Omega^\star_{\mathcal{X}}(\log \mathcal{D}))).$$

The de Rham cohomology classes that lie in the term $F^m H^m_{dR}(\mathcal{U})$ of the Hodge filtration can be realized by global forms with logarithmic poles, since there is an isomorphism

$$F^m H^m_{dR}(\mathcal{U}) = H^0(\mathcal{X}, \Omega^m_{\mathcal{X}}(\log \mathcal{D})). \tag{5.21}$$

This follows from the degeneration at the E_1 term of the Frölicher spectral sequence associated to the Hodge filtration (see Proposition 5.16 of [**67**], which we reproduce below as Proposition 5.16).

(5) **Griffiths-Schmid Theorem**, [**54**]: Let $\Omega_{\mathcal{C}^\infty(\mathcal{X})}(\log \mathcal{D})$ denote the $\mathcal{C}^\infty$-logarithmic de Rham complex, with $\mathcal{D}$ a simple normal crossings divisor in a complex smooth projective variety $\mathcal{X}$. Then there is an isomorphism of cohomologies $H^*_{dR}(\mathcal{U}) = H^*(\Omega_{\mathcal{C}^\infty(\mathcal{X})}(\log \mathcal{D}))$.

EXAMPLE 5.15. The hypercohomology in (5.20) cannot, in general, be replaced by cohomology of global sections, even assuming that $\mathcal{U}$ is affine. Let $\mathcal{U}$ be the complement of a finite set of n points in a smooth projective curve $\mathcal{X}$. Then $\dim H^1_{dR}(\mathcal{U}) = 2g + n - 1$, while by Riemann–Roch the space of global sections of the sheaf of logarithmic differentials has dimension $g + n - 1$.

Let $\mathcal{X}$ be a smooth projective variety of dimension $n = \dim \mathcal{X}$, with $\mathcal{D}$ a simple normal crossings divisor with affine complement $\mathcal{U} = \mathcal{X} \smallsetminus \mathcal{D}$. The Hodge filtration $F^p H^k_{dR}(\mathcal{U})$ is induced by the naive filtration on $\Omega^\star_{\mathcal{X}}(\log \mathcal{D})$.

The following fact is very well known, see [**43**]. We recall it here simply for the reader's convenience.

PROPOSITION 5.16. *The Hodge filtration satisfies*

$$F^n H^n_{dR}(\mathcal{U}) = H^0(\mathcal{X}, \Omega^n_{\mathcal{X}}(\log \mathcal{Z})). \tag{5.22}$$

PROOF. Given a complex $K^\star$ that comes from a double complex $K^{p,q}$, the spectral sequence of a filtration F with

$$F^p K^n = \oplus_{r \geq p, r+s=n} K^{r,s}$$

is given by

$$E_0^{p,q} = \mathrm{Gr}_p^F K^{p+q} = F^p K^{p+q}/F^{p+1}K^{p+q} = K^{p,q}$$
$$E_1^{p,q} = H^{p+q}(\mathrm{Gr}_p^F K^\star) = H^q(K^{p,\star})$$
$$E_\infty^{p,q} = \mathrm{Gr}_p^F H^{p+q}(K^\star).$$

In particular, the Frölicher spectral sequence associated to the Hodge filtration $F^p H^k_{dR}(\mathcal{U})$ has

$$E_1^{p,q} = H^q(\mathcal{X}, \Omega^p_{\mathcal{X}}(\log \mathcal{D}))$$
$$E_\infty^{p,q} = F^p H^{p+q}_{dR}(\mathcal{U})/F^{p+1}H^{p+q}_{dR}(\mathcal{U}).$$

In particular, $E_1^{n,0} = H^0(\mathcal{X}, \Omega^n_{\mathcal{X}}(\log \mathcal{D}))$ and $E_\infty^{n,0} = F^n H^n_{dR}(\mathcal{U})$. Deligne proved in [**43**] that, if $\mathcal{D}$ is normal crossings, then this spectral sequence degenerates at the E_1 term. This gives (5.22). □

5.7. Order of pole and Hodge filtration. It is in general difficult to estimate where a given class lies in the Hodge filtration. However, in the case of meromorphic forms, one can estimate the position in the Hodge fitration in terms of a filtration on the complex of meromorphic forms determined by the order of pole. More precisely, if $\mathcal{X}$ is a smooth projective variety and $\mathcal{D} \subset \mathcal{X}$ a simple normal crossings divisor, and $\mathcal{M}^\star_{\mathcal{D},\mathcal{X}}$ denotes the complex of meromorphic differential forms, one defines the polar filtration by taking $P^k\mathcal{M}^m_{\mathcal{D},\mathcal{X}}$ to consist of meromorphic m-forms with poles of order at most $m-k+1$, if $m-k \geq 0$ and zero otherwise. Deligne showed in §II.3, Proposition 3.13 of [**42**] and Proposition 3.1.11 of [**43**], that the natural morphism

$$(\Omega^\star_{\mathcal{X}}(\log \mathcal{D}), F^\star) \to (\mathcal{M}^\star_{\mathcal{D},\mathcal{X}}, P^\star)$$

is a filtered quasi-isomorphism.

THEOREM 5.17. *Let Γ be a primitive $\mathbb{C}$-log divergent graph in the sense of Definition 5.12. Then there is a meromorphic form with logarithmic poles $\beta^{(Z)}_\Gamma$ such that $\omega^{(Z)}_\Gamma$ can be replaced, up to coboundaries, by a form $\eta^{(Z)}_\Gamma = \epsilon\, \beta^{(Z)}_\Gamma \wedge \bar\beta^{(Z)}_\Gamma$, where ϵ is a sign.*

PROOF. We proceed as in Proposition 5.10. It suffices to check on a local model, with local coordinates. In $\mathbb{A}^{2m}$ with coordinates $\{x_1, \dots, x_{2m}\}$ consider a subspace of codimension k, given by $L = \{x_1 = \cdots = x_k = 0\}$, for some $k < m$. Consider a smooth differential form

$$\omega = \prod_{j\in E} \frac{1}{\|x_j\|^{2r}} \bigwedge_{i=1}^k dx_i \wedge d\bar{x}_i \wedge \theta_k,$$

where $\theta_k = dx_{k+1} \wedge d\bar{x}_{k+1} \wedge \cdots \wedge dx_m \wedge d\bar{x}_m$. The set E is a subset (with repetitions) of the coordinates x_i with $i = 1, \dots, k$. The form ω can be rewritten, for convenience, as a product $\omega = \epsilon\, \alpha \wedge \bar\alpha$, where ϵ is a sign and α is the meromorphic form

$$\alpha = \left(\prod_{j\in E} \frac{1}{x_j^r} \bigwedge_{i=1}^k dx_i \right) \wedge dx_{k+1} \wedge \cdots \wedge dx_m.$$

Let $\pi : \tilde{\mathbb{A}}^{2m} \to \mathbb{A}^{2m}$ be the blowup of $\mathbb{A}^{2m}$ along L. On $\tilde{\mathbb{A}}^{2m}$ we have local coordinates w_i with $w_i = x_i$ for $i \geq k$ and $w_i w_k = x_i$ for $i < k$, where $\{w_k = 0\}$ is the exceptional divisor of the blowup. The pullback gives

$$\pi^*(\bigwedge_{i=1}^{k} dx_i) = w_k^{k-1} \bigwedge_{i=1}^{k} dw_i$$

and $\pi^*(dx_{k+1} \wedge \cdots \wedge dx_m) = dw_{k+1} \wedge \cdots \wedge dw_m$, hence the pull back $\pi^*(\omega) = \epsilon\pi^*(\alpha) \wedge \overline{\pi^*(\alpha)}$ has a meromorphic form $\pi^*(\alpha)$ with a pole of order $re - k + 1$ along the exceptional divisor, where $e = \#E$. The form $\pi^*(\alpha)$, which is of degree m and pole of order $re - k + 1$ is therefore in the term P^ℓ of the polar filtration, with $\ell = m - (re - k + 1) + 1 = m - re + k$. This implies that its class is in the term of the Hodge filtration of order at least $m - re + k$.

When we apply this to the form $\omega_\Gamma^{(Z)}$ and the blowup along a diagonal $\Delta_\gamma^{(Z)}$ of codimension $k = D(N_\gamma - 1)$, with $N_\gamma = \#V_\gamma$, and with $r = D - 1$ and $e = e_\gamma = \#E_\gamma$, we obtain $\omega_\Gamma^{(Z)} = \epsilon\, \alpha_\Gamma^{(Z)} \wedge \bar{\alpha}_\Gamma^{(Z)}$, where $\alpha_\Gamma^{(Z)}$ is a meromorphic form of degree DN_Γ, with poles of order $(D-1)e_\gamma - D(N_\gamma - 1) + 1$ along the divisor $D_\gamma^{(Z)}$. Thus, the class of the form $\alpha_\Gamma^{(Z)}$ is in the term $F^{DN_\Gamma + D(N_\gamma - 1) - (D-1)e_\gamma} H^{DN_\Gamma}(\mathcal{U})$ of the Hodge filtration, unless the pole order is lowered by cancellations due to exact forms, which would result in a higher level in the filtration. In particular, if the graph Γ is $\mathbb{C}$-log divergent, then the subgraphs $\gamma \subseteq \Gamma$ satisfies $D(N_\gamma - 1) - (D-1)e_\gamma \geq 0$, see Remark 5.11 and Definition 5.12. Then the meromorphic form $\alpha_\Gamma^{(Z)}$ is in $F^{DN_\Gamma} H^{DN_\Gamma}(\mathcal{U})$. In this case, by the general result on the Hodge filtration and the Frölicher spectral sequence mentioned in (5.21) above, we can replace $\alpha_\Gamma^{(Z)}$ by a global DN_Γ-form $\beta_\Gamma^{(Z)}$ with logarithmic poles along $D_\gamma^{(Z)}$. One then checks that, if the meromorphic form $\alpha_\Gamma^{(Z)}$ can be replaced, up to a coboundary, by $\beta_\Gamma^{(Z)}$, then the closed form $\omega_\Gamma^{(Z)}$ can also be replaced, up to a coboundary, by $\eta_\Gamma^{(Z)} = \epsilon\beta_\Gamma^{(Z)} \wedge \bar{\beta}_\Gamma^{(Z)}$. Indeed, if $\alpha_\Gamma^{(Z)} = \beta_\Gamma^{(Z)} + d\xi_\Gamma^{(Z)}$ then $\omega_\Gamma^{(Z)} - \eta_\Gamma^{(Z)} = \epsilon_1 \beta_\Gamma \wedge d\bar{\xi}_\Gamma^{(Z)} + \epsilon_2 d\xi_\Gamma^{(Z)} \wedge \bar{\beta}_\Gamma^{(Z)} + \epsilon_3 d\xi_\Gamma^{(Z)} \wedge d\bar{\xi}_\Gamma^{(Z)}$, where the ϵ_i are signs. Since $d\alpha_\Gamma^{(Z)} = d\beta_\Gamma^{(Z)} = 0$, the expression above is a coboundary. □

5.8. Toy model regularization. Consider the simple toy model example of the differential form

$$\omega = \frac{dz \wedge d\bar{z}}{\|z\|^2} = \frac{dz}{z} \wedge \frac{d\bar{z}}{\bar{z}} \tag{5.23}$$

on $\mathbb{G}_m(\mathbb{C}) = \mathbb{C}^*$.

LEMMA 5.18. *A cutoff regularization of the integral on $\mathbb{C}^*$ of (5.23) has a logarithmic divergence in the cutoff scale, with a coefficient expressible in terms of a period integral of the form with logarithmic poles dz/z.*

PROOF. Using polar coordinates $z = re^{i\theta}$ we can write ω as

$$\omega = -2i\, \frac{dr}{r} \wedge d\theta.$$

The integral on $\mathbb{C}^*$ has divergences at both $r = 0$ and $r \to \infty$: integrating with cutoffs leads to

$$-2i \int_0^{2\pi} d\theta \int_\epsilon^\Lambda \frac{dr}{r} = -4\pi i \log(\frac{\Lambda}{\epsilon}).$$

The coefficient is an integer multiple of the period of the meromorphic form with logarithmic poles

$$2\pi i = \int_{S^1} \frac{dz}{z}. \tag{5.24}$$

□

Note that the divergent integral $\int_{\mathbb{C}^*} \omega$ also admits other forms of regularization:

(1) Gaussian cutoff:

$$\mathcal{I}_\epsilon(a) = \int_\epsilon^\infty \frac{dr}{r} e^{-ar} = \mathrm{Ei}(-a\epsilon) \sim \log(a\epsilon) + \gamma - a\epsilon + \frac{(a\epsilon)^2}{4} + \cdots,$$

with Ei the exponential integral function;

(2) Pauli–Villars regularization: $\mathcal{I}(a) - \mathcal{I}(b) = \int_0^\infty \frac{dr}{r}(e^{-ar} - e^{-br}) = \log(\frac{a}{b})$;

(3) Dimensional regularization:

$$\mathcal{I}_s(a) = \mu^{2s} \int_0^\infty \frac{dr}{r^{1-s}} e^{-ar} = -\Gamma(s)(\frac{a}{\mu^2})^{-s} \sim \frac{1}{s} + \gamma + \log\frac{a}{\mu^2} + \cdots$$

As a similar, higher dimensional, toy model, we consider the case of the $2n$-form

$$\omega = \frac{dz_1}{z_1} \wedge \frac{d\bar{z}_1}{\bar{z}_1} \wedge \cdots \wedge \frac{dz_n}{z_n} \wedge \frac{d\bar{z}_n}{\bar{z}_n} \tag{5.25}$$

on the torus $\mathbb{G}_m^n$ inside the affine space $\mathbb{A}^n$. As before, we can regularize the integral of ω on $\mathbb{G}_m^n$ with a cutoff regularization (or with one of the other regularization methods listed above).

LEMMA 5.19. *A cutoff regularization of the integral on $\mathbb{G}_m(\mathbb{C})^n$ of* (5.25) *has a logarithmic divergence in the cutoff scale, with a coefficient expressible in terms of a period integral of an n-form with logarithmic poles η, integrated on an n-dimensional locus Σ (a real torus) inside the Leray coboundary $\mathcal{L}(\mathcal{D}) = \cup_i \mathcal{L}(D_i)$ of the divisor $\mathcal{D} = \cup_i D_i$ of coordinate hyperplanes $D_i = \{z_i = 0\}$.*

PROOF. One simply applies the previous lemma coordinatewise and obtains a product integral with one logarithmic divergence in each variable. The coefficient of this divergence is (up to a sign) of the form $2^n \int_\Sigma \eta$, where $\eta = \frac{dz_1}{z_1} \wedge \cdots \wedge \frac{dz_n}{z_n}$ and $\Sigma = T^n = S^1 \times \cdots \times S^1$. We view $\Sigma \subset \cup_i \mathcal{L}(D_i)$ where $\mathcal{L}(D_i)$ is the boundary of a unit disk tubular neighborhood of D_i. □

With this toy model case in mind, we see how one can approach the computation of a regularized integral of the form $\beta_\Gamma^{(Z)} \wedge \bar{\beta}_\Gamma^{(Z)}$ obtained in Theorem 5.17, with the period (5.24) replaced by period integrals of iterated Poincaré residues of the form with logarithmic poles $\beta_\Gamma^{(Z)}$, on components and intersections of components of the divisor $\mathcal{D}_\Gamma^{(Z)}$ of the compactification, or equivalently integrals of the form $\beta_\Gamma^{(Z)}$ along iterated Leray coboundaries. We will discuss this procedure in the coming sections. For the remaining of this section, we discuss how one can proceed, in the case of graphs that are not $\mathbb{C}$-log divergent, to use forms with logarithmic poles.

5.9. Cellular structures, torifications, and constructible torifications. We now sketch a possible approach for the more general case, where the graph Γ is not necessarily $\mathbb{C}$-log divergent. In this case, unlike the previous result of Theorem 5.17, we cannot appeal to the Hodge filtration result to replace the meromorphic form $\pi_\Gamma^*(\alpha_\Gamma^{(Z)})$ in the pullback $\pi_\Gamma^*(\omega_\Gamma^{(Z)}) = \epsilon\pi_\Gamma^*(\alpha_\Gamma^{(Z)}) \wedge \pi_\Gamma^*(\bar{\alpha}_\Gamma^{(Z)})$ with a meromorphic form $\beta_\Gamma^{(Z)}$ with logarithmic poles. Thus, we need to proceed in a different way. A useful observation, for this case, is the "Stein open sets" method mentioned in §5.6.1, based on Lemma 2.5 of [**34**]. More precisely, we will show that if a smooth projective variety $\mathcal{X}$ has a decomposition into cells (cellular structure), or more generally into tori (torifications, or the certain weaker structures of constructible torifications) then the restriction of a meromorphic form α to each top dimensional set in the decomposition can be replaced by a form β with logarithmic poles. We will then use results of [**15**], [**64**] on the existence of (constructible) torifications on the compactifications $\overline{\mathrm{Conf}}_\Gamma(X)$, under the assumption that X has a cell decomposition.

As before, suppose given a smooth projective variety $\mathcal{X}$ with a normal crossings divisor $\mathcal{D}$, and a form $\omega = \alpha \wedge \bar{\alpha}$, where α is a meromorphic form on $\mathcal{X}$ with poles along $\mathcal{D}$. The first observation (see Proposition 5.8 of [**67**]) is that, if $\mathcal{X}$ has a cellular decomposition, with top dimensional cells $\mathcal{X}_k$, then by Lemma 2.5 of [**34**] the restrictions $\alpha|_{\mathcal{X}_k}$ can be replaced with cohomologous meromorphic forms β_k on $\mathcal{X}_k$ with logaritmic poles along $\mathcal{D}$. These forms do not necessarily match on the boundaries of the cells, due to the existence of nontrivial Čech cocycles: the hypercohomology cannot be replaced globally by cohomology, but it is possible locally in the Stein open sets $\mathcal{X}_k$. The regularized integral of ω on $\mathcal{X} \smallsetminus \mathcal{D}$ can then be replaced by a sum of regularized integrals $\sum_k \int_{X_k \smallsetminus \mathcal{D}} \omega|_{X_k}$. After replacing $\omega|_{X_k}$ with $\beta_k \wedge \bar{\beta}_k$, one can apply, within each cell X_k the regularization procedure modeled on the toy model case of §5.8. This results in a logarithmic divergence with coefficient a period integral $\int_{X_k \cap \Sigma} \beta_k$. The nature of this period now depends on how the divisor $\mathcal{D}$ and its Leray coboundary $\mathcal{L}(\mathcal{D})$ intersect each cell X_k and the nature of the motives $\mathfrak{m}(\mathcal{D} \cap X_k)$.

In order to apply this method to our situation we need a further step. Indeed, the compactifications $\overline{\mathrm{Conf}}_\Gamma(X)$ do not necessarily admit a cellular structure, even assuming that the variety X does. However, an argument similar to the one outlined above applies to other decompositions of the variety.

A *torification* of a variety $\mathcal{X}$ is a decomposition of $\mathcal{X}$ into a disjoint union of tori $T_j = \mathbb{G}_m^{n_j}$ (see [**63**] for more details). By the same argument as above, if T_k denotes the subcollection of top dimensional tori, the restrictions $\alpha|_{T_k}$ can be replaced by meromorphic forms with logarithmic poles along $\mathcal{D}$, again by Lemma 2.5 of [**34**]. As in the previous case, the forms with logarithmic poles β_k on T_k do not necessarily extend to a single form η on all of $\mathcal{X} \smallsetminus \mathcal{D}$. One can proceed, as in the case of cellular decompositions, to compute regularized integrals on each T_k. In each case the coefficient of the logarithmic divergence will be a period integral that depends on the nature of the motive $\mathfrak{m}(\mathcal{D} \cap T_k)$.

A further level of generality is the notion of constructible torification introduced in [**64**]. In this case, one only requires that the variety $\mathcal{X}$ admits a decomposition into a disjoint union of pieces that are constructible sets obtained, starting from $\mathbb{G}_m$ by repeated applications of products, disjoint unions, and complements. Roughly speaking, in a constructible torification the building blocks are complements of tori inside other tori.

Under the assumption that X is a smooth projective variety with a cellular decomposition (in particular, for a projective space $X = \mathbb{P}^D$), Lemma 8.1 of [**64**] shows that the blowup of X^{V_Γ} along a diagonal Δ_γ admits a torification. When one iterates the blowup construction, one obtains a (constructible) torification on $\overline{\mathrm{Conf}}_\Gamma(X)$. The top dimensional pieces C_k of this decompositions are, up to sets of measure zero, complements of hypersurfaces in top dimensional tori, to which Lemma 2.5 of [**34**] can still be applied. The resulting periods, which appear as coefficients of the logarithmic divergences in the regularized integrals on the top dimensional pieces of this decomposition, depend on the nature of the motives $\mathfrak{m}(\mathcal{D} \cap C_k)$.

6. Regularization and residue integration

In this section, we describe a regularization of the Feynman integral

$$\int_{\tilde{\sigma}_\Gamma^{(Z,y)} \smallsetminus \mathcal{D}_\Gamma} \eta_\Gamma^{(Z)}, \tag{6.1}$$

which is based on a generalization of the toy model case of §5.8, applied to the form with logarithmic poles $\eta_\Gamma^{(Z)} = \beta_\Gamma^{(Z)} \wedge \bar{\beta}_\Gamma^{(Z)}$ described in the previous section. We present the regularization in distributional terms, using the theory of principal values and residue currents. This will show that one can express the resulting residue integrals in the regularization procedure in terms of the iterated Poincaré residues, or equivalently in terms of an integration along a locus in a Leray coboundary of an intersections of divisors $D_\gamma^{(Z)}$.

As we will discuss more in detail in the proof of Proposition 6.1 below, the Leray coboundary of a hypersurface is dual to the Poincaré residue map and assigns to a k-dimensional Σ in the hypersurface a $k+1$-dimensional chain in the hypersurface complement that is a circle bundle over Σ. The Leray coboundary of a transverse intersection of r hypersurfaces is dual to the iterated residue map and assigns to a Σ in the intersection of the hypersurfaces a T^r-torus bundle over Σ. The precise definition of iterated residues and Leray conboundaries and their main properties are also recalled in the proof of Proposition 6.1. For a general treatment of Leray cobondaries and their properties, we refer the reader to [**13**].

6.1. Iterated Poincaré residues. One can associate to a holomorphic differential form β_Γ with logarithmic poles along $\mathcal{D}_\Gamma$ a Poincaré residue on each non-empty intersection of a collection of divisors $D_\gamma^{(Z)}$ that corresponds to a $\mathcal{G}_\Gamma$-nest $\mathcal{N} = \{\gamma_1, \ldots, \gamma_r\}$.

PROPOSITION 6.1. *Let $N_\Gamma = \# V_\Gamma$. Let β_Γ be a meromorphic DN_Γ-form with logarithmic poles along $\mathcal{D}_\Gamma$. For every $\mathcal{G}_\Gamma$-nest $\mathcal{N} = \{\gamma_1, \ldots, \gamma_r\}$, there is a Poincaré residue $\mathcal{R}_\mathcal{N}(\beta_\Gamma)$, which defines a cohomology class in $H^{DN_\Gamma - r}(V_\mathcal{N})$, on the complete intersection $V_\mathcal{N}^{(Z)} = D_{\gamma_1}^{(Z)} \cap \cdots \cap D_{\gamma_r}^{(Z)}$. The pairing of $\mathcal{R}_\mathcal{N}(\eta_\Gamma)$ with an $(DN_\Gamma - r)$-cycle $\Sigma_\mathcal{N}$ in $V_\mathcal{N}^{(Z)}$ is equal to*

$$\int_{\Sigma_\mathcal{N}} \mathcal{R}_\mathcal{N}(\beta_\Gamma) = \frac{1}{(2\pi i)^r} \int_{\mathcal{L}_\mathcal{N}(\Sigma_\mathcal{N})} \beta_\Gamma, \tag{6.2}$$

where $\mathcal{L}_\mathcal{N}(\Sigma_\mathcal{N})$ is the DN_Γ-cycle in $Z[\Gamma]$ given by an iterated Leray coboundary of $\Sigma_\mathcal{N}$, which is a T^r-torus bundle over $\Sigma_\mathcal{N}$. Under the assumption that the variety X is a mixed Tate motive, the integrals (6.2) *are periods of mixed Tate motives.*

PROOF. As shown in Proposition 5.5, the divisors $D_\gamma^{(Z)}$ in $Z[\Gamma]$ have the property that

$$V_{\mathcal{N}}^{(Z)} = D_{\gamma_1}^{(Z)} \cap \cdots \cap D_{\gamma_r}^{(Z)} \neq \emptyset \Leftrightarrow \{\gamma_1, \ldots, \gamma_r\} \quad \text{is a } \mathcal{G}_\Gamma - \text{nest}, \tag{6.3}$$

with transverse intersections.

Consider the first divisor $D_{\gamma_1}^{(Z)}$ in the $\mathcal{G}_\Gamma$-nest $\mathcal{N}$, and a tubular neighborhood $N_{\Gamma,\gamma_1} = N_{Z[\Gamma]}(D_{\gamma_1}^{(Z)})$ of $D_{\gamma_1}^{(Z)}$ in N_{Γ,γ_1}. This is a unit disk bundle over $D_{\gamma_1}^{(Z)}$ with projection $\pi : N_{\Gamma,\gamma_1} \to D_{\gamma_1}^{(Z)}$ and with $\sigma : D_{\gamma_1}^{(Z)} \hookrightarrow N_{\Gamma,\gamma_1}$ the zero section. The Gysin long exact sequence in homology gives

$$\cdots \to H_k(N_{\Gamma,\gamma_1} \smallsetminus D_{\gamma_1}^{(Z)}) \stackrel{\iota_*}{\to} H_k(N_{\Gamma,\gamma_1})$$

$$\stackrel{\sigma^!}{\to} H_{k-2}(D_{\gamma_1}^{(Z)}) \stackrel{\pi^!}{\to} H_{k-1}(N_{\Gamma,\gamma_1} \smallsetminus D_{\gamma_1}^{(Z)}) \to \cdots$$

where $\pi^!$ is the Leray coboundary map, which assigns to a chain Σ in $D_{\gamma_1}^{(Z)}$ the homology class in $N_{\Gamma,\gamma_1} \smallsetminus D_{\gamma_1}^{(Z)}$ of the boundary $\partial\pi^{-1}(\Sigma)$ of the disk bundle $\pi^{-1}(\Sigma)$ over Σ, which is an S^1-bundle over Σ. Its dual is a morphism

$$\mathcal{R}_{\gamma_1} : H^{k+1}(N_{\Gamma,\gamma_1} \smallsetminus D_{\gamma_1}^{(Z)}) \to H^k(D_{\gamma_1}^{(Z)}),$$

which is the residue map. The iterated residue map is obtained by considering the complements $\mathcal{U}_0 = N_{\Gamma,\gamma_1} \smallsetminus D_{\gamma_1}^{(Z)}$ and

$$\mathcal{U}_1 = D_{\gamma_1}^{(Z)} \smallsetminus \bigcup_{1<k\leq r} D_{\gamma_k}^{(Z)},$$

$$\mathcal{U}_2 = (D_{\gamma_1}^{(Z)} \cap D_{\gamma_2}^{(Z)}) \smallsetminus \bigcup_{2<k\leq r} D_{\gamma_k}^{(Z)},$$

and so on. One obtains a sequence of maps

$$H^k(\mathcal{U}_0) \stackrel{\mathcal{R}_{\gamma_1}}{\to} H^{k-1}(\mathcal{U}_1) \stackrel{\mathcal{R}_{\gamma_2}}{\to} H^{k-2}(\mathcal{U}_2) \to \cdots \stackrel{\mathcal{R}_{\gamma_r}}{\to} H^{k-r}(V_{\mathcal{N}}^{(Z)}).$$

The composition $\mathcal{R}_{\mathcal{N}} = \mathcal{R}_{\gamma_r} \circ \cdots \circ \mathcal{R}_{\gamma_1}$ is the iterated residue map. Because the residue map is dual to Leray coboundary, under the pairing of homology and cohomology one obtains

$$\langle \mathcal{R}_{\mathcal{N}}(\beta), \Sigma \rangle = \langle \beta, \mathcal{L}_{\mathcal{N}}(\Sigma) \rangle,$$

where $\mathcal{L}_{\mathcal{N}} = \mathcal{L}_{\gamma_1} \circ \cdots \circ \mathcal{L}_{\gamma_r}$ is the compositions of the Leray coboundary maps of the divisors $D_{\gamma_k}^{(Z)}$. The resulting $\mathcal{L}_{\mathcal{N}}(\Sigma)$ is therefore, by construction, a T^r-torus bundle over Σ. At the level of differential forms, the residue map $\mathcal{R}_{\gamma_1}$ is given by integration along the circle fibers of the S^1-bundle $\partial\pi^{-1}(\Sigma) \to \Sigma$. Thus, the pairing $\langle \mathcal{R}_{\mathcal{N}}(\beta), \Sigma \rangle$ contains a $2\pi i$ factor, coming from the integration of a form df/f, with f the local defining equation of the hypersurface, along the circle fibers. This means that, when writing the pairings in terms of differential forms, one obtains (6.2). As shown in [**36**], if $\mathfrak{m}(X)$ is mixed Tate, the divisors D_γ and their intersections in $\overline{\mathrm{Conf}}_\Gamma(X)$ are mixed Tate motives, and so are the $D_\gamma^{(Z)}$ and their intersections $V_{\mathcal{N}}^{(Z)}$ in $Z[\Gamma]$. $\square$

Given a $\mathcal{G}_\Gamma$-nest $\mathcal{N} = \{\gamma_1, \dots, \gamma_r\}$ let $\mathcal{V}_\mathcal{N} = D_{\gamma_1} \cap \cdots \cap D_{\gamma_r}$ be the intersection of the corresponding divisors in $\overline{\mathrm{Conf}}_\Gamma(X)$. The residues $\mathcal{R}_\mathcal{N}(\beta_\Gamma)$ of Proposition 6.1 pair with a $DN_\Gamma - r$-dimensional cycles $\Sigma_\mathcal{N} \subset \mathcal{V}_\mathcal{N} \times \{y\} \subset \mathcal{V}_\mathcal{N}^{(Z)}$, with $N_\Gamma = \# V_\Gamma$,

$$\langle \mathcal{R}_\mathcal{N}(\beta_\Gamma), \Sigma_\mathcal{N} \rangle = \int_{\Sigma_\mathcal{N}} \mathcal{R}_\mathcal{N}(\beta_\Gamma) = \frac{1}{(2\pi i)^r} \int_{\mathcal{L}^r(\Sigma_\mathcal{N})} \beta_\Gamma. \tag{6.4}$$

We can then reinterpret and extend the toy model case of §5.8 in the following way. We write $\eta_\Gamma = \beta_\Gamma \wedge \bar{\beta}_\Gamma$ (ignoring a possible sign), with β_Γ holomorphic with logarithmic poles along $\mathcal{D}_\Gamma$. We will also ignore some rational constant factors in the computation, as they will not affect the nature of the resulting periods.

COROLLARY 6.2. *Suppose given a $\mathcal{G}_\Gamma$-nest $\mathcal{N} = \{\gamma_1, \dots, \gamma_r\}$ as above, with $\mathcal{V}_\mathcal{N} = D_{\gamma_1} \cap \cdots \cap D_{\gamma_r}$. There is a regularization of the integral* (6.1), *such that, its restriction to a neighborhood of $\mathcal{V}_\mathcal{N}$ is an iterated Poincaré residue.*

PROOF. When integrating the $2DN_\Gamma$-form η_Γ in a neighborhood $\mathcal{U}$ of $\mathcal{V}_\mathcal{N}$, we have (up to a sign)

$$\int_\mathcal{U} \eta_\Gamma = \int_\mathcal{U} \eta \wedge \beta_\mathcal{N} \wedge \bar{\beta}_\mathcal{N},$$

where we can locally write the form in coordinates

$$\beta_\mathcal{N} = \frac{df_1}{f_1} \wedge \cdots \wedge \frac{df_r}{f_r},$$

with f_j the function that locally defines the divisor D_{γ_j}, and with η a $(2DN_\Gamma - r)$-form which is regular along $\mathcal{V}_\mathcal{N}$. The integral of $\beta_\mathcal{N} \wedge \bar{\beta}_\mathcal{N}$ can be modeled as in the toy model case of §5.8. Thus, up to a rational factor, the integral of $\beta_\mathcal{N} \wedge \bar{\beta}_\mathcal{N}$ can be regularized as in §5.8, so that it results in a periods integral

$$\int_{T^r} \beta_\mathcal{N},$$

where the torus T^r is the fiber of the Leray coboundary $\mathcal{L}^r(\mathcal{V}_\mathcal{N})$, identified with a T^r-bundle over $\mathcal{V}_\mathcal{N}$, while the remaining term η is integrated along $\mathcal{V}_\mathcal{N}$-directions. Indeed, the chain of integration $\tilde{\sigma}_\Gamma^{(Z,y)} = \overline{\mathrm{Conf}}_\Gamma(X) \times \{y\}$ intersects the loci $\mathcal{V}_\mathcal{N}^{(Z)} = \mathcal{V}_\mathcal{N} \times X^{V_\Gamma}$ along $\mathcal{V}_\mathcal{N} \times \{y\}$, where the $\mathcal{V}_\mathcal{N}$ are the intersections $\mathcal{V}_\mathcal{N} = D_{\gamma_1} \cap \cdots \cap D_{\gamma_r}$ of the divisors D_{γ_k} in $\overline{\mathrm{Conf}}_\Gamma(X)$. Thus, $\Sigma = \mathcal{V}_\mathcal{N} \times \{y\}$ defines a $2DN_\Gamma - r$-dimensional cycle in $\mathcal{V}_\mathcal{N}^{(Z)}$, with $N_\Gamma = \# V_\Gamma$, which can be paired with the form η of degree $2DN_\Gamma - r$ on $\mathcal{V}_\mathcal{N}^{(Z)}$. Thus, we write the integral equivalently (up to a rational factor) as a residue integral

$$\int_{\mathcal{L}^r(\mathcal{V}_\mathcal{N})} \eta \wedge \beta_\mathcal{N}.$$

□

REMARK 6.3. The reason for passing to the wonderful compactification $Z[\Gamma]$ and pulling back the form $\omega_\Gamma^{(Z)}$ along the projection $\pi_\Gamma : Z[\Gamma] \to Z^{V_\Gamma}$ is in order to pass to a setting where the locus of divergence is described by divisors D_γ intersecting transversely in $\overline{\mathrm{Conf}}_\Gamma(X)$, while the diagonals Δ_γ in X^{V_Γ} have higher codimensions and do not in general intersect transversely. These transversality issues are discussed in more detail in [**36**]. There is a generalization of the theory of forms with logarithmic poles and Poincaré residues [**74**], that extends the case of [**42**] of normal crossings divisors to more general classes of divisors, but in this more general setting the Poincaré residue gives meromorphic instead of holomorphic forms.

6.2. Current-regularization of complexified amplitudes. We review here briefly some well known facts about residue and principal value currents and we apply them to reinterpret the regularization procedure described above for the integral (6.1) in terms of currents and principal values.

6.2.1. *Residue currents and Mellin transforms.* Recall that, for a single smooth hypersurface defined by an equation $\{f = 0\}$, the residue current $[Z_f]$, supported on the hypersurface, is defined as

$$[Z_f] = \frac{1}{2\pi i}\bar{\partial}[\frac{1}{f}] \wedge df := \frac{1}{2\pi i}\bar{\partial}\partial \log |f|^2.$$

This is known as the Poincaré–Lelong formula. It can also be seen as a limit

$$\int_{Z_f} \varphi = \lim_{\epsilon \to 0} \frac{1}{2\pi i} \int_{|f|=\epsilon} \frac{df}{f} \wedge \varphi.$$

A generalization is given by the Coleff–Herrera residue current [**39**], associated to a collection of functions $\{f_1, \dots, f_r\}$. Under the assumption that these define a complete intersection $V = \{f_1 = \dots = f_r = 0\}$, the residue current

$$\mathcal{R}_f = \bar{\partial}[\frac{1}{f_1}] \wedge \dots \wedge \bar{\partial}[\frac{1}{f_r}] \tag{6.5}$$

is obtained as a limit

$$\mathcal{R}_f(\varphi) = \lim_{\delta \to 0} \int_{T_{\epsilon(\delta)}(f)} \frac{\varphi}{f_1 \cdots f_r},$$

with $T_{\epsilon(\delta)}(f) = \{|f_k| = \epsilon_k(\delta)\}$, with the limit taken over "admissible paths" $\epsilon(\delta)$, which satisfy the properties

$$\lim_{\delta \to 0} \epsilon_r(\delta) = 0, \qquad \lim_{\delta \to 0} \frac{\epsilon_k(\delta)}{(\epsilon_{k+1}(\delta))^\ell} = 0,$$

for $k = 1, \dots, r$ and any positive integer ℓ. The test form φ is a $(2n-r)$-form of type $(n, n-r)$, where $2n$ is the real dimension of the ambient space, and the residual current obtained in this way is a $(0, r)$-current. For more details, see §3 of [**18**] and [**84**]. Notice that, while in general one cannot take products of distributions, the Coleff–Herrera product (6.5) is well defined for residue currents, as well as between residue and principal value currents.

Moreover, the Mellin transform

$$\Gamma_f^\varphi(\lambda) = \int_{\mathbb{R}_+^r} \mathcal{I}_f^\varphi(\epsilon)\, \epsilon^{\lambda - I}\, d\epsilon, \tag{6.6}$$

with

$$\mathcal{I}_f^\varphi(\epsilon) = \int_{T_\epsilon(f)} \frac{\varphi}{f_1 \cdots f_r}$$

and with

$$\epsilon^{\lambda - I}\, d\epsilon = \epsilon_1^{\lambda_1 - 1} \cdots \epsilon_r^{\lambda_r - 1}\, d\epsilon_1 \wedge \dots \wedge d\epsilon_r,$$

can also be written, as in [**18**], [**84**], as

$$\Gamma_f^\varphi(\lambda) = \frac{1}{(2\pi i)^r} \int_{\mathcal{X}} |f|^{2(\lambda - I)} \overline{df} \wedge \varphi, \tag{6.7}$$

where the integration is on the ambient variety $\mathcal{X}$ and where

$$|f|^{2(\lambda-I)} = |f_1|^{2(\lambda_1-1)} \cdots |f_r|^{2(\lambda_r-1)}, \quad \text{and} \quad \overline{df} = \overline{df_1} \wedge \cdots \wedge \overline{df_r}.$$

When $\{f_1, \ldots, f_r\}$ define a complete intersection, the function $\lambda_1 \cdots \lambda_r \Gamma_f^{\varphi}(\lambda)$ is holomorphic in a neighborhood of $\lambda = 0$ and the value at zero is given by the residue current ([**18**], [**84**])

$$\mathcal{R}_f(\varphi) = \lambda_1 \cdots \lambda_r \Gamma_f^{\varphi}(\lambda)|_{\lambda=0}. \tag{6.8}$$

Equivalently, (6.7) and (6.8) can also be written as

$$\lim_{\lambda\to 0} \lambda \Gamma_f^{\varphi}(\lambda) = \lim_{\lambda\to 0} \frac{1}{(2\pi i)^r} \int_{\mathcal{X}} \frac{\bar\partial |f_r|^{2\lambda_r} \wedge \cdots \wedge \bar\partial |f_1|^{2\lambda_1}}{f_r \cdots f_1} \wedge \varphi, \tag{6.9}$$

where the factor λ on the left-hand-side stands for $\lambda_1 \cdots \lambda_r$ as in (6.8).

The Poincaré–Lelong formula, in this more general case of a complete intersection defined by a collection $\{f_1, \ldots, f_r\}$, expresses the integration current Z_f as

$$[Z_f] = \frac{1}{(2\pi i)^r} \bar\partial [\frac{1}{f_r}] \wedge \cdots \wedge \bar\partial [\frac{1}{f_1}] \wedge df_1 \wedge \cdots \wedge df_r. \tag{6.10}$$

The correspondence between residue currents and the Poincaré residues on complete intersections, discussed above in §6.1, is given for instance in Theorem 4.1 of [**4**].

6.2.2. *Principal value current.* The principal value current $[1/f]$ of a single holomorphic function f can be computed as [**56**], [**76**]

$$\langle [\frac{1}{f}], \phi \rangle = \lim_{\epsilon\to 0} \int_{|f|>\epsilon} \frac{\phi \, d\zeta \wedge d\bar\zeta}{f}, \tag{6.11}$$

where ϕ is a test function. More generally, for $\{f_1, \ldots, f_r\}$ as above, the principal value current is given by

$$\langle [\frac{1}{f}], \phi \rangle = \lim_{\epsilon\to 0} \int_{N_\epsilon(f)} \frac{\phi}{f_r \cdots f_1}, \tag{6.12}$$

with ϕ a test form and with

$$N_\epsilon(f) = \{|f_k| > \epsilon_k\}. \tag{6.13}$$

More generally we will use the following notation.

DEFINITION 6.4. Given a (p,q)-form η on an m-dimensional smooth projective variety $\mathcal{X}$, with poles along an effective divisor $\mathcal{D} = D_1 \cup \cdots D_r$, where the components D_k are smooth hypersurfaces defined by equations $f_k = 0$, the principal value current $PV(\eta)$ is defined by

$$\langle PV(\eta), \phi \rangle = \lim_{\epsilon\to 0} \int_{N_\epsilon(f)} \eta \wedge \phi, \tag{6.14}$$

for an $(m-p, m-q)$ test form ϕ, with $N_\epsilon(f)$ defined as in (6.13).

The following simple Lemma describes how the principal value current and the residue currents are related.

LEMMA 6.5. *When the test form ϕ is modified to $\phi + \bar{\partial}\psi$, the principal value current satisfies*

$$\langle [\frac{1}{f}], \phi + \bar{\partial}\psi \rangle = \langle [\frac{1}{f}], \phi \rangle - \langle \bar{\partial}[\frac{1}{f}], \psi \rangle, \tag{6.15}$$

where $\bar{\partial}[1/f]$ is the residue current $\mathcal{R}_f$ of (6.5).

PROOF. By Stokes theorem, we have

$$\langle [\frac{1}{f}], \bar{\partial}\psi \rangle = \lim_{\epsilon \to 0} \int_{|f|>\epsilon} \frac{\bar{\partial}\psi}{f} = \lim_{\epsilon \to 0} - \int_{|f|=\epsilon} \frac{\psi}{f} = -\langle \bar{\partial}[\frac{1}{f}], \psi \rangle.$$

□

We now return to the case of the complexified amplitudes (3.9) and describe the corresponding regularization and ambiguities.

6.2.3. *Principal value and complexified amplitude.* We can regularize the complexified amplitude given by the integral (6.1), interpreted in the distributional sense, as in §3.3.4, using the principal value current.

DEFINITION 6.6. The principal value regularization of the amplitude (6.1) is given by the current $PV(\eta_\Gamma^{(Z)})$ defined as in (6.14),

$$\langle PV(\eta_\Gamma^{(Z)}), \varphi \rangle = \lim_{\epsilon \to 0} \int_{N_\epsilon(f)} \varphi \, \eta,$$

for a test function φ.

We can also write the regularized integral in the following form.

LEMMA 6.7. *The regularized integral satisfies*

$$\langle PV(\eta_\Gamma^{(Z)}), \varphi \rangle = \lim_{\lambda \to 0} \int_{\tilde{\sigma}_\Gamma^{(Z,y)}} |f_n|^{2\lambda_n} \cdots |f_1|^{2\lambda_1} \eta_\Gamma^{(Z,y)} \, \varphi$$

where $n = n_\Gamma$ is the cardinality $n_\Gamma = \#\mathcal{G}_\Gamma$ of the building set $\mathcal{G}_\Gamma$ and the f_k are the defining equations of the $D_\gamma^{(Z)}$ in $\mathcal{G}_\Gamma$

PROOF. The form $\eta_\Gamma^{(Z)}$ has poles along the divisor $\mathcal{D}_\Gamma = \cup_{\Delta_\gamma^{(Z)} \in \mathcal{G}_\Gamma} D_\gamma^{(Z)}$. Thus, if we denote by f_k, with $k = 1, \ldots, n$ with $n = \#\mathcal{G}_\Gamma$ the defining equations of the $D_\gamma^{(Z)}$, we can write the principal value current in the form

$$\lim_{\lambda \to 0} \int_{\tilde{\sigma}_\Gamma^{(Z,y)}} \frac{|f_n|^{2\lambda_n} \cdots |f_1|^{2\lambda_1}}{f_n \cdots f_1} h \, \varphi,$$

with h an algebraic from without poles and φ is a test function. □

6.2.4. *Pseudomeromorphic currents.* If $\{f_1, \ldots, f_n\}$ define a complete intersection $V = \{f_1 = \cdots = f_r = 0\}$ in a smooth projective variety $\mathcal{X}$, an *elementary pseudomeromorphic current* is a current of the form

$$C_{r,n} := [\frac{1}{f_n}] \wedge \cdots \wedge [\frac{1}{f_{r+1}}] \wedge \bar{\partial}[\frac{1}{f_r}] \wedge \cdots \wedge \bar{\partial}[\frac{1}{f_1}], \tag{6.16}$$

for some $1 \leq r \leq n$, where the products of principal value and residue currents are well defined Coleff–Herrera products and the resulting current is commuting in the principal value factors and anticommuting in the residue factors. These distributions also have a Mellin transform formulation (see [**18**]) as

$$\langle C_{r,n}, \phi \rangle = \lim_{\lambda \to 0} \frac{1}{(2\pi i)^r} \int_{\mathcal{X}} \prod_{k=r+1}^{n} \frac{|f_k|^{2\lambda_k}}{f_k} \bigwedge_{j=1}^{r} \bar{\partial}\left(\frac{|f_j|^{2\lambda_j}}{f_j}\right) \wedge \phi. \tag{6.17}$$

6.3. Ambiguities of regularized integrals. We can use the formalism of residue currents recalled above to describe the ambiguities in the principal value regularization of the complexified amplitudes of Definition 6.6.

6.3.1. *Complexified amplitude and residue currents.* As in §6.1, consider a $\mathcal{G}_\Gamma$-nest $\{\gamma_1, \dots, \gamma_r\}$ and the associated intersection $V_{\mathcal{N}}^{(Z)} = D_{\gamma_1}^{(Z)} \cap \cdots \cap D_{\gamma_r}^{(Z)}$. Also let $n_\Gamma = \#\mathcal{G}_\Gamma$ and let f_k, for $k = 1, \dots, n_\Gamma$ be the defining equations for the $D_\gamma^{(Z)}$, for γ ranging over the subgraphs defining the building set $\mathcal{G}_\Gamma$. For $\epsilon = (\epsilon_k)$, we define

$$\tilde{\sigma}_{\Gamma,\epsilon}^{(Z,y)} := \tilde{\sigma}_{\Gamma}^{(Z,y)} \cap N_\epsilon(f), \tag{6.18}$$

with $N_\epsilon(f)$ defined as in (6.13). The principal value regularization of Definition 6.6 can then be written as

$$\langle PV(\eta_\Gamma^{(Z,y)}), \varphi \rangle = \lim_{\epsilon \to 0} \int_{\tilde{\sigma}_{\Gamma,\epsilon}^{(Z,y)}} \varphi\, \eta_\Gamma^{(Z,y)},$$

where the limit is taken over admissible paths.

Similarly, given a $\mathcal{G}_\Gamma$-nest $\mathcal{N} = \{\gamma_1, \dots, \gamma_r\}$, we introduce the notation

$$\tilde{\sigma}_{\Gamma,\mathcal{N},\epsilon}^{(Z,y)} := \tilde{\sigma}_{\Gamma}^{(Z,y)} \cap T_{\mathcal{N},\epsilon}(f) \cap N_{\mathcal{N},\epsilon}(f), \tag{6.19}$$

where $T_{\mathcal{N},\epsilon}(f) = \{|f_k| = \epsilon_k,\ k = 1, \dots, r\}$ and $N_{\mathcal{N},\epsilon}(f) = \{|f_k| > \epsilon,\ k = r+1, \dots, n\}$, where we have ordered the n subgraphs γ in $\mathcal{G}_\Gamma$ so that the first r belong to the nest $\mathcal{N}$.

PROPOSITION 6.8. *For a $\mathcal{G}_\Gamma$-nest $\mathcal{N} = \{\gamma_1, \dots, \gamma_r\}$, as above, the limit*

$$\lim_{\epsilon \to 0} \int_{\tilde{\sigma}_{\Gamma,\mathcal{N},\epsilon}^{(Z,y)}} \varphi\, \eta_\Gamma^{(Z,y)} \tag{6.20}$$

determines a pseudomeromorphic current, whose residue part is an iterated residue supported on $V_{\mathcal{N}}^{(Z)} = D_{\gamma_1}^{(Z)} \cap \cdots \cap D_{\gamma_r}^{(Z)}$, obtained as in Corollary 6.2.

PROOF. Let f_k, with $k = 1, \dots, n$ with $n = \#\mathcal{G}_\Gamma$, be the defining equations of the $D_\gamma^{(Z)}$. We assume the subgraphs in $\mathcal{G}_\Gamma$ are ordered so that the first r belong to the given $\mathcal{G}_\Gamma$-nest $\mathcal{N}$. We can then write the current (6.20) in the form $\langle C_{r,n}, h\varphi \rangle$, where $C_{r,n}$ is the elementary pseudomeromorphic current of (6.16) and h is algebraic without poles. Using a local coordinates model, as in Corollary 6.2 one can identify the resulting residue current with the regularization computed there. □

6.3.2. *Residue currents as ambiguities.* With the same setting as in Proposition 6.8, we then have the following characterization of the ambiguities of the principal value regularization.

PROPOSITION 6.9. *The ambiguities in the current-regularization $PV(\eta_\Gamma^{(Z,y)})$ are given by iterated residues supported on the intersections $V_{\mathcal{N}}^{(Z)} = D_{\gamma_1}^{(Z)} \cap \cdots \cap D_{\gamma_r}^{(Z)}$, of divisors corresponding to $\mathcal{G}_\Gamma$-nests $\mathcal{N} = \{\gamma_1, \dots, \gamma_r\}$.*

PROOF. As above, we have

$$\langle PV(\eta_\Gamma^{(Z,y)}),\varphi\rangle = \lim_{\epsilon\to 0}\int_{\tilde\sigma^{(Z,y)}_{\Gamma,\epsilon}} \varphi\,\eta_\Gamma^{(Z,y)} = \lim_{\lambda\to 0}\int_{\tilde\sigma^{(Z,y)}_{\Gamma}} \frac{|f_n|^{2\lambda_n}\cdots|f_1|^{2\lambda_1}}{f_n\cdots f_1}\,h\,\varphi$$

If we replace the form $h\,\varphi$ with a form $h\varphi+\bar\partial_{\mathcal N}\psi$, where $\mathcal N$ is a $\mathcal G_\Gamma$-nest and the notation $\bar\partial_{\mathcal N}\psi$ means a form

$$\bar\partial_{\mathcal N}\psi := \psi_n\cdots\psi_{r+1}\,\bar\partial\psi_r\wedge\cdots\wedge\bar\partial\psi_1,$$

for test functions ψ_k, $k=1,\dots,n$, we obtain a pseudomeromorphic current

$$\langle PV(\eta_\Gamma^{(Z,y)}),\bar\partial_{\mathcal N}\psi\rangle = \langle[\frac{1}{f_n}]\wedge\cdots\wedge[\frac{1}{f_{r+1}}]\wedge\bar\partial[\frac{1}{f_r}]\wedge\cdots\wedge\bar\partial[\frac{1}{f_1}],\psi\rangle,$$

with $\psi=\psi_n\cdots\psi_1$. Notice then that the residue part

$$\langle\bar\partial[\frac{1}{f_r}]\wedge\cdots\wedge\bar\partial[\frac{1}{f_1}],\psi\rangle = \mathcal R_{\mathcal N}(\psi) \tag{6.21}$$

is an iterated residue supported on $V^{(Z)}_{\mathcal N}=D^{(Z)}_{\gamma_1}\cap\cdots\cap D^{(Z)}_{\gamma_r}$. □

By the results of §6.1, and the relation (6.21) between residue currents and iterated Poincaré residues (see [**4**]), when evaluated on algebraic test forms on the varieties $V^{(Z)}_{\mathcal N}$, these ambiguities can be expressed in terms of periods of mixed Tate motives. In other words, the mixed Tate nature of the period is not changed by the presence of these ambiguities, as these are themselves expressible in terms of mixed Tate periods of intersections of components of the boundary divisor.

7. Other regularization methods

We now discuss a regularization method for the evaluation of the integrals of the complexified amplitudes,

$$\int_{\tilde\sigma_\Gamma^{(Z,y)}} \pi_\Gamma^*(\omega_\Gamma^{(Z)}),$$

with the pullback $\pi_\Gamma^*(\omega_\Gamma^{(Z)})$ to $Z[\Gamma]$ as in Corollary 5.13. The chain of integration $\tilde\sigma_\Gamma^{(Z,y)}$ is as in (5.17), obtained from the form $\omega_\Gamma^{(Z)}$ and the chain $\sigma_\Gamma^{(Z)}$ of Definition 3.6. The geometric method of regularization we describe in this section is based on the *deformation to the normal cone.*

A general method of regularization consists of deforming the chain of integration so that it no longer intersects the locus of divergences. We first describe briefly why this cannot be done directly within the space $Z[\Gamma]$ considered above, and then we introduce a simultaneous deformation of the form and of the space where integration happens, so that the integral can be regularized according to the general method mentioned above.

7.1. Deformations and linking. To illustrate where the problem arises, if one tries to deform the chain of integration away from the locus of divergence in $Z[\Gamma]$, consider the local problem near a point $z\in D_\gamma^{(Z)}$ in the intersection of $\tilde\sigma_\Gamma^{(Z,y)}$ with one of the divisors in the divergence locus of the form $\pi_\Gamma^*(\omega_\Gamma)$. Near this point, the locus of divergence is a product $D_\gamma\times X^{V_\Gamma}$. We look at the intersection of the integration chain $\tilde\sigma_\Gamma^{(Z,y)}$ with a small tubular neighborhood T_ϵ of $D_\gamma\times X^{V_\Gamma}$. We have

$$\tilde\sigma_\Gamma^{(Z,y)}\cap\partial T_\epsilon = \partial\pi_\epsilon^{-1}(D_\gamma)\times\{y\},$$

with $\pi_\epsilon : T_\epsilon(D_\gamma) \to D_\gamma$ the projection of the 2-disc bundle and $\partial\pi_\epsilon^{-1}(D_\gamma)$ a circle bundle, locally isomorphic to $D_\gamma \times S^1$. Thus, locally, $\tilde{\sigma}_\Gamma^{(Z,y)} \cap T_\epsilon$ looks like a ball $B^{2D|\mathbf{V}_\Gamma|} \times \{0\}$ inside a ball B^{4DN_Γ}, with $N_\Gamma = \#V_\Gamma$. Locally, we can think of the problem of deforming the chain of integration in a neighborhood of the divergence locus as the question of deforming a ball $B^{2DN_\Gamma} \times \{0\}$ leaving fixed the boundary $S^{2DN_\Gamma - 1} \times \{0\}$ inside a ball B^{4DN_Γ} so as to avoid the locus $\{0\} \times B^{2DN_\gamma}$ that lies in the divergence locus, with $N_\gamma = \#V_\gamma$. However, we find the following simple topological fact.

LEMMA 7.1. *The spheres $S^{2DN_\Gamma-1} \times \{0\}$ and $\{0\} \times S^{2DN_\gamma-1}$ are linked inside the sphere $S^{4DN_\Gamma-1}$.*

PROOF. This can be seen, for instance, by computing their Gauss linking integral (see [**78**])

$$\mathrm{Lk}(M,N) = \frac{1}{\mathrm{Vol}(S)} \int_{M\times N} \frac{\Omega_{k,\ell}(\alpha)}{\sin^n(\alpha)} [x, dx, y, dy] \tag{7.1}$$

with $M = S^k \times \{0\}$, $N = \{0\} \times S^\ell$, $S = S^n$, and with $k = \ell = 2D|\mathbf{V}_\Gamma| - 1$ and $n = 4D|\mathbf{V}_\Gamma| - 1$, where

$$\Omega_{k,\ell}(\alpha) := \int_{\theta=\alpha}^{\pi} \sin^k(\theta - \alpha)\sin^\ell(\theta) d\theta,$$

$$\alpha(x,y) := \mathrm{dist}_{S^n}(x,y), \quad x \in M,\ y \in N,$$

$$[x, dx, y, dy] := \det(x, \frac{\partial x}{\partial s_1}, \dots, \frac{\partial x}{\partial s_k}, y, \frac{\partial y}{\partial t_1}, \dots, \frac{\partial y}{\partial t_\ell})\, ds\, dt,$$

with x, y the embeddings of S^k and S^ℓ in S^n and s, t the local coordinates on S^k and S^ℓ. Then one can see (§4 of [**78**]) that in S^n with $n = k + \ell + 1$ the linking number is $\mathrm{Lk}(S^k \times \{0\}, \{0\} \times S^\ell) = 1$. □

This type of problem can be easily avoided by introducing a simultaneous deformation of the form $\pi_\Gamma^*(\omega_\Gamma)$ and of the space $Z[\Gamma]$ as we show in the following.

7.2. Form extension. We first embed the configuration space Z^{V_Γ} as the fiber over zero in a one parameter family $Z^{V_\Gamma} \times \mathbb{P}^1$. We extend the form $\omega_\Gamma^{(Z)}$ to $Z^{V_\Gamma} \times \mathbb{P}^1$ using the additional coordinate $\zeta \in \mathbb{P}^1$ to alter the differential form in a suitable way.

DEFINITION 7.2. The extension of the complexified amplitude $\omega_\Gamma^{(Z)}$ on the space $Z^{V_\Gamma} \times \mathbb{P}^1$ is given by

$$\tilde{\omega}_\Gamma^{(Z)} = \prod_{e \in E_\Gamma} \frac{1}{(\|x_{s(e)} - x_{t(e)}\|^2 + |\zeta|^2)^{D-1}} \bigwedge_{v \in V_\Gamma} dx_v \wedge d\bar{x}_v \wedge d\zeta \wedge d\bar{\zeta}, \tag{7.2}$$

where ζ is the local coordinate on $\mathbb{P}^1$.

LEMMA 7.3. *The divergent locus $\{\tilde{\omega}_\Gamma^{(Z)} = \infty\}$ of the form (7.2) on $Z^{\mathbf{V}_\Gamma} \times \mathbb{P}^1$ is given by the locus $\cup_{e \in \mathbf{E}_\Gamma} \Delta_e^{(Z)} \subset Z^{\mathbf{V}_\Gamma} \times \{0\}$.*

PROOF. The locus of divergence is the intersection of $\{\zeta = 0\}$ and the union of the products $\Delta_e^{(Z)} \times \mathbb{P}^1 = \{x_{s(e)} - x_{t(e)} = 0\}$. □

Notice that we have introduced in the form (7.2) an additional variable of integration, $d\zeta \wedge d\bar{\zeta}$. The reason for shifting the degree of the form will become clear later in this section (see §7.6 below), where we see that, when using the deformation to the normal cone, the chain of integration $\sigma_\Gamma^{(Z,y)}$ is also extended by an additional complex dimension to $\sigma_\Gamma^{(Z,y)} \times \mathbb{P}^1$, of which one then takes a proper transform and deforms it inside the deformation to the normal cone. In terms of the distributional interpretation of the complexified amplitudes of §3.3.4, the relation between the form (7.2) and the original amplitude (3.9) can be written as

$$\omega_\Gamma^{(Z)} = \int \prod_{e \in \mathbf{E}_\Gamma} \frac{\delta(\zeta = 0)}{(\|x_{s(e)} - x_{t(e)}\|^2 + |\zeta|^2)^{D-1}} \bigwedge_{v \in \mathbf{V}_\Gamma} dx_v \wedge d\bar{x}_v \wedge d\zeta \wedge d\bar{\zeta}, \tag{7.3}$$

where the distributional delta constraint can be realized as a limit of normalized integrations on small tubular neighborhoods of the central fiber $\zeta = 0$ in the trivial fibration $Z^{\mathbf{V}_\Gamma} \times \mathbb{P}^1$.

7.3. Deformation to the normal cone. The deformation to the normal cone is the natural algebro-geometric replacement for tubular neighborhoods in smooth geometry, see [**46**]. We use it here to extend the configuration space Z^{V_Γ} to a trivial fibration $Z^{V_\Gamma} \times \mathbb{P}^1$ and then replacing the fiber over $\{0\} \in \mathbb{P}^1$ with the wonderful compactification $Z[\Gamma]$. This will allow us to simultaneously regularize the form and the chain of integration. For simplicity we illustrate the construction for the case where the graph Γ is itself biconnected.

PROPOSITION 7.4. *Let Γ be a biconnected graph. Starting with the product $Z^{V_\Gamma} \times \mathbb{P}^1$, a sequence of blowups along loci parameterized by the $\Delta_\gamma^{(Z)} \times \{0\}$, with γ induced biconnected subgraphs yields a variety $\mathcal{D}(Z[\Gamma])$ fibered over $\mathbb{P}^1$ such that the fiber over all points $\zeta \in \mathbb{P}^1$ with $\zeta \neq 0$ is still equal to Z^{V_Γ}, while the fiber over $\zeta = 0$ has a component equal to the wonderful compactification $Z[\Gamma]$ and other components given by projectivizations $\mathbb{P}(C \oplus 1)$ with C the normal cone of the blowup locus.*

PROOF. We start with the product $Z^{V_\Gamma} \times \mathbb{P}^1$. We then perform the first blowup of the iterated sequence of §5.2.1 on the fiber over $\zeta = 0$ namely we blowup the locus $\Delta_\Gamma^{(Z)} \times \{0\}$, with $\Delta_\Gamma^{(Z)}$ the deepest diagonal, inside $Z^{\mathbf{V}_\Gamma} \times \mathbb{P}^1$. (Note that this is where we are using the biconnected hypothesis on Γ, otherwise the first blowup may be along induced biconnected subgraphs with a smaller number of vertices.) The blowup $\mathrm{Bl}_{\Delta_\Gamma^{(Z)} \times \{0\}}(Z^{V_\Gamma} \times \mathbb{P}^1)$ is equal to $Z^{V_\Gamma} \times (\mathbb{P}^1 \smallsetminus \{0\})$ away from $\zeta = 0$, while over the point $\zeta = 0$ it has a fiber with two components. One of the components is isomorphic to the blowup of Z^{V_Γ} along $\Delta_\Gamma^{(Z)}$, that is, $\mathrm{Bl}_{\Delta_\Gamma^{(Z)}}(Z^{V_\Gamma}) = Y_1$, with the notation of §5.2.1. The other component is equal to $\mathbb{P}(C_{Z^{V_\Gamma}}(\Delta_\Gamma^{(Z)}) \oplus 1)$ where $C_{Z^{V_\Gamma}}(\Delta_\Gamma^{(Z)})$ is the normal cone of $\Delta_\Gamma^{(Z)}$ in Z^{V_Γ}. Since $\Delta_\Gamma^{(Z)} \simeq X \times X^{V_\Gamma}$ is smooth, the normal cone is the normal bundle of $\Delta_\Gamma^{(Z)}$ in Z^{V_Γ}. The two Cartier divisors Y_1 and $\mathbb{P}(C_{Z^{V_\Gamma}}(\Delta_\Gamma^{(Z)}) \oplus 1)$ meet along $\mathbb{P}(C_{Z^{V_\Gamma}}(\Delta_\Gamma^{(Z)}))$. We can then proceed to blow up the further loci $\Delta_\gamma^{(Z)}$ with $\gamma \in \mathcal{G}_{n-1,\Gamma}$ inside the special fiber $\tilde{\pi}^{-1}(0)$ in $\mathrm{Bl}_{\Delta_\Gamma^{(Z)} \times \{0\}}(Z^{V_\Gamma} \times \mathbb{P}^1)$, where

$$\tilde{\pi} : \mathrm{Bl}_{\Delta_\Gamma^{(Z)} \times \{0\}}(Z^{V_\Gamma} \times \mathbb{P}^1) \to Z^{V_\Gamma} \times \mathbb{P}^1$$

is the projection. These loci lie in the intersection of the two components of the special fiber $\tilde{\pi}^{-1}(0)$. Thus, at the next stage we obtain a variety that again agrees

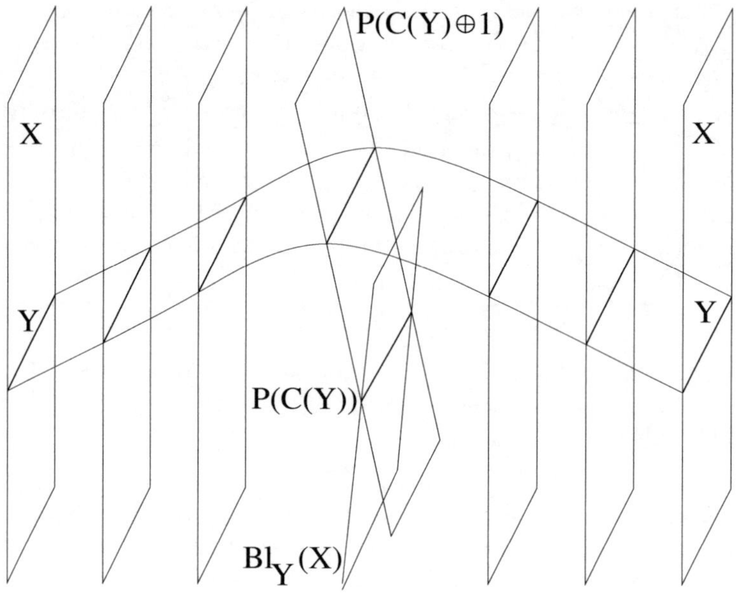

FIGURE 1. Deformation to the normal cone.

with $Z^{V_\Gamma} \times (\mathbb{P}^1 \smallsetminus \{0\})$ away from the central fiber, while over $\zeta = 0$ it now has a component equal to Y_2 and further components coming from the normal cone after this additional blowup. After iterating this process as in §5.2.1 we obtain a variety that has fiber Z^{V_Γ} over all points $\zeta \neq 0$ and over $\zeta = 0$ it has a component equal to the wonderful compactification $Z[\Gamma]$ and other components coming from normal cones. $\square$

Notice that one can also realize the iterated blowup of §5.2.1 as a single blowup over a more complicated locus and perform the deformation to the normal cone for that single blowup. We proceed as in Proposition 7.4, as it will be easier in this way to follow the effect that this deformation has on the motive.

The main reason for introducing the deformation to the normal cone, as we discuss more in detail in §7.6 below, is the fact that it will provide us with a natural mechanism for deforming the chain of integration away from the locus of divergences. The key idea is depicted in Figure 1, where one considers a variety $\mathcal{X}$ and the deformation $\mathrm{Bl}_{\mathcal{Y}\times\{0\}}(\mathcal{X} \times \mathbb{P}^1)$. If $\pi : \mathrm{Bl}_{\mathcal{Y}\times\{0\}}(\mathcal{X} \times \mathbb{P}^1) \to \mathbb{P}^1$ denotes the projection, the special fiber $\pi^{-1}(0)$ has two components, one given by the blowup $\mathrm{Bl}_{\mathcal{Y}}(\mathcal{X})$ of $\mathcal{X}$ along $\mathcal{Y}$ and the other is the normal cone $\mathbb{P}(C_{\mathcal{X}}(\mathcal{Y}) \oplus 1)$ of $\mathcal{Y}$ inside $\mathcal{X}$. The two components meet along $\mathbb{P}(C_{\mathcal{X}}(\mathcal{Y}))$. As shown in §2.6 of [**47**], one can use the deformation to the normal cone to deform $\mathcal{Y}$ to the zero section of the normal cone. Thus, given a subvariety $\mathcal{Z} \subset \mathcal{Y}$ the proper transform $\overline{\mathcal{Z} \times \mathbb{P}^1}$ in $\mathrm{Bl}_{\mathcal{Y}\times\{0\}}(\mathcal{X} \times \mathbb{P}^1)$ gives a copy of $\mathcal{Z}$ inside the special fiber $\pi^{-1}(0)$ lying in the normal cone component, see Figure 1.

7.4. Deformation and the motive. We check that passing from the space Z^{V_Γ} to the deformation $\mathcal{D}(Z[\Gamma])$ described in Proposition 7.4 does not alter the nature of the motive.

It is easy to see that this is the case at the level of virtual motives, that is, classes in the Grothendieck ring of varieties.

PROPOSITION 7.5. *If the class $[X]$ in the Grothendieck ring of varieties $K_0(\mathcal{V})$ is a virtual mixed Tate motive, that is, it lies in the subring $\mathbb{Z}[\mathbb{L}]$ generated by the Lefschetz motive $\mathbb{L} = [\mathbb{A}^1]$, then the class of $\mathcal{D}(Z[\Gamma])$ is also in $\mathbb{Z}[\mathbb{L}]$.*

PROOF. As shown in Proposition 7.4, the space $\mathcal{D}(Z[\Gamma])$ is a fibration over $\mathbb{P}^1$, which is a trivial fibration over $\mathbb{P}^1 \smallsetminus \{0\}$ with fiber Z^{V_Γ}. By inclusion-exclusion, we can write the class $[\mathcal{D}(Z[\Gamma])]$ in $K_0(\mathcal{V})$ as a sum of the class of the fibration over $\mathbb{P}^1 \smallsetminus \{0\}$, which is

$$[Z^{V_\Gamma} \times (\mathbb{P}^1 \smallsetminus \{0\})] = [X]^{2N_\Gamma}\, \mathbb{L},$$

with $N_\Gamma = \# V_\Gamma$, and the class $[\pi^{-1}(0)]$ of the fiber over $\zeta = 0$, with $\pi : \mathcal{D}(Z[\Gamma]) \to \mathbb{P}^1$ the fibration. The component $[X]^{2N_\Gamma}\mathbb{L}$ is in $\mathbb{Z}[\mathbb{L}]$ if the class $[X] \in \mathbb{Z}[\mathbb{L}]$ as we are assuming, so we need to check that the class $[\pi^{-1}(0)]$ is also in $\mathbb{Z}[\mathbb{L}]$. The locus $\pi^{-1}(0)$ is constructed in a sequence of steps as shown in Proposition 7.4. At the first step, we are dealing with the deformation to the normal cone $\mathrm{Bl}_{\Delta_\Gamma^{(Z)}}(Z^{V_\Gamma} \times \mathbb{P}^1)$ and the fiber over zero is the union of Y_1 and $\mathbb{P}(C_{Z^{V_\Gamma}}(\Delta_\Gamma^{(Z)}) \oplus 1)$, intersecting along $\mathbb{P}(C_{Z^{V_\Gamma}}(\Delta_\Gamma^{(Z)}))$. Since $\Delta_\Gamma^{(Z)} \simeq X \times X^{V_\Gamma}$ is smooth and a Tate motive, $\mathbb{P}(C_{Z^{V_\Gamma}}(\Delta_\Gamma^{(Z)}) \oplus 1)$ is a projective bundle over a Tate motive so it is itself a Tate motive. So is $\mathbb{P}(C_{Z^{V_\Gamma}}(\Delta_\Gamma^{(Z)}))$, for the same reason. So is also Y_1 because of the blowup formula for Grothendieck classes [**19**],

$$[Y_1] = [Z^{V_\Gamma}] + \sum_{k=1}^{\mathrm{codim}(\Delta_\Gamma^{(Z)} \times \{0\}) - 1} [\Delta_\Gamma^{(Z)}]\, \mathbb{L}^k.$$

By the inclusion-exclusion relations in the Grothendieck ring, it then follows that if the two components of the fiber over zero are in $\mathbb{Z}[\mathbb{L}]$ and the class of their intersection also is, then so is also the class of the union, which is the class of the fiber itself. At the next step the fiber over zero is blown up again, this time along the (dominant transforms of) $\Delta_\gamma^{(Z)}$ with $\gamma \in \mathcal{G}_{n-1,\Gamma}$. Each of these is a blowup of a variety whose class is a virtual mixed Tate motive along a locus whose class is also a virtual mixed Tate motive, hence repeated application of the blowup formula in the Grothendieck ring and an argument analogous to the one used in the first step shows that the Grothendieck class of the fiber over zero is also in $\mathbb{Z}[\mathbb{L}]$. $\square$

We can then, with a similar technique, improve the result from the level of Grothendieck classes to the level of motives.

PROPOSITION 7.6. *If the motive $\mathfrak{m}(X)$ of the variety X is mixed Tate, then the motive $\mathfrak{m}(\mathcal{D}(Z[\Gamma]))$ of the deformation $\mathcal{D}(Z[\Gamma])$ is also mixed Tate.*

PROOF. As in the case of the Grothendieck classes, it suffices to check that, at each step in the construction of $\mathcal{D}(Z[\Gamma])$, the result remains inside the category of mixed Tate motives. It is clear that, if $\mathfrak{m}(X)$ is mixed Tate, then $\mathfrak{m}(Z)$, $\mathfrak{m}(Z^{V_\Gamma})$ and $\mathfrak{m}(Z^{V_\Gamma} \times \mathbb{P}^1)$ also are. At the next step, we use the blowup formula for Voevodsky motives (Proposition 3.5.3 of [**88**]) and we obtain

$$\mathfrak{m}(\mathrm{Bl}_{\Delta_\Gamma^{(Z)} \times \{0\}}(Z^{V_\Gamma} \times \mathbb{P}^1)) =$$

$$\mathfrak{m}(Z^{V_\Gamma} \times \mathbb{P}^1)) \oplus \bigoplus_{k=1}^{\operatorname{codim}(\Delta_\Gamma^{(Z)} \times \{0\})-1} \mathfrak{m}(\Delta_\Gamma^{(Z)})(k)[2k].$$

This implies that $\mathfrak{m}(\mathrm{Bl}_{\Delta_\Gamma^{(Z)} \times \{0\}}(Z^{V_\Gamma} \times \mathbb{P}^1))$ is mixed Tate if $\mathfrak{m}(X)$ is. The successive steps are again obtained by blowing up loci $\Delta_\gamma^{(Z)}$ whose motive $\mathfrak{m}(\Delta_\gamma^{(Z)})$ is mixed Tate, inside a variety whose motive is mixed Tate by the previous step, hence repeated application of the blowup formula for motives yields the result. □

The analog of Remark 5.8 also holds for the motive $\mathfrak{m}(\mathcal{D}(Z[\Gamma])$.

7.5. Amplitude form on the deformation. Let $\tilde{\omega}_\Gamma^{(Z)}$ be the regularized form defined in (7.2). In order to allow room for a regularization of the chain of integration, we pull it back to the deformation to the normal cone described above.

DEFINITION 7.7. The regularization of the form $\omega_\Gamma^{(Z)}$ on the deformation space $\mathcal{D}(Z[\Gamma])$ is the pullback

$$\tilde{\pi}_\Gamma^*(\tilde{\omega}_\Gamma^{(Z)}), \tag{7.4}$$

where $\tilde{\pi}_\Gamma : \mathcal{D}(Z[\Gamma]) \to Z^{V_\Gamma} \times \mathbb{P}^1$ is the projection and $\tilde{\omega}_\Gamma^{(Z)}$ is the extension (7.2) of the complexified amplitude to $Z^{V_\Gamma} \times \mathbb{P}^1$.

The locus of divergence $\{\tilde{\pi}_\Gamma^*(\tilde{\omega}_\Gamma^{(Z)}) = \infty\}$ inside the deformation space $\mathcal{D}(Z[\Gamma])$ is then given by the following.

LEMMA 7.8. *The locus of divergence of the regularized Feynman amplitude $\tilde{\pi}_\Gamma^*(\tilde{\omega}_\Gamma^{(Z)})$ on the space $\mathcal{D}(Z[\Gamma])$ is a union of divisors inside the central fiber,*

$$\bigcup_{\Delta_\gamma^{(Z)} \in \mathcal{G}_\Gamma} D_\gamma^{(Z)} \subset \pi^{-1}(0), \tag{7.5}$$

where $\pi : \mathcal{D}(Z[\Gamma]) \to \mathbb{P}^1$ is the projection of the fibration.

PROOF. When pulling back the regularized form $\tilde{\omega}_\Gamma^{(Z)}$ from $Z^{V_\Gamma} \times \mathbb{P}^1$ to $\mathcal{D}(Z[\Gamma])$, the poles of $\tilde{\omega}_\Gamma^{(Z)}$ along the diagonals $\Delta_\gamma^{(Z)} \times \{0\}$ yield (as in Proposition 5.10 and Corollary 5.13) poles along the divisors $D_\gamma^{(Z)}$, contained in the central fiber $\pi^{-1}(0)$ at $\zeta = 0$ of $\mathcal{D}(Z[\Gamma])$. □

7.6. Deformation of the chain of integration. We now describe a regularization of the chain of integration, based on the deformation to the normal cone.

PROPOSITION 7.9. *The proper transform of the chain $\sigma_\Gamma^{(Z,y)} \times \mathbb{P}^1$ inside $\mathcal{D}(Z[\Gamma])$ gives a deformation of the chain of integration, which does not intersect the locus of divergences of the form $\tilde{\pi}_\Gamma^*(\tilde{\omega}_\Gamma^{(Z)})$.*

PROOF. Consider the chain $\sigma_\Gamma^{(Z,y)} = X^{V_\Gamma} \times \{y\}$ of (3.10), inside Z^{V_Γ}. Extend it to a chain $\sigma_\Gamma^{(Z,y)} \times \mathbb{P}^1$ inside $Z^{V_\Gamma} \times \mathbb{P}^1$. Let $\overline{\sigma_\Gamma^{(Z,y)} \times \mathbb{P}^1}$ denote the proper transform in the blowup $\mathcal{D}(Z[\Gamma])$. Then, as illustrated in Figure 1, we obtain a deformation of $\sigma_\Gamma^{(Z,y)}$ inside the normal cone component of the special fiber $\pi^{-1}(0)$ in $\mathcal{D}(Z[\Gamma])$ that is separated from the intersection with the component given by the blowup $Z[\Gamma]$. □

7.7. Regularized integral. Using the deformation of the chain of integration and of the form, one can regularize the integral by

$$\int_{\Sigma_\Gamma^{(Z,y)}} \tilde{\pi}_\Gamma^*(\tilde{\omega}_\Gamma^{(Z)}), \tag{7.6}$$

where $\Sigma_\Gamma^{(Z,y)}$ denotes the $(2N_\Gamma + 2)$-chain on $\mathcal{D}(Z[\Gamma])$, with $N_\Gamma = \#V_\Gamma$, obtained as in Proposition 7.9. As in (7.3), one also has a corresponding integral on the intersection of the deformed chain $\Sigma^{(Z,y)}$ with the central fiber, which we can write as

$$\int_{\Sigma_\Gamma^{(Z,y)}} \delta(\pi^{-1}(0))\, \tilde{\pi}_\Gamma^*(\tilde{\omega}_\Gamma^{(Z)}).$$

7.7.1. *Behavior at infinity.* The regularization (7.6) described above avoids divergences along the divisors $D_\gamma^{(Z)}$ in $Z[\Gamma]$. It remains to check the behavior at infinity, both in the $\mathbb{P}^1$-direction added in the deformation construction, and along the locus $\mathcal{D}_\infty$ in $\mathcal{D}(Z[\Gamma])$ defined, in the intersection of each fiber $\pi^{-1}(\zeta)$ with the chain of integration $\Sigma_\Gamma^{(Z,y)}$, by $\Delta_{\Gamma,\infty} := X^{V_\Gamma} \smallsetminus \mathbb{A}^{DN_\Gamma}$, with $N_\Gamma = \#V_\Gamma$.

PROPOSITION 7.10. *The integral* (7.6) *is convergent at infinity when* $D > 2$.

PROOF. For the behavior of (7.6) when $\zeta \to \infty$ in $\mathbb{P}^1$, we see that the form behaves like $r^{-2D+2} r dr$, where $r = |\zeta|$. This gives a convergent integral for $2D - 3 > 1$. For the behavior at $\Delta_{\Gamma,\infty}$, consider first the case where a single radial coordinate $r_v = |x_v| \to \infty$. In polar coordinates, we then have a radial integral $r_v^{-(2D-2)\#E_{\Gamma,v}} r^{D-1} dr$, where $E_{\Gamma,v} = \{e \in E_\Gamma \,|\, v \in \partial(e)\}$ is the valence $v(v)$ of the vertex v. This gives a convergent integral when $(2D-2)v(v) - D + 1 > 1$. Since $v(v) \geq 1$ and $2D - 2 \geq 0$, we have $(2D-2)\#E_{\Gamma,v} - D + 1 \geq D - 1$, so the condition is satisfied whenever $D > 2$. More generally, one can have several $r_v \to \infty$. The strongest constraint comes from the case that behaves like $r^{-(2D-2)\sum_v v(v)} r^{DN_\Gamma - 1}$, with $N_\Gamma = \#V_\Gamma$. In this case the convergence condition is given by $(2D-2)v_\Gamma - DN_\Gamma > 0$, where $v_\Gamma = \sum_{v \in V_\Gamma} v(v)$. Again we have $v_\Gamma \geq N_\Gamma$, and we obtain

$$(2D-2)v_\Gamma - DN_\Gamma \geq (D-2)N_\Gamma > 0,$$

whenever $D > 2$. In this case the condition for convergence at $|\zeta| \to \infty$ is also satisfied. □

8. Appendix: Amplitudes and Bochner–Martinelli kernels

In this Appendix we present some additional complementary material. Using the relation between the Feynman amplitude ω_Γ of (3.2) and the Green functions of the Laplacian, we compute $d\omega_\Gamma$ in terms of an integral kernel associated to the graph Γ and the affine space $\mathbb{A}^{DN_\Gamma} \subset X^{V_\Gamma}$, with $N_\Gamma = \#V_\Gamma$. We also discuss the relation between the complexified amplitude $\omega_\Gamma^{(Z)}$ of (3.9) and the Bochner–Martinelli kernel.

8.1. Real Green functions and differentials. As before, we consider the Green function of the real Laplacian on $\mathbb{A}^D(\mathbb{R})$, with $D = 2\lambda + 2$, given by

$$G_{\mathbb{R}}(x,y) = \frac{1}{\|x-y\|^{2\lambda}}. \tag{8.1}$$

Consider then the differential form $\omega = G_{\mathbb{R}}(x,y)\, dx \wedge dy$. This corresponds to the Feynman amplitude (3.2) in the case of the graph consisting of a single edge, with configuration space $(X \times X) \smallsetminus \Delta$, with $\Delta = \{(x,y)\,|\, x = y\}$ the diagonal.

LEMMA 8.1. *The form $\omega = G_{\mathbb{R}}(x,y)\, dx \wedge dy$ is not closed. Its differential is given by*

$$d\omega = -\lambda \sum_{k=1}^{D} \frac{(x_k - y_k)}{\|x-y\|^D} \left(dx \wedge dy \wedge d\bar{x}_k - dx \wedge dy \wedge d\bar{y}_k \right). \tag{8.2}$$

PROOF. We have

$$d\omega = \bar{\partial}\omega = \sum_{k=1}^{D} \frac{\partial G_{\mathbb{R}}}{\partial \bar{x}_k} dx \wedge dy \wedge d\bar{x}_k + \sum_{k=1}^{D} \frac{\partial G_{\mathbb{R}}}{\partial \bar{y}_k} dx \wedge dy \wedge d\bar{y}_k.$$

We then see that

$$\frac{\partial \|x-y\|^{-2\lambda}}{\partial \bar{x}_k} = -\lambda \frac{(x_k - y_k)}{\|x-y\|^D}, \quad \text{and} \quad \frac{\partial \|x-y\|^{-2\lambda}}{\partial \bar{y}_k} = \lambda \frac{(x_k - y_k)}{\|x-y\|^D},$$

so that we obtain (8.2). □

8.2. Feynman amplitudes and integral kernels on graphs. We first consider the form defining the Feynman amplitude ω_Γ of (3.2). It is not a closed form. In fact, we compute here explicitly its differential $d\omega_\Gamma$ in terms of some integral kernels associated to graphs.

Recall that, given a graph Γ and a vertex $v \in V_\Gamma$, the graph $\Gamma \smallsetminus \{v\}$ has

$$V_{\Gamma \smallsetminus \{v\}} = V_\Gamma \smallsetminus \{v\}, \quad \text{and} \quad E_{\Gamma \smallsetminus \{v\}} = E_\Gamma \smallsetminus \{e \in E_\Gamma \,|\, v \in \partial(e)\}, \tag{8.3}$$

that is, one removes a vertex along with its star of edges.

DEFINITION 8.2. Suppose given a graph Γ and a vertex $v \in V_\Gamma$. Let $N_\Gamma = \# V_\Gamma$. Let $\epsilon_v = (-1)^{N_\Gamma} \epsilon'_v$, where ϵ'_v is the sign determined by

$$\epsilon'_v \,(\bigwedge_{w \neq v} dx_w) \wedge dx_v = \bigwedge_{v' \in \mathbf{V}_\Gamma} dx_{v'}.$$

Let $S_{\Gamma,v}$ and $T_{\Gamma,v}$ be the expressions defined as

$$S_{\Gamma,v} = \prod_{e \in E_\Gamma : v = s(e)} \frac{1}{\|x_v - x_{t(e)}\|^{2\lambda}}, \qquad T_{\Gamma,v} = \prod_{e \in E_\Gamma : v = t(e)} \frac{1}{\|x_{s(e)} - x_v\|^{2\lambda}}. \tag{8.4}$$

Also denote by $\kappa_{v,w,k}$ the expression

$$\kappa_{v,w,k} := \frac{(x_{v,k} - x_{w,k})}{\|x_v - x_w\|^D}. \tag{8.5}$$

We define the differential form

$$\kappa^{\mathbb{R}}_{\Gamma,v} := \epsilon_v \sum_{k=1}^{D} \left(S_{\Gamma,v} \sum_{e: v = s(e)} T_{\Gamma \smallsetminus e, v} \kappa_{v,t(e),k} - T_{\Gamma,v} \sum_{e : v = t(e)} S_{\Gamma \smallsetminus e, v} \kappa_{s(e),v,k} \right) dx_v \wedge d\bar{x}_{v,k}, \tag{8.6}$$

in the coordinates $x = (x_v \,|\, v \in V_\Gamma)$. Given an oriented graph Γ, we then consider the integral kernel

$$\mathcal{K}_{\mathbb{R},\Gamma} = \lambda \sum_{v \in \mathbf{V}_\Gamma} \omega_{\Gamma \smallsetminus \{v\}} \wedge \kappa^{\mathbb{R}}_{\Gamma,v}, \tag{8.7}$$

where $\omega_{\Gamma\smallsetminus\{v\}}$ is the form (3.2) for the graph (8.3).

PROPOSITION 8.3. *The differential $d\omega_\Gamma$ is the integral kernel $\mathcal{K}_{\mathbb{R},\Gamma}$ of* (8.7).

PROOF. First observe that, for ω_Γ the Feynman amplitude of (3.2), we have $d\omega_\Gamma = \bar\partial\omega_\Gamma$, that is,

$$d\omega_\Gamma = \sum_{v\in V_\Gamma}\sum_{k=1}^{D}(-1)^{N_\Gamma}\frac{\partial}{\partial\bar x_{v,k}}\omega_\Gamma\wedge d\bar x_{v,k},$$

with $N_\Gamma = \#V_\Gamma$. We introduce the notation

$$\upsilon_{\Gamma,v} := \prod_{e:v\notin\partial(e)}\frac{1}{\|x_{s(e)}-x_{t(e)}\|^{2\lambda}}. \tag{8.8}$$

We can then write, for any given $v\in V_\Gamma$,

$$\prod_{e\in E_\Gamma}\frac{1}{\|x_{s(e)}-x_{t(e)}\|^{2\lambda}} = \upsilon_{\Gamma,v}S_{\Gamma,v}T_{\Gamma,v},$$

and we obtain

$$\partial_{\bar x_{v,k}}\left(\prod_e\frac{1}{\|x_{s(e)}-x_{t(e)}\|^{2\lambda}}\right) = \upsilon_{\Gamma,v}S_{\Gamma,v}\partial_{\bar x_{v,k}}T_{\Gamma,v}+\upsilon_{\Gamma,v}T_{\Gamma,v}\partial_{\bar x_{v,k}}S_{\Gamma,v}.$$

We also have

$$\frac{\partial}{\partial\bar x_{v,k}}T_{\Gamma,v} = \lambda\sum_{e:v=t(e)}T_{\Gamma\smallsetminus e,v}\frac{-(x_{s(e),k}-x_{v,k})}{\|x_{s(e)}-x_v\|^D}$$

$$\frac{\partial}{\partial\bar x_{v,k}}S_{\Gamma,v} = \lambda\sum_{e:v=s(e)}S_{\Gamma\smallsetminus e,v}\frac{(x_{v,k}-x_{t(e),k})}{\|x_v-x_{t(e)}\|^D},$$

where we used the fact that, for $z,w\in\mathbb{A}^D$, one has

$$\frac{\partial}{\partial\bar w_k}\frac{1}{\|z-w\|^{2\lambda}} = -\frac{\lambda\,(w_k-z_k)}{\|z-w\|^D}.$$

This gives

$$\partial_{\bar x_{v,k}}\left(\prod_e\frac{1}{\|x_{s(e)}-x_{t(e)}\|^{2\lambda}}\right) = \lambda\upsilon_{\Gamma,v}\left(\sum_{v=s(e)}S_{\Gamma\smallsetminus e,v}T_{\Gamma,v}\kappa_{v,t(e),k} - \sum_{v=t(e)}S_{\Gamma,v}T_{\Gamma\smallsetminus e,v}\kappa_{s(e),v,k}\right).$$

Thus, we find

$$d\omega_\Gamma = \lambda(-1)^{\mathbf{V}_\Gamma}\sum_{v\in\mathbf{V}_\Gamma}\upsilon_{\Gamma,v}\bigwedge_{w\neq v}dx_w\wedge\kappa^{\mathbb{R}}_{\Gamma,v}.$$

We then identify the term $\upsilon_{\Gamma,v}\wedge_{w\neq v}dx_w$ with the form $\omega_{\Gamma\smallsetminus\{v\}}$. □

It is easy to see that this recovers (8.2) in the case of the graph consisting of two vertices and a single edge between them.

8.3. Complex Green functions and the Bochner–Martinelli kernel. On $\mathbb{A}^D(\mathbb{C}) \subset X$ the complex Laplacian

$$\Delta = \sum_{k=1}^{D} \frac{\partial^2}{\partial x_k \partial \bar{x}_k}.$$

has a fundamental solution of the form ($D > 1$)

$$G_{\mathbb{C}}(x,y) = \frac{-(D-2)!}{(2\pi i)^D \|x-y\|^{2D-2}}. \tag{8.9}$$

The Bochner–Martinelli kernel is given by

$$\mathcal{K}_{\mathbb{C}}(x,y) = \frac{(D-1)!}{(2\pi i)^D} \sum_{k=1}^{D} (-1)^{k-1} \frac{\bar{x}_k - \bar{y}_k}{\|x-y\|^{2D}} d\bar{x}_{[k]} \wedge dx, \tag{8.10}$$

where we write

$$dx = dx_1 \wedge \cdots \wedge dx_D \quad \text{and} \quad d\bar{x}_{[k]} = d\bar{x}_1 \wedge \cdots \wedge \widehat{d\bar{x}_k} \wedge \cdots \wedge d\bar{x}_D,$$

with the k-th factor removed. The following facts are well known (see [**53**], §3.2 and [**60**]):

$$\begin{aligned} \mathcal{K}_{\mathbb{C}}(x,y) &= \sum_{k=1}^{D} (-1)^{k-1} \frac{\partial G_{\mathbb{C}}}{\partial x_k} d\bar{x}_{[k]} \wedge dx \\ &= (-1)^{D-1} \partial_x G_{\mathbb{C}} \wedge \sum_{k=1}^{D} d\bar{x}_{[k]} \wedge dx_{[k]}. \end{aligned} \tag{8.11}$$

where ∂_x and $\bar{\partial}_x$ denote the operators ∂ and $\bar{\partial}$ in the variables $x = (x_k)$. For fixed y, the coefficients of $\mathcal{K}_{\mathbb{C}}(x,y)$ are harmonic functions on $\mathbb{A}^D \smallsetminus \{y\}$ and $\mathcal{K}_{\mathbb{C}}(x,y)$ is closed, $d_x \mathcal{K}(x,y) = 0$. Moreover, the Bochner–Martinelli integral formula holds: for a bounded domain Σ with piecewise smooth boundary $\partial\Sigma$, a function $f \in \mathcal{C}^2(\overline{\Sigma})$ and for $y \in \Sigma$,

$$f(y) = \int_{\partial\Sigma} f(x) \mathcal{K}_{\mathbb{C}}(x,y) + \int_{\Sigma} \Delta(f)(x)\, G_{\mathbb{C}}(x,y) d\bar{x} \wedge dx - \int_{\partial\Sigma} G_{\mathbb{C}}(x,y)\, \mu_f(x), \tag{8.12}$$

where $\mu_f(x)$ is the form

$$\mu_f(x) = \sum_{k=1}^{D} (-1)^{D+k-1} \frac{\partial f}{\partial \bar{x}_k} dx_{[k]} \wedge d\bar{x}. \tag{8.13}$$

The integral (8.12) vanishes when $y \notin \Sigma$. A related Bochner–Martinelli integral, which can be derived from (8.12) (see Thm 1.3 of [**60**]) is of the form

$$f(y) = \int_{\partial\Sigma} f(x)\, \mathcal{K}_{\mathbb{C}}(x,y) - \int_{\Sigma} \bar{\partial} f \wedge \mathcal{K}_{\mathbb{C}}(x,y), \tag{8.14}$$

for $y \in \Sigma$ and $f \in \mathcal{C}^1(\overline{\Sigma})$, with $\bar{\partial} f = \sum_k \partial_{\bar{x}_k} f\, d\bar{x}_k$.

Similarly, one can consider Green forms associated to the Laplacians on $\Omega^{p,q}$ forms and related Bochner–Martinelli kernels, see [**82**].

8.4. Complexified amplitudes and the Bochner–Martinelli kernels. We now consider the Feynman amplitude $\hat{\omega}_\Gamma$ of (3.12) in the complexified case discussed in §3.3.3. We first introduce a Bochner–Martinelli (BM) kernel for graphs.

8.4.1. *Bochner–Martinelli kernel for graphs.* We define Bochner–Martinelli kernels for graphs in the following way.

Suppose given an oriented graph Γ and vertices $v, w \in V_\Gamma$. In the coordinates $x_v, x_w \in X^{V_\Gamma}$, consider the forms

$$\kappa^{\mathbb{C}}_{v,w} = \sum_{k=1}^{D} (-1)^{k-1} \frac{\bar{x}_{v,k} - \bar{x}_{w,k}}{\|x_v - x_w\|^{2D}} \, dx_v \wedge d\bar{x}_{v,[k]}, \tag{8.15}$$

$$\kappa^{\mathbb{C},*}_{v,w} = \sum_{k=1}^{D} (-1)^{k-1} \frac{x_{v,k} - x_{w,k}}{\|x_v - x_w\|^{2D}} \, dx_{v,[k]} \wedge d\bar{x}_v. \tag{8.16}$$

We construct two kinds of integral kernels, in terms of the $\kappa^{\mathbb{C}}_{v,w}$ and $\kappa^{\mathbb{C},*}_{v,w}$ and of expressions

$$S^{\mathbb{C}}_{\Gamma,v} = \prod_{\substack{e \in E_\Gamma \\ v = s(e)}} \frac{1}{\|x_v - x_{t(e)}\|^{2D-2}}, \qquad T^{\mathbb{C}}_{\Gamma,v} = \prod_{\substack{e \in E_\Gamma \\ v = t(e)}} \frac{1}{\|x_{s(e)} - x_v\|^{2D-2}}. \tag{8.17}$$

which are the analogs of (8.4) in the complexified case.

DEFINITION 8.4. For an oriented graph Γ and a vertex $v \in V_\Gamma$, we set

$$\kappa^{\mathbb{C}}_{\Gamma,v} = \sum_{e: v = s(e)} S^{\mathbb{C}}_{\Gamma,v} T^{\mathbb{C}}_{\Gamma \smallsetminus e, v} \kappa^{\mathbb{C}}_{v,t(e)} - \sum_{e: v = t(e)} S^{\mathbb{C}}_{\Gamma \smallsetminus e, v} T^{\mathbb{C}}_{\Gamma,v} \kappa^{\mathbb{C}}_{s(e),v}. \tag{8.18}$$

$$\kappa^{\mathbb{C},*}_{\Gamma,v} = \sum_{e: v = s(e)} S^{\mathbb{C}}_{\Gamma,v} T^{\mathbb{C}}_{\Gamma \smallsetminus e, v} \kappa^{\mathbb{C},*}_{v,t(e)} - \sum_{e: v = t(e)} S^{\mathbb{C}}_{\Gamma \smallsetminus e, v} T^{\mathbb{C}}_{\Gamma,v} \kappa^{\mathbb{C},*}_{s(e),v}. \tag{8.19}$$

We also define a simpler BM kernel for graphs, of the form

$$\kappa^{BM,\mathbb{C}}_{\Gamma,v} = \sum_{e: v \in \partial(e)} \epsilon_e \, \kappa^{\mathbb{C}}_{s(e),t(e)} \, dx_v \wedge d\bar{x}_{v,[k]} \tag{8.20}$$

and

$$\kappa^{BM,\mathbb{C},*}_{\Gamma,v} = \sum_{e: v \in \partial(e)} \epsilon_e \, \kappa^{\mathbb{C},*}_{s(e),t(e)} \, dx_{v,[k]} \wedge d\bar{x}_v, \tag{8.21}$$

where the sign ϵ_e is ± 1 depending on whether $v = s(e)$ or $v = t(e)$.

8.4.2. *Complexified amplitude and Bochner–Martinelli kernel.* The Bochner–Martinelli kernel of graphs defined above is related to the Feynman amplitude (3.12) by the following.

PROPOSITION 8.5. *Let $\hat{\omega}_\Gamma$ be the Feynman amplitude* (3.12). *Then*

$$\partial \hat{\omega}_\Gamma = \sum_{v \in V_\Gamma} \epsilon_v \, \hat{\omega}_{\Gamma \smallsetminus \{v\}} \wedge \kappa^{\mathbb{C}}_{\Gamma,v} \tag{8.22}$$

$$\bar{\partial} \hat{\omega}_\Gamma = \sum_{v \in V_\Gamma} \epsilon_v \, \hat{\omega}_{\Gamma \smallsetminus \{v\}} \wedge (-1)^{D-1} \kappa^{\mathbb{C},*}_{\Gamma,v}, \tag{8.23}$$

where the sign ϵ_v is defined by

$$\epsilon_v \left(\bigwedge_{w \neq v} \sum_k (-1)^{k-1} dx_{w,[k]} \wedge d\bar{x}_{w,[k]} \right) \wedge \left(\sum_k (-1)^{k-1} dx_{v,[k]} \wedge d\bar{x}_{v,[k]} \right) =$$

$$\bigwedge_{v'\in V_\Gamma} \sum_k (-1)^{k-1} dx_{v',[k]} \wedge d\bar{x}_{v',[k]}.$$

PROOF. The argument is analogous to Proposition 8.3. □

8.5. Bochner–Martinelli integral on graphs. There is an analog of the classical Bochner–Martinelli integral (8.14) for the kernel (8.18) of graphs.

We first recall some well known facts about the Laplacian on graphs, see e.g. [**17**]. Given a graph Γ, one defines the exterior differential δ from functions on V_Γ to functions on E_Γ by

$$(\delta h)(e) = h(s(e)) - h(t(e))$$

and the δ^* operator from functions on edges to functions on vertices by

$$(\delta^*\xi)(v) = \sum_{e:v=s(e)} \xi(e) - \sum_{e:v=t(e)} \xi(e).$$

Thus, the Laplacian $\Delta_\Gamma = \delta^*\delta$ on Γ is given by

$$(\Delta_\Gamma f)(v) = \sum_{e:v=s(e)} (h(v) - h(t(e))) - \sum_{e:v=t(e)} (h(s(e)) - h(v))$$

$$= N_v\, h(v) - \sum_{e:v\in\partial(e)} h(v_e),$$

where N_v is the number of vertices connected to v by an edge, and v_e is the other endpoint of e (we assume as usual that Γ has no looping edges). Thus, a harmonic function h on a graph is a function on V_Γ satisfying

$$h(v) = \frac{1}{N_v} \sum_{e:v\in\partial(e)} h(v_e). \tag{8.24}$$

Motivated by the usual notion of graph Laplacian Δ_Γ and the harmonic condition (8.24) for graphs recalled here above, we introduce an operator

$$(\Delta_{\Gamma,v} f)(x) = \sum_{e:v\in\partial(e)} f(x_{v_e}), \tag{8.25}$$

which assigns to a complex valued function f defined on $\mathbb{A}^D \subset X$ a complex valued function $\Delta_{\Gamma,v} f$ defined on X^{V_Γ}.

We then have the following result.

PROPOSITION 8.6. *Let f be a complex valued function defined on $\mathbb{A}^D \subset X$. Also suppose given a bounded domain Σ with piecewise smooth boundary $\partial\Sigma$ in $\mathbb{A}^D$ and assume that f is C^1 on $\overline{\Sigma}$. For a given $v \in \mathbf{V}_\Gamma$ consider the set of $x = (x_w) \in \mathbb{A}^{D|\mathbf{V}_\Gamma|} \subset X^{\mathbf{V}_\Gamma}$, such that $x_{v_e} \in \Sigma$ for all $v_e \neq v$ endpoints of edges e with $v \in \partial(e)$. For such $x = (x_w)$ we have*

$$(\Delta_{\Gamma,v} f)(x) = \frac{(D-1)!}{(2\pi i)^D N_v} \left(\int_{\partial\Sigma} f(x_v) \kappa^{BM,\mathbb{C}}_{\Gamma,v}(x) - \int_\Sigma \bar{\partial}_{x_v} f(x_v) \wedge \kappa^{BM,\mathbb{C}}_{\Gamma,v}(x) \right), \tag{8.26}$$

with $\kappa^{BM,\mathbb{C}}_{\Gamma,v}$ and $\kappa^{BM,\mathbb{C}}_{\Gamma,v}$ as in (8.20) and (8.21), where the integration on Σ and $\partial\Sigma$ is in the variable x_v and $\Delta_{\Gamma,v} f$ is defined as in (8.25).

PROOF. We have

$$\int_{\partial\Sigma} f(x_v)\kappa^{BM,\mathbb{C}}_{\Gamma,v}(x) = \sum_{e:v\in\partial(e)} \epsilon_e \int_{\partial\Sigma} f(x_v)\sum_{k=1}^{D}(-1)^{k-1}\frac{(\bar{x}_{s(e),k}-\bar{x}_{t(e),k})}{\|x_{s(e)}-x_{t(e)}\|^{2D}}\,\eta_v$$

where

$$\eta_v = dx_v \wedge d\bar{x}_{v,[k]}.$$

We write the integral as

$$\sum_{e:v=s(e)} \int_{\partial\Sigma} f(x_v)\sum_{k=1}^{D}(-1)^{k-1}\frac{(\bar{x}_{v,k}-\bar{x}_{v_e,k})}{\|x_v-x_{v_e}\|^{2D}}\,\eta_v$$

$$-\sum_{e:v=t(e)} \int_{\partial\Sigma} f(x_v)\sum_{k=1}^{D}(-1)^{k-1}\frac{(\bar{x}_{v_e,k}-\bar{x}_{v,k})}{\|x_v-x_{v_e}\|^{2D}}\,\eta_v$$

$$=\sum_{e:v\in\partial(e)} \int_{\partial\Sigma} f(x_v)\sum_{k=1}^{D}(-1)^{k-1}\frac{(\bar{x}_{v,k}-\bar{x}_{v_e,k})}{\|x_v-x_{v_e}\|^{2D}}\,\eta_v.$$

The case of the integral on Σ is analogous. We then apply the classical result (8.14) about the Bochner–Martinelli integral and we obtain

$$\int_{\partial\Sigma} f(x)\kappa^{BM,\mathbb{C}}_{\Gamma,v}(x) - \int_{\Sigma} \bar{\partial}_{x_v} f(x)\wedge\kappa^{BM,\mathbb{C}}_{\Gamma,v}(x) = \frac{(2\pi i)^D}{(D-1)!}\sum_{e:v\in\partial(e)} f(x_{v_e}).$$

□

8.6. Bochner–Martinelli kernel and forms with logarithmic poles. In view of our discussion about replacing the complexified amplitude by an algebraic form with logarithmic poles, we mention here a well known property of the Bochner–Martinelli kernel, which provides a completely explicit toy model model case for the kind of substitution we aim for in the more general case of the complexified amplitude.

Consider then the Bochner–Martinelli form, for $x, w \in \mathbb{C}^D$

$$\Psi(x,y) = \sum_{k=1}^{D}(-1)^{k-1}\frac{(\bar{x}_k-\bar{y}_k)}{\|x-y\|^{2D}}\,dx\wedge d\bar{x}_{[k]} \tag{8.27}$$

Let Ω^{D-1} be the pullback, under the projection $p:\mathbb{C}^D\smallsetminus\{0\}\to\mathbb{P}^{D-1}(\mathbb{C})$, of the Kähler volume form of the Fubini–Study metric on $\mathbb{P}^{D-1}(\mathbb{C})$, and let $\pi:\mathrm{Bl}_y(\mathbb{C}^D)\to\mathbb{C}^D$ denote the projection of the blowup of $\mathbb{C}^D$ at a point y. We then have the following result.

LEMMA 8.7. *The pullback $\pi^*(\Psi)$ of the Bochner–Martinelli form (8.27) is*

$$\pi^*(\Psi) = C\,\theta\wedge\Omega^{D-1}, \tag{8.28}$$

where Ω is as above and $\theta = d\log(\lambda)$ on the fibers $\{\lambda x\}_{\lambda\in\mathbb{C}}$, and with the constant $C\in\mathbb{Q}(2\pi i)$.

PROOF. The result follows as in [**53**], pp. 371-373. One checks with a direct computation that the constant is given by $C=(2\pi i)^{D-1}/(D-1)!$. □

LEMMA 8.8. *The form*

$$\hat{\omega}_{v,w} = \|x_v - x_w\|^{-(2D-2)} \sum_{k=1}^{D} (-1)^{k-1} dx_{v,[k]} \wedge d\bar{x}_{v,[k]} \tag{8.29}$$

satisfies

$$\partial \hat{\omega}_{v,w} = \Psi(x_v, x_w). \tag{8.30}$$

where the ∂*-operator is taken in the* x_v *variables.*

PROOF. The form $\partial\hat{\omega}_{v,w}$ is given by

$$\sum_{j=1}^{D} \frac{\partial}{\partial x_{v,j}} \left(\frac{1}{\|x_w - x_v\|^{2D-2}} \right) dx_{v,j} \wedge \sum_{k=1}^{D} (-1)^{k-1} dx_{v,[k]} \wedge d\bar{x}_{v,[k]} = \Psi(x_v, x_w).$$

□

Note that the form $\hat{\omega}_{v,w}$ of (8.29) is the form $\hat{\omega}_\Gamma$ for the graph Γ consisting of two vertices v, w and a single edge between them. We then have the following result, which shows that the form $\hat{\omega}_{v,w}$ can be replaced by an algebraic form with logarithmic poles, up to a coboundary.

PROPOSITION 8.9. *Let* π *be the projection of the blowup along* $\Delta = \{x_v = x_w\}$. *The form* $\hat{\omega}_{v,w}$ *satisfies, for some form* ξ,

$$\pi^*(\hat{\omega}_{v,w}) = C \log(\lambda)\, \Omega^{D-1} + d\xi, \tag{8.31}$$

with $C = (2\pi i)^{D-1}/(D-1)!$.

PROOF. By (8.28) and (8.30), we have $\partial\pi^*(\hat{\omega}_{v,w}) = C\theta\wedge\Omega^{D-1}$ and $\bar{\partial}\pi^*(\hat{\omega}_{v,w}) = 0$. Thus, we have

$$d(\pi^*(\hat{\omega}_{v,w}) - C \log(\lambda)\, \Omega^{D-1}) = 0.$$

The cohomology of the blowup is given additively by (see [**53**], p. 605)

$$H^*(\mathrm{Bl}_\Delta(\mathbb{P}^{D-1} \times \mathbb{P}^{D-1})) = \pi^* H^*(\mathbb{P}^{D-1} \times \mathbb{P}^{D-1}) \oplus H^*(E) / \pi^* H^*(\Delta).$$

This has no odd dimensional cohomology, hence for the above form of degree $2D-1$ we obtain that $\pi^*(\hat{\omega}_{v,w})$ and $C \log(\lambda)\, \Omega^{D-1}$ differ by a coboundary, with C as in Lemma 8.7. □

In particular, as a direct consequence of the previous result, in the case of a tree we obtain the following.

COROLLARY 8.10. *If the graph* Γ *is a tree, then*

$$\pi^*(\hat{\omega}_\Gamma) = \wedge_{e \in E_\Gamma} \pi^*(\hat{\omega}_{s(e),t(e)})$$

is cohomologous to a differential form with logarithmic poles.

References

[1] Jakob Ablinger, Johannes Blümlein, and Carsten Schneider, *Harmonic sums and polylogarithms generated by cyclotomic polynomials*, J. Math. Phys. **52** (2011), no. 10, 102301, 52, DOI 10.1063/1.3629472. MR2894586

[2] I. A. Aĭzenberg and A. P. Yuzhakov, *Integral representations and residues in multidimensional complex analysis*, Translations of Mathematical Monographs, vol. 58, American Mathematical Society, Providence, RI, 1983. Translated from the Russian by H. H. McFaden; Translation edited by Lev J. Leifman. MR735793

[3] L.A. Aizenberg, A.K. Tsikh, A.P. Yuzhakov, *Multidimensional residues and applications*, in "Several Complex Variables, II", Encyclopedia of Mathematical Sciences, Vol.8, Springer Verlag, 1994.

[4] A.G. Aleksandrov, A.K. Tsikh, *Multi-logarithmic differential forms on complete intersections*, J. Siberian Fed. Univ. Math and Phys. 2 (2008) 105–124.

[5] Sigitas Ališauskas, *Coupling coefficients of* $\mathrm{SO}(n)$ *and integrals involving Jacobi and Gegenbauer polynomials*, J. Phys. A **35** (2002), no. 34, 7323–7345, DOI 10.1088/0305-4470/35/34/307. MR1945516

[6] Paolo Aluffi and Matilde Marcolli, *Feynman motives of banana graphs*, Commun. Number Theory Phys. **3** (2009), no. 1, 1–57, DOI 10.4310/CNTP.2009.v3.n1.a1. MR2504753

[7] Paolo Aluffi and Matilde Marcolli, *Parametric Feynman integrals and determinant hypersurfaces*, Adv. Theor. Math. Phys. **14** (2010), no. 3, 911–963. MR2801414

[8] Paolo Aluffi and Matilde Marcolli, *Feynman motives and deletion-contraction relations*, Topology of algebraic varieties and singularities, Contemp. Math., vol. 538, Amer. Math. Soc., Providence, RI, 2011, pp. 21–64, DOI 10.1090/conm/538/10594. MR2777815

[9] Paolo Aluffi and Matilde Marcolli, *Graph hypersurfaces and a dichotomy in the Grothendieck ring*, Lett. Math. Phys. **95** (2011), no. 3, 223–232, DOI 10.1007/s11005-011-0461-5. MR2775124

[10] Paolo Aluffi and Matilde Marcolli, *Algebro-geometric Feynman rules*, Int. J. Geom. Methods Mod. Phys. **8** (2011), no. 1, 203–237, DOI 10.1142/S0219887811005099. MR2782886

[11] Paolo Aluffi and Matilde Marcolli, *Graph hypersurfaces and a dichotomy in the Grothendieck ring*, Lett. Math. Phys. **95** (2011), no. 3, 223–232, DOI 10.1007/s11005-011-0461-5. MR2775124

[12] Tom M. Apostol and Thiennu H. Vu, *Dirichlet series related to the Riemann zeta function*, J. Number Theory **19** (1984), no. 1, 85–102, DOI 10.1016/0022-314X(84)90094-5. MR751166

[13] V. I. Arnol′d, S. M. Guseĭn-Zade, and A. N. Varchenko, *Singularities of differentiable maps. Vol. I*, Monographs in Mathematics, vol. 82, Birkhäuser Boston, Inc., Boston, MA, 1985. The classification of critical points, caustics and wave fronts; Translated from the Russian by Ian Porteous and Mark Reynolds. MR777682

[14] Arthur Erdélyi, Wilhelm Magnus, Fritz Oberhettinger, and Francesco G. Tricomi, *Higher transcendental functions. Vols. I, II*, McGraw-Hill Book Company, Inc., New York-Toronto-London, 1953. Based, in part, on notes left by Harry Bateman. MR0058756

[15] Dori Bejleri and Matilde Marcolli, *Quantum field theory over* $\mathbb{F}_1$, J. Geom. Phys. **69** (2013), 40–59, DOI 10.1016/j.geomphys.2013.03.002. MR3040951

[16] Prakash Belkale and Patrick Brosnan, *Matroids, motives, and a conjecture of Kontsevich*, Duke Math. J. **116** (2003), no. 1, 147–188, DOI 10.1215/S0012-7094-03-11615-4. MR1950482

[17] Itai Benjamini and László Lovász, *Harmonic and analytic functions on graphs*, J. Geom. **76** (2003), no. 1-2, 3–15, DOI 10.1007/s00022-033-1697-8. Combinatorics, 2002 (Maratea). MR2005527

[18] Carlos A. Berenstein, Roger Gay, Alekos Vidras, and Alain Yger, *Residue currents and Bezout identities*, Progress in Mathematics, vol. 114, Birkhäuser Verlag, Basel, 1993. MR1249478

[19] Franziska Bittner, *The universal Euler characteristic for varieties of characteristic zero*, Compos. Math. **140** (2004), no. 4, 1011–1032, DOI 10.1112/S0010437X03000617. MR2059227

[20] James D. Bjorken and Sidney D. Drell, *Relativistic quantum fields*, McGraw-Hill Book Co., New York-Toronto-London-Sydney, 1965. MR0187642

[21] S. Bloch, Leonidas Alaoglu Memorial Lecture, Caltech, May 2012.

[22] Spencer Bloch, Hélène Esnault, and Dirk Kreimer, *On motives associated to graph polynomials*, Comm. Math. Phys. **267** (2006), no. 1, 181–225, DOI 10.1007/s00220-006-0040-2. MR2238909

[23] Spencer Bloch, Matt Kerr, and Pierre Vanhove, *A Feynman integral via higher normal functions*, Compos. Math. **151** (2015), no. 12, 2329–2375, DOI 10.1112/S0010437X15007472. MR3433889

[24] Spencer Bloch and Dirk Kreimer, *Mixed Hodge structures and renormalization in physics*, Commun. Number Theory Phys. **2** (2008), no. 4, 637–718, DOI 10.4310/CNTP.2008.v2.n4.a1. MR2492196

[25] Spencer Bloch and Pierre Vanhove, *The elliptic dilogarithm for the sunset graph*, J. Number Theory **148** (2015), 328–364, DOI 10.1016/j.jnt.2014.09.032. MR3283183

[26] J. Blümlein, D. J. Broadhurst, and J. A. M. Vermaseren, *The multiple zeta value data mine*, Comput. Phys. Comm. **181** (2010), no. 3, 582–625, DOI 10.1016/j.cpc.2009.11.007. MR2578167

[27] Christian Bogner and Francis Brown, *Feynman integrals and iterated integrals on moduli spaces of curves of genus zero*, Commun. Number Theory Phys. **9** (2015), no. 1, 189–238, DOI 10.4310/CNTP.2015.v9.n1.a3. MR3339854

[28] D. J. Broadhurst and D. Kreimer, *Association of multiple zeta values with positive knots via Feynman diagrams up to 9 loops*, Phys. Lett. B **393** (1997), no. 3-4, 403–412, DOI 10.1016/S0370-2693(96)01623-1. MR1435933

[29] Francis Brown, *Mixed Tate motives over* $\mathbb{Z}$, Ann. of Math. (2) **175** (2012), no. 2, 949–976, DOI 10.4007/annals.2012.175.2.10. MR2993755

[30] F. Brown, D. Doryn, *Framings for graph hypersurfaces*, arXiv:1301.3056.

[31] Francis Brown and Oliver Schnetz, *A K3 in* ϕ^4, Duke Math. J. **161** (2012), no. 10, 1817–1862, DOI 10.1215/00127094-1644201. MR2954618

[32] Francis Brown and Oliver Schnetz, *Modular forms in quantum field theory*, Commun. Number Theory Phys. **7** (2013), no. 2, 293–325, DOI 10.4310/CNTP.2013.v7.n2.a3. MR3164866

[33] Francis Brown, Oliver Schnetz, and Karen Yeats, *Properties of* c_2 *invariants of Feynman graphs*, Adv. Theor. Math. Phys. **18** (2014), no. 2, 323–362. MR3273316

[34] Francisco J. Castro-Jiménez, Luis Narváez-Macarro, and David Mond, *Cohomology of the complement of a free divisor*, Trans. Amer. Math. Soc. **348** (1996), no. 8, 3037–3049, DOI 10.1090/S0002-9947-96-01690-X. MR1363009

[35] Ö. Ceyhan, in preparation.

[36] Özgür Ceyhan and Matilde Marcolli, *Feynman integrals and motives of configuration spaces*, Comm. Math. Phys. **313** (2012), no. 1, 35–70, DOI 10.1007/s00220-012-1484-1. MR2928218

[37] Özgür Ceyhan and Matilde Marcolli, *Algebraic renormalization and Feynman integrals in configuration spaces*, Adv. Theor. Math. Phys. **18** (2014), no. 2, 469–511. MR3273319

[38] K. G. Chetyrkin, A. L. Kataev, and F. V. Tkachov, *New approach to evaluation of multiloop Feynman integrals: the Gegenbauer polynomial x-space technique*, Nuclear Phys. B **174** (1980), no. 2-3, 345–377, DOI 10.1016/0550-3213(80)90289-8. MR591619

[39] Nicolas R. Coleff and Miguel E. Herrera, *Les courants résiduels associés à une forme méromorphe* (French), Lecture Notes in Mathematics, vol. 633, Springer, Berlin, 1978. MR492769

[40] Alain Connes and Matilde Marcolli, *Renormalization and motivic Galois theory*, Int. Math. Res. Not. **76** (2004), 4073–4091, DOI 10.1155/S1073792804143122. MR2109986

[41] Alain Connes and Matilde Marcolli, *Noncommutative geometry, quantum fields and motives*, American Mathematical Society Colloquium Publications, vol. 55, American Mathematical Society, Providence, RI; Hindustan Book Agency, New Delhi, 2008. MR2371808

[42] Pierre Deligne, *Équations différentielles à points singuliers réguliers* (French), Lecture Notes in Mathematics, Vol. 163, Springer-Verlag, Berlin-New York, 1970. MR0417174

[43] Pierre Deligne, *Théorie de Hodge. II* (French), Inst. Hautes Études Sci. Publ. Math. **40** (1971), 5–57. MR498551

[44] Dzmitry Doryn, *On one example and one counterexample in counting rational points on graph hypersurfaces*, Lett. Math. Phys. **97** (2011), no. 3, 303–315, DOI 10.1007/s11005-011-0501-1. MR2826814

[45] Pedro Freitas, *Integrals of polylogarithmic functions, recurrence relations, and associated Euler sums*, Math. Comp. **74** (2005), no. 251, 1425–1440, DOI 10.1090/S0025-5718-05-01747-3. MR2137010

[46] William Fulton, *Intersection theory*, 2nd ed., Ergebnisse der Mathematik und ihrer Grenzgebiete. 3. Folge. A Series of Modern Surveys in Mathematics [Results in Mathematics and Related Areas. 3rd Series. A Series of Modern Surveys in Mathematics], vol. 2, Springer-Verlag, Berlin, 1998. MR1644323

[47] William Fulton, *Introduction to intersection theory in algebraic geometry*, CBMS Regional Conference Series in Mathematics, vol. 54, Published for the Conference Board of the Mathematical Sciences, Washington, DC; by the American Mathematical Society, Providence, RI, 1984. MR735435

[48] William Fulton and Robert MacPherson, *A compactification of configuration spaces*, Ann. of Math. (2) **139** (1994), no. 1, 183–225, DOI 10.2307/2946631. MR1259368

[49] J. Gallier, *Notes on spherical harmonics and linear representations of Lie groups*, preprint, 2009.

[50] A.B. Goncharov, *Periods and mixed motives*, arXiv.math/0202154v2

[51] A. B. Goncharov and Yu. I. Manin, *Multiple ζ-motives and moduli spaces $\overline{\mathcal{M}}_{0,n}$*, Compos. Math. **140** (2004), no. 1, 1–14, DOI 10.1112/S0010437X03000125. MR2004120

[52] Phillip A. Griffiths, *On the periods of certain rational integrals. I, II*, Ann. of Math. (2) 90 (1969), 460-495; ibid. (2) **90** (1969), 496–541, DOI 10.2307/1970746. MR0260733

[53] Phillip Griffiths and Joseph Harris, *Principles of algebraic geometry*, Wiley-Interscience [John Wiley & Sons], New York, 1978. Pure and Applied Mathematics. MR507725

[54] Phillip Griffiths and Wilfried Schmid, *Recent developments in Hodge theory: a discussion of techniques and results*, Discrete subgroups of Lie groups and applicatons to moduli (Internat. Colloq., Bombay, 1973), Oxford Univ. Press, Bombay, 1975, pp. 31–127. MR0419850

[55] A. Grothendieck, *On the de Rham cohomology of algebraic varieties*, Inst. Hautes Études Sci. Publ. Math. **29** (1966), 95–103. MR199194

[56] M. Herrera and D. Lieberman, *Residues and principal values on complex spaces*, Math. Ann. **194** (1971), 259–294, DOI 10.1007/BF01350129. MR296352

[57] C. Itzykson, J.B. Zuber, *Quantum Field Theory*, Dover Publications, 2006

[58] Dieter Jungnickel, *Graphs, networks and algorithms*, 3rd ed., Algorithms and Computation in Mathematics, vol. 5, Springer, Berlin, 2008. MR2363884

[59] Georg Junker, *Explicit evaluation of coupling coefficients for the most degenerate representations of* SO(n), J. Phys. A **26** (1993), no. 7, 1649–1661. MR1219686

[60] Alexander M. Kytmanov, *The Bochner-Martinelli integral and its applications*, Birkhäuser Verlag, Basel, 1995. Translated from the Russian by Harold P. Boas and revised by the author. MR1409816

[61] Li Li, *Chow motive of Fulton-MacPherson configuration spaces and wonderful compactifications*, Michigan Math. J. **58** (2009), no. 2, 565–598, DOI 10.1307/mmj/1250169077. MR2595554

[62] Li Li, *Wonderful compactification of an arrangement of subvarieties*, Michigan Math. J. **58** (2009), no. 2, 535–563, DOI 10.1307/mmj/1250169076. MR2595553

[63] Javier López Peña and Oliver Lorscheid, *Torified varieties and their geometries over* $\mathbb{F}_1$, Math. Z. **267** (2011), no. 3-4, 605–643, DOI 10.1007/s00209-009-0638-0. MR2776051

[64] Yuri I. Manin and Matilde Marcolli, *Moduli operad over* $\mathbb{F}_1$, Absolute arithmetic and $\mathbb{F}_1$-geometry, Eur. Math. Soc., Zürich, 2016, pp. 331–361. MR3701904

[65] Matilde Marcolli, *Feynman motives*, World Scientific Publishing Co. Pte. Ltd., Hackensack, NJ, 2010. MR2604634

[66] Matilde Marcolli, *Motivic renormalization and singularities*, Quanta of maths, Clay Math. Proc., vol. 11, Amer. Math. Soc., Providence, RI, 2010, pp. 409–458. MR2732060

[67] Matilde Marcolli and Xiang Ni, *Rota-Baxter algebras, singular hypersurfaces, and renormalization on Kausz compactifications*, J. Singul. **15** (2016), 80–117. MR3562856

[68] N. Martin, *Compactification d'espaces de configurations*, Mémoire de Master Recherche, Centre de Mathématiques Laurent Schwartz, École Polytechnique, 2013.

[69] L. J. Mordell, *On the evaluation of some multiple series*, J. London Math. Soc. **33** (1958), 368–371, DOI 10.1112/jlms/s1-33.3.368. MR100181

[70] Mitsuo Morimoto, *Analytic functionals on the sphere*, Translations of Mathematical Monographs, vol. 178, American Mathematical Society, Providence, RI, 1998. MR1641900

[71] N. Nakanishi, *Graph Theory and Feynman Integrals*, Gordon and Breach, 1971.

[72] N.M. Nikolov, R. Stora, I. Todorov, *Configuration space renormalization of massless QFT as an extension problem for associate homogeneous distributions*, IHES preprints 2011: IHES/P/11/07.

[73] C. Sabbah, private communication, June 2013.

[74] Kyoji Saito, *Theory of logarithmic differential forms and logarithmic vector fields*, J. Fac. Sci. Univ. Tokyo Sect. IA Math. **27** (1980), no. 2, 265–291. MR586450

[75] Oliver Schnetz, *Quantum field theory over* $\mathbb{F}_q$, Electron. J. Combin. **18** (2011), no. 1, Paper 102, 23. MR2811071

[76] Laurent Schwartz, *Division par une fonction holomorphe sur une variété analytique complexe* (French), Summa Brasil. Math. **3** (1955), 181–209 (1955). MR139937

[77] Didier Sébilleau, *On the computation of the integrated products of three spherical harmonics*, J. Phys. A **31** (1998), no. 34, 7157–7168, DOI 10.1088/0305-4470/31/34/017. MR1641160

[78] Clayton Shonkwiler and David Shea Vela-Vick, *Higher-dimensional linking integrals*, Proc. Amer. Math. Soc. **139** (2011), no. 4, 1511–1519, DOI 10.1090/S0002-9939-2010-10603-2. MR2748445

[79] Richard P. Stanley, *Acyclic orientations of graphs*, Discrete Math. **5** (1973), 171–178, DOI 10.1016/0012-365X(73)90108-8. MR317988

[80] Elias M. Stein and Guido Weiss, *Introduction to Fourier analysis on Euclidean spaces*, Princeton University Press, Princeton, N.J., 1971. Princeton Mathematical Series, No. 32. MR0304972

[81] John R. Stembridge, *Counting points on varieties over finite fields related to a conjecture of Kontsevich*, Ann. Comb. **2** (1998), no. 4, 365–385, DOI 10.1007/BF01608531. MR1774975

[82] N. N. Tarkhanov and A. A. Shlapunov, *Green's formulas in complex analysis*, J. Math. Sci. (N.Y.) **120** (2004), no. 6, 1868–1900, DOI 10.1023/B:JOTH.0000020708.24761.bb. Complex analysis. MR2085500

[83] Leonard Tornheim, *Harmonic double series*, Amer. J. Math. **72** (1950), 303–314, DOI 10.2307/2372034. MR34860

[84] August Tsikh and Alain Yger, *Residue currents*, J. Math. Sci. (N.Y.) **120** (2004), no. 6, 1916–1971, DOI 10.1023/B:JOTH.0000020710.57247.b7. Complex analysis. MR2085502

[85] Pierre Vanhove, *The physics and the mixed Hodge structure of Feynman integrals*, String-Math 2013, Proc. Sympos. Pure Math., vol. 88, Amer. Math. Soc., Providence, RI, 2014, pp. 161–194, DOI 10.1090/pspum/088/01455. MR3330287

[86] D. A. Varshalovich, A. N. Moskalev, and V. K. Khersonskiĭ, *Quantum theory of angular momentum*, World Scientific Publishing Co., Inc., Teaneck, NJ, 1988. Irreducible tensors, spherical harmonics, vector coupling coefficients, $3nj$ symbols; Translated from the Russian. MR1022665

[87] N. Ja. Vilenkin, *Special functions and the theory of group representations*, Translated from the Russian by V. N. Singh. Translations of Mathematical Monographs, Vol. 22, American Mathematical Society, Providence, R. I., 1968. MR0229863

[88] Vladimir Voevodsky, *Triangulated categories of motives over a field*, Cycles, transfers, and motivic homology theories, Ann. of Math. Stud., vol. 143, Princeton Univ. Press, Princeton, NJ, 2000, pp. 188–238. MR1764202

[89] Yu-lin Xu, *Fast evaluation of the Gaunt coefficients*, Math. Comp. **65** (1996), no. 216, 1601–1612, DOI 10.1090/S0025-5718-96-00774-0. MR1361813

Faculté des Sciences, de la Technologie et de la Communication, University of Luxembourg, 6 rue Richard Coudenhove-Kalergi, L-1359 Luxembourg

Email address: ozgur.ceyhan@uni.lu

Department of Mathematics, Division of Physics, Mathematics and Astronomy, Mail Code 253-37, California Institute of Technology, 1200 E. California Blvd. Pasadena, California 91125

Email address: matilde@caltech.edu

Clay Mathematics Proceedings
Volume **21**, 2020
https://doi.org/10.1090/cmip/21/478

Introductory course on ℓ-adic sheaves and their ramification theory on curves

Lars Kindler and Kay Rülling

Abstract. These are the notes accompanying 13 lectures given by the authors at the Clay Mathematics Institute Summer School 2014 in Madrid. They give an introduction to the theory of ℓ-adic sheaves with emphasis on their ramification theory on curves.

Contents

2010 *Mathematics Subject Classification.* Primary 11S15.
The first author was supported by the ERC Advanced Grant 226257.
The second author was supported by the ERC Advanced Grant 226257 and by the DFG Heisenberg Grant RU 1412/2-1.

1. Introduction

These are the notes accompanying 13 lectures given by the authors at the Clay Mathematics Institute Summer School 2014 in Madrid. The goal of this lecture series is to introduce the audience to the theory of ℓ-adic sheaves with emphasis on their ramification theory. Ideally, the lectures and these notes will equip the audience with the necessary background knowledge to read current literature on the subject, particularly [**16**], which is the focus of a second series of lectures at the same summer school. We do not attempt to give a panoramic exposition of recent research in the subject.

Before giving an outline of this document, the authors wish to stress that there is *no original mathematical content in these notes, and that inaccuracies, omissions and errors are solely their responsibility.*

To introduce ideas, let p be a prime number and U a smooth, connected curve over the finite field $\mathbb{F}_p$. If $\ell \neq p$ is a second prime number, an ℓ-adic sheaf can be understood as a continuous representation $\rho : \pi_1^{\text{ét}}(U,u) \to \mathrm{GL}_r(E)$, where E is a finite extension of the field of ℓ-adic numbers $\mathbb{Q}_\ell$, and where $\pi_1^{\text{ét}}(U,u)$ is the étale fundamental group of U with respect to the base point u. This group is an algebraic variant of the usual fundamental group of a topological space; additionally, it carries a topology. We summarize its construction and properties in Section 6.

A representation ρ as above has a geometric interpretation (hence the word "sheaf"), which is explained in Section 7. If X is the unique smooth projective curve over $\mathbb{F}_p$ containing U as an open dense subvariety, then one might ask whether ρ

extends to a representation of $\pi^{\text{ét}}(X,u)$, i.e. whether ρ factors through the canonical map $\pi_1^{\text{ét}}(U,u) \to \pi_1^{\text{ét}}(X,u)$. If it does, ρ is called unramified. If it does not, then it is natural to ask whether it can be measured "how bad" the ramification of ρ is. The two important definitions in this context are *tame ramification* and *wild ramification.* Tamely ramified ℓ-adic sheaves are much better understood than wildly ramified ones and they behave similarly to regular singular local systems on Riemann surfaces.

One of the main results presented in these notes is the construction and analysis of an invariant of ρ, locally at the finitely many closed points of $X \smallsetminus U$, which measures how wild the ramification of ρ is; this is the so called Swan conductor. In the course of the construction, we give a proof of the Hasse-Arf theorem and we show that the Swan conductor arises from the character of a projective $\mathbb{Z}_\ell$-representation, the Swan representation. See Sections 3 and 4.

The second main result gives a global, cohomological interpretation of the Swan conductor: This is the formula of Grothendieck-Ogg-Shafarevich, see Section 10.

The final sections survey generalizations of these notions to higher dimensional varieties. One approach based on ideas of Wiesend, further developed by Kerz, Schmidt, Drinfeld and Deligne is presented in Section 11. The idea is to study an ℓ-adic sheaf via its "skeleton": If U is an algebraic variety over $\mathbb{F}_p$, consider the family $\{\varphi_C : C \hookrightarrow U\}$ of all curves lying on U, and let $\varphi_C^N : C^N \to C \hookrightarrow U$ be the composition with the normalization of C. If ρ is an ℓ-adic sheaf on U, then, roughly, its skeleton is the family of ℓ-adic sheaves $(\varphi_C^N)^*\rho$ on C^N. Using the ramification theory on curves one obtains in this way a ramification theory for ℓ-adic sheaves in higher dimensions. Recently, Deligne proved that on a smooth connected scheme over a finite field, there are only finitely many irreducible lisse $\bar{\mathbb{Q}}_\ell$-representations with bounded rank and ramification, at least up to twist with a character of the Weil group of the ground field, see Theorem 11.19. One of the aims of these notes is to give the background necessary to understand the statement of this theorem. Notice, however, that its proof lies far beyond the scope of the material presented here and we refer the reader to [**16**] for details.

In the final section we give a very brief outlook on the higher-dimensional generalizations of the Grothedieck-Ogg-Shafarevich formula due to Kato-Saito. There are further generalizations of this formula due to Abbes and Saito who also develop a ramification theory in higher dimensions. We can say nothing about this, but give some references for further reading at the end of Section 12.

Part 1. Ramification theory of local fields

2. Infinite Galois theory

In this section we first define the notion of a profinite group and then briefly summarize Galois theory for infinite algebraic extensions. We will give more details in the more general situation of Section 6.

2.1. Profinite groups. We begin by recalling the notion of a projective limit; we mainly follow [**10**, Ch. V].

DEFINITION 2.1. (a) A *directed set* is a set I together with a partial ordering $\leq$, such that for every pair $i,j \in I$ there exists $n \in I$ such that $i \leq n$ and $j \leq n$.

(b) If $\mathcal{C}$ is a category, then a *projective system in $\mathcal{C}$ indexed by a directed set* I consists of the following data: For every $i \in I$ an object X_i of $\mathcal{C}$ and for every $j \in I$ with $i \leq j$, a morphism $\varphi_{ji} : X_j \to X_i$, such that if $i \leq j \leq n$, the diagram

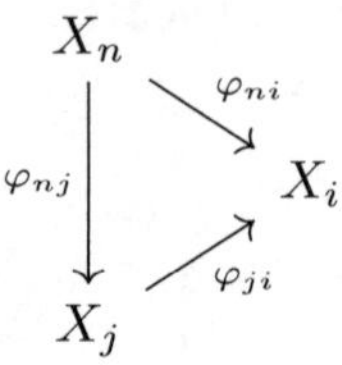

commutes, and such that $\varphi_{ii} = \mathrm{id}_{X_i}$.

(c) If $\{(X_i)_{i\in I}, (\varphi_{ji})_{i\leq j\in I}\}$ is a projective system, let $\mathcal{C}'$ be the following category: Its objects are tuples $(X, (q_i)_{i\in I})$, where X is an object of $\mathcal{C}$ and $q_i : X \to X_i$ morphisms in $\mathcal{C}$, such that for every $i \leq j \in I$, the diagram

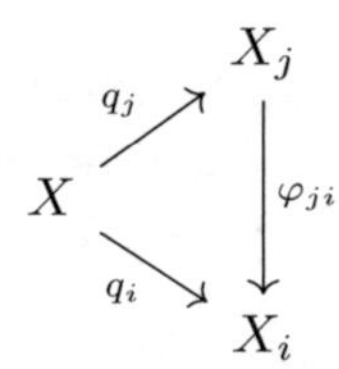

commutes. A morphism $(X, (q_i)_{i\in I}) \to (X', (q_i')_{i\in I})$ in $\mathcal{C}'$ is a morphism $f : X \to X'$ in $\mathcal{C}$, such that $q_i = q_i' f$ for all $i \in I$.

If $\mathcal{C}'$ has a final object, then it is unique up to unique isomorphism and this final object is called the *projective limit* of the projective system $\{(X_i)_{i\in I}, (\varphi_{ji})_{i\leq j\in I}\}$, and usually denoted $(\varprojlim_I X_i, (\mathrm{pr}_i)_{i\in I})$. Often the morphisms pr_i are omitted from the notation.

EXAMPLE 2.2. (a) The simplest example of a projective system is a constant projective system: If X is an object of $\mathcal{C}$, and I a directed set, consider the projective system given by $X_i = X$, $\varphi_{ji} = \mathrm{id}_X$. Clearly, $\varprojlim_I X_i \cong X$, and $\mathrm{pr}_i = \mathrm{id}$.

(b) If $\mathcal{C}$ is the category of groups, let $I = \mathbb{N}$ with its usual order, p a prime and $G_n := \mathbb{Z}/p^n\mathbb{Z}$. If $m \geq n$, then the projection map $\varphi_{mn} : \mathbb{Z}/p^m \to \mathbb{Z}/p^n$ makes $\{G_n, \{\varphi_{mn}\}_{n\leq m\in N}\}$ into a projective system, and the projective limit $\varprojlim_{\mathbb{N}} G_n$ of this system is an abelian group, the *p-adic integers* $\mathbb{Z}_p$.

Moreover, as the maps φ_{mn} are maps of rings, the group $\mathbb{Z}_p$ also carries a ring structure, which makes it a projective limit in the category of commutative rings.

(c) Let $\mathcal{C}$ be the category of groups and equip $I = \mathbb{N}$ with the partial order defined by $(n \text{ “}\leq\text{” } m) :\Leftrightarrow n|m$. If n, m are integers such that $n|m$, then again we have a projection morphism $\varphi_{mn} : \mathbb{Z}/m \twoheadrightarrow \mathbb{Z}/n$, and $\{(\mathbb{Z}/n)_{n\in N}, (\varphi_{mn})_{n|m\in\mathbb{N}}\}$ is a projective system. Its projective limit is an abelian group denoted $\widehat{\mathbb{Z}}$. It is not difficult to check that $\widehat{\mathbb{Z}} \cong \prod_{p \text{ prime}} \mathbb{Z}_p$.

As in the previous example, $\widehat{\mathbb{Z}}$ is also a projective limit in the category of commutative rings.

(d) More generally: If $\mathcal{C}$ is the category of groups, I a directed set and $\{(G_i)_{i\in I}, (\varphi_{ji})_{j\geq i\in I})\}$ a projective system in $\mathcal{C}$, then $\varprojlim_I G_i$ exists and it is a subgroup of the product $\prod_{i\in I} G_i$:

$$\varprojlim_I G_i = \left\{ (g_i) \in \prod_{i\in I} G_i \,\middle|\, \forall i \le j : \varphi_{ji}(g_j) = g_i \right\} \subseteq \prod_{i\in I} G_i. \tag{2.1}$$

The morphisms $\mathrm{pr}_j : \varprojlim_I G_i \to G_j$ belonging to the datum of the projective limit are induced by the projection maps $\prod_{i\in I} G_i \to G_j$.

We are now interested in the category **TopGroup** of topological groups.

PROPOSITION 2.3. *The projective limit of a projective system of topological groups exists. If the groups in the projective system are Hausdorff and quasi-compact, then so is the inverse limit of the system.*

Indeed, this follows from the description (2.1).

If G is a finite group, we consider it as a topological group by equipping it with the discrete topology.

PROPOSITION 2.4. *Let I be a directed set and $\{(G_i)_{i\in I}, (\varphi_{ji})_{i\le j\in I}\}$ a projective system of finite groups. Its projective limit $\varprojlim_I G_i$ computed in the category of abstract groups exists and it is a subgroup of the product $\prod_{i\in I} G_i$. If we equip each G_i with the discrete topology and $\varprojlim_I G_i$ with the topology induced by $\prod_{i\in I} G_i$, then $\varprojlim_I G_i$ is also a projective limit of the system $\{(G_i)_{i\in I}, (\varphi_{ji})_{i\le j\in I}\}$ computed in the category of topological groups.*

Moreover, the topology on $\varprojlim_I G_i$ is the coarsest topology such that the projections pr_i are continuous.

In other words: We get the same result if we equip the abstract group $\varprojlim_I G_i$ with the topology induced by the product, or if we compute the projective limit in the category of topological groups, by considering the G_i as discrete groups.

DEFINITION 2.5. A topological group G is called *profinite*, if it is isomorphic to the projective limit of a projective system of finite (discrete) groups in the category of topological groups.

Similarly, if p is a prime number, G is called *pro-p-group* if G is isomorphic to the projective limit of a projective system of finite p-groups.

EXAMPLE 2.6.
(a) The groups $\mathbb{Z}_p$ and $\widehat{\mathbb{Z}}$ from Example 2.2 are profinite groups.
(b) The group $\mathbb{Z}_p$ is a pro-p-group.
(c) If G is an abstract group, then we write

$$\widehat{G} := \varprojlim_{\substack{H \triangleleft G \\ (G:H)<\infty}} G/H,$$

where H runs through the directed set of normal subgroups of finite index of G. The profinite group $\widehat{G}$ is called the *profinite completion of G*.
(d) Similarly, if p is a prime number, we write

$$\widehat{G}^{(p)} := \varprojlim_{\substack{H \triangleleft G \\ (G:H)=p\text{-power}}} G/H,$$

where H runs through the directed set of normal subgroups of finite p-power index. The pro-p-group $\widehat{G}^{(p)}$ is called the *pro-p-completion of G*.

In fact, there is a purely topological description of profinite groups:

PROPOSITION 2.7 ([**10**, V, §1.4]). *A topological group is profinite if and only if it is compact and totally disconnected.*

EXERCISE 2.8. If G is a topological group and U an open subgroup, show that U is also closed. Moreover, if G is compact, show that U has finite index in G.

COROLLARY 2.9. *If G is profinite, then*

$$G \cong \varprojlim G/U,$$

where U runs through the set of open normal subgroups of G.

COROLLARY 2.10. *If G is a profinite group and $H \subseteq G$ a closed normal subgroup, then H and G/H are both profinite.*

More precisely,

$$H \cong \varprojlim H/H \cap U \text{ and } G/H \cong \varprojlim G/UH,$$

where in both cases U runs through the set of open normal subgroups of G.

REMARK 2.11. Some caution is required: If G is a profinite group, then it is not always true that the canonical map $G \to \widehat{G}$ of G into its profinite completion is an isomorphism. In other words, it is not true that every finite index subgroup is open. However, if G is a topologically finitely generated profinite group (i.e. if there exists a finitely generated dense subgroup of G), then every subgroup of finite index is open ([**42**]).

For example, if p is a prime, consider the projective system $((\mathbb{Z}/p\mathbb{Z})^n, \varphi_{ji})$ indexed by $\mathbb{N}$, where for $j \geq i$, $\varphi_{ji} : (\mathbb{Z}/p\mathbb{Z})^j \to (\mathbb{Z}/p\mathbb{Z})^i$ is the projection onto the first i factors. Then $\varprojlim_n ((\mathbb{Z}/p\mathbb{Z})^n, \varphi_{ji})$ is the product of countably many copies of $\mathbb{Z}/p\mathbb{Z}$. This is a profinite group which has infinitely many finite index subgroups which are not open. This can be used to show that there are finite index subgroups of $\mathrm{Gal}(\overline{\mathbb{Q}}/\mathbb{Q})$, which are not open. In fact this is true for the subfield of $\overline{\mathbb{Q}}$ spanned by $\sqrt{-1}$ and $\{\sqrt{p} \mid p \text{ prime}\}$ (exercise or [**40**, §7.]).

2.2. The Galois correspondence for infinite algebraic extensions. In this section we recall the basic statements of Galois theory for infinite extensions, as it can be found, for example, in [**41**], [**40**] or [**52**]. For a more general treatment see Section 6.

We assume that the reader is familiar with Galois theory of finite extensions.

Let K be a field and L an algebraic extension of K. The extension L/K is called Galois if $L^{\mathrm{Aut}_K(L)} = K$. In this case we also write $\mathrm{Gal}(L/K) := \mathrm{Aut}_K(L)$. Equivalently, L/K is Galois if and only if the minimal polynomial of every element $\alpha \in L$ splits into distinct linear factors over L. In particular, Galois extensions are separable.

PROPOSITION 2.12. *If L/K is a Galois extension of fields, write*

$$\mathcal{G}_{L/K} := \{F \subseteq L \mid F/K \text{ is a finite Galois extension}\}.$$

Then $L = \bigcup_{F \in \mathcal{G}_{L/K}} F$, and the inclusion relation makes $\mathcal{G}_{L/K}$ into a directed set. If $F_1 \subseteq F_2 \in \mathcal{G}_{L/K}$, then restriction of automorphisms induces a homomorphism

$\mathrm{Gal}(F_2/K) \to \mathrm{Gal}(F_1/K)$. *This makes the set of groups* $\mathrm{Gal}(F/K)$, $F \in \mathcal{G}_{L/K}$, *into a projective system, and we have*

$$\mathrm{Gal}(L/K) = \varprojlim_{F \in \mathcal{G}_{L/K}} \mathrm{Gal}(F/K).$$

In particular, $\mathrm{Gal}(L/K)$ *is a profinite group.*

EXERCISE 2.13. (a) Let p be a prime, $n \in \mathbb{N}$, $q = p^n$ and $\mathbb{F}_q$ the field with q elements. Fix an algebraic closure $\overline{\mathbb{F}_q}$. Show that the profinite group $\mathrm{Gal}(\overline{\mathbb{F}}_q/\mathbb{F}_q)$ is isomorphic to $\widehat{\mathbb{Z}}$.

(b) Let $\mathbb{Q}^{\mathrm{cyc}} \subseteq \mathbb{C}$ be the smallest subfield containing all roots of unity. Is $\mathbb{Q}^{\mathrm{cyc}}$ a Galois extension of $\mathbb{Q}$? If so, what is $\mathrm{Gal}(\mathbb{Q}^{\mathrm{cyc}}/\mathbb{Q})$?

THEOREM 2.14 (Galois correspondence). *Let* L/K *be a Galois extension and* $\mathrm{Gal}(L/K)$ *its Galois group. There is an order reversing bijection*

$$\begin{array}{ccc} \{\textit{subextensions of } L/K\} & \xrightarrow{\cong} & \{\textit{closed subgroups of } \mathrm{Gal}(L/K)\} \\ F & \longmapsto & \mathrm{Gal}(L/F) \end{array}$$

Under this correspondence,

(a) *finite subextensions of* L/K *correspond to open subgroups,*

(b) *subextensions which are Galois over* K *correspond to closed normal subgroups.*

REMARK 2.15. In Section 6 we will see a vast geometric generalization of this correspondence. If F/K is a finite separable extension, then the associated morphism $\mathrm{Spec}\, F \to \mathrm{Spec}\, K$ will be interpreted as a (nontrivial!) covering space. The Galois group of F/K will then act as the group of deck transformations.

3. Ramification groups and the theorem of Hasse-Arf

In this rather lengthy section we summarize the ramification theory of finite separable extensions of a complete discretely valued field. Our main references are [**50**] and [**41**]. If L/K is such an extension, assume that the associated extension of residue fields is also separable (which is automatic if the residue field of K is perfect). In this case, if L/K is Galois, we will construct two descending filtrations on the Galois group $G := \mathrm{Gal}(L/K)$: the *lower numbering filtration* $\{G_u\}_{u \in \mathbb{Z}_{\geq -1}}$ and the *upper numbering* $\{G^v\}_{v \in \mathbb{R}_{\geq -1}}$. The subgroups appearing in both filtrations are the same, but the lower numbering is adapted to taking subgroups, while the upper numbering is adapted to taking quotients of G, i.e. if $H \triangleleft G$ is a normal subgroup, then $H_u = G_u \cap H$, and $(G/H)^v = G^v/(H \cap G^v)$.

The numbers $\lambda \in \mathbb{R}_{\geq -1}$, for which $G^\lambda \neq G^{\lambda+\varepsilon}$ for all $\varepsilon > 0$, are called *jumps* or *breaks* of the filtration. The theorem of Hasse-Arf (Theorem 3.63) states that the jumps are integers if G is abelian. This theorem is a crucial ingredient for the ramification theory of ℓ-adic representations developed in the following sections.

3.1. Discretely valued fields and discrete valuation rings.

We first recall some basic material on discrete valuations.

DEFINITION 3.1. Let K be a field. A *discrete valuation* is a surjective homomorphism of abelian groups

$$v : K^\times \to \mathbb{Z}$$

such that $v(x+y) \geq \min\{v(x), v(y)\}$, for all $x, y \in K^\times$ with $x+y \neq 0$. We extend v to K, with the convention that $v(0) = \infty$.

If K is equipped with a discrete valuation v, then (K, v) is called a *discretely valued field*.

REMARK 3.2. Some authors do not require a valuation to be surjective, but only nontrivial.

EXERCISE 3.3. Let (K, v) be a discretely valued field and $x, y \in K$. Show that $v(x+y) = \min\{v(x), v(y)\}$ if $v(x) \neq v(y)$.

If (K, v) is a discretely valued field, then the subset $A := A_K := \{x \in K | v(x) \geq 0\}$ is a commutative ring with unique maximal ideal $\mathfrak{m} := \mathfrak{m}_K := \{x \in A | v(x) > 0\}$. The local ring A is called *valuation ring of* K, and it is in fact a *discrete valuation ring*, i.e. a local principal ideal domain, which is not a field. A generator π of $\mathfrak{m}_K$ is called *uniformizer of* A*, or uniformizer of* K. With our definitions, we always have $v(\pi) = 1$. The *residue field* $A/\mathfrak{m}_K$ is also said to be the residue field of K.

EXAMPLE 3.4. The three main examples to keep in mind are as follows:

(a) Let $K = \mathbb{Q}$ be the field of rational numbers and p a prime number. An nonzero integer $a \in \mathbb{Z}$ can be uniquely written as $a = a'p^r$, with $a' \in \mathbb{Z}$ prime to p, and $r \in \mathbb{Z}_{\geq 0}$. We define $v_p(a) := r$, and for $\frac{a}{b} \in \mathbb{Q}^\times$, $v_p(\frac{a}{b}) := v_p(a) - v_p(b)$. This defines a discrete valuation on $\mathbb{Q}$, which is called the *p-adic valuation.* The valuation ring associated with this valuation is $\mathbb{Z}_{(p)}$: the ring of rational numbers with denominators prime to p, and clearly p is a uniformizer of $(\mathbb{Q}, v_p)$. The residue field of $(\mathbb{Q}, v_p)$ is $\mathbb{F}_p$, the finite field with p elements.

(b) Let k be a field and C a connected, affine, normal, algebraic curve over k. This means that C is of the form $C = \operatorname{Spec} A$, with A a normal, finite type k-algebra of Krull dimension 1. Let $K := k(C)$ be its function field, i.e., the local ring of C at its generic point, which amounts to saying that K is the field of fractions of A. If $c \in C$ is a closed point, we can associated with it a discrete valuation v_c on K: c corresponds to a maximal ideal $\mathfrak{m}_c \subseteq A$, and if $f \in A \setminus \{0\}$, there exists an integer $r \in \mathbb{Z}_{\geq 0}$, such that $f \in \mathfrak{m}_c^r \setminus \mathfrak{m}_c^{r+1}$. We define $v_c(f) := r$, and for $\frac{f}{g} \in K^\times$, $v_c(\frac{f}{g}) = v_c(f) - v_c(g)$. The discrete valuation ring associate with v_c is $A_{\mathfrak{m}_c} = \mathcal{O}_{C,c}$, and its residue field $k(c) = A_{\mathfrak{m}_c}/\mathfrak{m}_c$ is a finite extension of k.

If C is a normal, proper curve over k, the converse is also true: Every discrete valuation of K which is trivial on k is of the shape v_c for some closed point $c \in C$ (valuative criterion of properness).

(c) More generally, if k is a field and X a connected, finite type, normal k-scheme, to every codimension 1 point η of X (i.e. to every point η such that the closure of $\{\eta\}$ has codimension 1), one can attach a discrete valuation v_η in the same way as in the previous example. However, if $\dim X > 1$, there are discrete valuations on the function field $K = k(X)$ which do not arise from codimension 1 points of X[1].

The globalization of the notion of a discrete valuation ring is called *Dedekind domain.*

[1]There are even discrete valuations which do not arise from codimension 1 points on a different model of $k(X)$, see [**35**, Ch. 8, Thm. 3.26].

DEFINITION 3.5. A *Dedekind domain* A is a noetherian integral domain, which is not a field, such that the localization at each maximal ideal is a discrete valuation ring. This is equivalent to A being normal, noetherian and 1-dimensional (see [**50**, Ch. I, Prop. 4]).

Of course $\mathbb{Z}$ is a Dedekind domain, and so is the ring A from Example 3.4, (b). In fact, geometrically, one can think of Dedekind domains as rings of functions on a smooth curve.

3.2. Completion. A discretely valued field (K, v) is equipped with a natural topology, given by the "ultrametric" norm $\|\cdot\|_v$ defined by $\|x\|_v = \exp(-v(x))$. In fact, one could also take $\|x\| := a^{v(x)}$ for a any real number in $(0,1)$, but the induced topology on K is independent of the choice of a. We will mostly be interested in discretely valued fields (K, v) which are complete with respect to this topology. Such fields are, naturally, called *complete discretely valued fields*. If (K, v) is complete, then its valuation ring A is complete in the $\mathfrak{m}_K$-adic topology, i.e. $A \cong \varprojlim_n A/\mathfrak{m}_K^n$.

There is a standard process of completing a discretely valued field (K, v): Let K_v denote the ring of Cauchy sequences in K modulo the maximal ideal of sequences converging to 0. The valuation v can be extended uniquely to K_v according to the exercise below, so that (K_v, v) becomes a complete discretely valued field. If A is the valuation ring of K with maximal ideal $\mathfrak{m}_K$, then K_v is the fraction field of the complete discrete valuation ring $\varprojlim_n A/\mathfrak{m}^n$.

EXERCISE 3.6. Let $(a_n)_{n\in\mathbb{N}}$ be a Cauchy sequence in K not converging to 0. Show that the sequence of integers $(v(a_n))_{n\in\mathbb{N}}$ becomes stationary.

EXAMPLE 3.7. We extend Example 3.4:

(a) If $K = \mathbb{Q}$ is given the p-adic valuation v_p, then its completion is the field of p-adic numbers $\mathbb{Q}_p$. Its valuation ring is the ring of p-adic integers $\mathbb{Z}_p$. Note that the additive group underlying $\mathbb{Z}_p$ coincides with the pro-p-completion of the abelian group $\mathbb{Z}$.

(b) If k is a field and C a normal, connected, algebraic curve over k, and $c \in C(k)$ a rational point, then the choice of a uniformizer t of $\mathcal{O}_{C,c}$ gives rise to a canonical isomorphism between the completion (K_c, v_c) of $k(C)$ with respect to v_c and the field $k((t))$ of Laurent series over k. Geometrically, K_c should be thought of as an infinitesimally small, punctured disc around c.

3.3. Extensions of discretely valued fields. If (K, v) is a complete discretely valued field with valuation ring A, and L/K a finite extension of fields, then there exists a unique discrete valuation v_L on L and an integer $e > 0$, such that for all $x \in K$, $v_L(x) = ev(x)$, and L is complete with respect to the discrete valuation v_L. Moreover, the integral closure B of A in L is a complete discrete valuation ring and a free A-module of rank $[L : K]$ ([**50**, Ch. II, Prop. 3]).

DEFINITION 3.8. Keep the notations from the previous paragraph.

- By abuse of notation we say that v_L *extends* v.
- The integer e is called the *ramification index of* L *over* K.
- The extension L/K induces a finite extension of their residue fields. Its degree is denoted by f and called the *residue degree of* L *over* K.
- We say that L/K is *totally ramified* if $f = 1$.

- If $e = 1$ and if the extension of residue fields is separable, then L/K is called *unramified*.

In the situation of the definition, if $n = [L : K]$, then $n = ef$.

We globalize: Let (K, v) be a discretely valued field (not necessarily complete) with valuation ring A and L/K a finite extension. Denote by B the integral closure of A in L, then according to the Krull-Akizuki theorem ([**41**, Ch. I, Thm. 12.8]) B is a Dedekind domain, integral over A, but not necessarily local. In other words: The discrete valuation v can be extended to L, but there are finitely many distinct valuations $w_1, \dots, w_r$ which extend it.

LEMMA 3.9. *The Dedekind domain B is a finitely generated A-algebra, and hence a finitely generated A-module, in each of the following situations:*

(a) *A is a complete discrete valuation ring (see the discussion before Definition 3.8).*
(b) *L/K is separable* ([**50**, Ch. I, Prop. 8]).
(c) *A is the localization of a finitely generated algebra over a field*[2].

In our applications, B will always be a finitely generated, hence finite, A-algebra, so from now on we tacitly assume this is the case.

PROPOSITION 3.10 ([**50**, Ch. II, Thm. 1]). *Let (K, v) be a discretely valued field with valuation ring A and L/K a finite extension of degree n. Assume that the integral closure B of A in L is a finite A-algebra. Write $w_1, \dots, w_r$ for the valuations of L extending v.*

(a) The canonical map $L \otimes_K K_v \to \prod_{i=1}^r L_{w_i}$ is an isomorphism.
(b) L_{w_i}/K_v is a finite extension.

Thus, for each i we obtain numbers n_i, e_i, and f_i: the degree, ramification index and residue degree of the extension L_{w_i}/K_v of complete discretely valued fields. As before we have $n_i = e_i f_i$ and

$$n = \sum_{i=1}^{r} n_i = \sum_{i=1}^{r} e_i f_i. \tag{3.1}$$

There is also an ideal theoretic interpretation of these numbers. The Dedekind domain B is a finite extension of the discrete valuation ring A, and the discrete valuations $w_1, \dots, w_r$ correspond to the finitely many nonzero prime ideals $\mathfrak{P}_i$ of B, i.e. $\mathfrak{P}_i = \{b \in B | w_i(b) > 0\}$. Since B is a Dedekind domain, it has the pleasant property that any ideal I can be uniquely written as a product

$$I = \mathfrak{P}_1^{s_1} \cdot \ldots \cdot \mathfrak{P}_r^{s_r}.$$

In particular, if $\mathfrak{m}$ is the maximal ideal of A, then $\mathfrak{m}B$ is a proper ideal in B, as the extension is integral, and hence $\mathfrak{m}B$ can be written as such a product of prime ideals. Indeed,

$$\mathfrak{m}B = \mathfrak{P}_1^{e_1} \cdot \ldots \cdot \mathfrak{P}_r^{e_r},$$

where the e_i are precisely the ramification indices from above. Moreover, since B is 1-dimensional, the (nonzero) prime ideals $\mathfrak{P}_i$ are maximal ideals, so we get finite extensions $k(v) = A/\mathfrak{m} \subseteq B/\mathfrak{P}_i$, and these extensions are precisely the degree f_i extensions of the residue fields of L_{w_i}/K_v.

[2] or more generally, if A is a *Japanese/N-2 ring*, see [**37**, (31.H)].

DEFINITION 3.11. Keep the notation from the previous paragraph.

(a) If $r = 1$ and $f_1 = 1$, i.e. if there is only one valuation w_1 on L extending v and the residue extension is trivial, then L/K is called *totally ramified*.
(b) If $e_i = 1$ and if the residue extension $A/\mathfrak{m}_K \subseteq B/\mathfrak{P}_i$ of L_{w_i}/K_{v_K} is separable, then L/K is called *unramified at* $\mathfrak{P}_i$ *or at* w_i.
(c) If L/K is unramified at $w_1, \dots, w_r$, then L/K is called *unramified*.

3.4. Discriminant and different. We start with a definition.

DEFINITION 3.12. Let L/K be a finite extension of fields, or, more generally, let K be a ring and L a K-algebra which is free of finite rank as a K-module.

- For $x \in L$ define $N_{L/K}(x) \in K$ as the determinant of the K-linear endomorphism of L given by multiplication with x. The map $N_{L/K} : L^\times \to K^\times$ is called the *norm of* L/K.
- Similarly we define the *trace map of* L/K, $\mathrm{Tr}_{L/K} : L \to K$, by defining $\mathrm{Tr}_{L/K}(x)$ to be the trace of the K-linear endomorphism of L given by multiplication with x.

An easy computation shows the following standard facts.

PROPOSITION 3.13. *Let L/K be a finite separable extension, and fix an algebraic closure $\overline{K}$ of K. Let $\sigma_1, \dots, \sigma_r$ be the K-embeddings of L into $\overline{K}$. Then for $x \in L$, we have*

$$N_{L/K}(x) = \prod_{i=1}^{r} \sigma_i(x)$$

and

$$\mathrm{Tr}_{L/K}(x) = \sum_{i=1}^{r} \sigma_i(x).$$

Moreover if K is complete with respect to the discrete valuation v, and v_L discrete valuation on L extending v, then for $x \in L$,

$$v_L(x) = \frac{1}{f} v(N_{L/K}(x)),$$

where f is the degree of the residue extension of L/K.

Recall the following fact.

PROPOSITION 3.14 ([**7**, IX, §2, Prop. 5]). *If R is a finite K-algebra, then R is a product of finite separable field extensions of K if and only if $R \times R \to R$, $(x, y) \mapsto \mathrm{Tr}_{L/K}(xy)$ is a nondegenerate bilinear form.*

DEFINITION 3.15. Let A be a Dedekind domain with fraction field K and L a finite separable extension. Write B for the integral closure of A in L. Define the B-module

$$B^\vee := \{x \in L \mid \mathrm{Tr}_{L/K}(xy) \in A \text{ for all } y \in B\}.$$

PROPOSITION 3.16. *We have $B \subseteq B^\vee$, and $B^\vee$ is a finitely generated B-submodule of L, i.e. a fractional ideal of B. The map*

$$B^\vee \to \mathrm{Hom}_A(B, A), \quad x \mapsto (y \mapsto \mathrm{Tr}_{L/K}(xy))$$

is an isomorphism of B-modules. The inverse of the fractional ideal $B^\vee$ is an ideal in B.

PROOF. Since L/K is separable, Proposition 3.14 shows that the map of K-vector spaces

$$L \to \mathrm{Hom}_K(L, K), \quad x \mapsto (y \mapsto \mathrm{Tr}_{L/K}(xy)) \tag{3.2}$$

is an isomorphism. On the other hand, every A-morphism $B \to A$ extends uniquely to a K-morphism $L \to K$, so $\mathrm{Hom}_A(B, A)$ is a sub-A-module of $\mathrm{Hom}_K(L, K)$, and it is precisely the image of $B^\vee$ under (3.2). Since B is a finitely generated A-module, the same is true for $\mathrm{Hom}_A(B, A)$ and hence for $B^\vee$.

From $\mathrm{Tr}_{L/K}(B) \subseteq A$, it follows that $B^\vee \supseteq B$, so $B = B^\vee \cdot (B^\vee)^{-1} \supseteq (B^\vee)^{-1}$. This finishes the proof. □

Here is another characterization of $B^\vee$:

LEMMA 3.17. *With the notations from above, $B^\vee$ is the largest sub-B-module of L with the property that $\mathrm{Tr}_{L/K}(B^\vee) \subseteq A$. In particular, if $\mathfrak{b} \subseteq L$ and $\mathfrak{a} \subseteq K$ are fractional ideals of B, A, then $\mathrm{Tr}_{L/K}(\mathfrak{b}) \subseteq \mathfrak{a}$ if and only if $\mathfrak{b} \subseteq \mathfrak{a}B^\vee$.*

PROOF. If E is a sub-B-module of L such that $\mathrm{Tr}_{L/K}(E) \subseteq A$, then $E \subseteq B^\vee$, so the maximality statement is clear.

For the second statement, note that $\mathrm{Tr}_{L/K}(\mathfrak{b}) \subseteq \mathfrak{a}$ if and only if $\mathrm{Tr}_{L/K}(\mathfrak{a}^{-1}\mathfrak{b}) \subseteq A$, if and only if $\mathfrak{b} \subseteq \mathfrak{a}B^\vee$. □

DEFINITION 3.18. Keeping the notations from Definition 3.15, we write $\mathfrak{D}_{B/A} := (B^\vee)^{-1}$. This ideal of B is called the *different of B/A*.

The ideal $N_{L/K}(\mathfrak{D}_{B/A}) := \mathfrak{d}_{B/A}$ of A is called the *discriminant of B/A*.

REMARK 3.19. Here the norm of an ideal is defined as follows: In a Dedekind domain, every ideal can be uniquely written as a product of prime ideals. If $\mathfrak{P}$ is a nonzero prime ideal of B, then we define $N_{L/K}(\mathfrak{P}) := \prod_{\mathfrak{P}|\mathfrak{p}} \mathfrak{p}^{f_\mathfrak{p}}$, where $f_\mathfrak{p}$ is the degree of the residue extension $(B/\mathfrak{P})/(A/\mathfrak{p})$. Since we will only use these definitions in the case where L/K is an extension of *complete* discretely valued fields, we will only care about Dedekind rings which are discrete valuation rings, so we can forget the more complicated definition from above, and take the usual norm.

PROPOSITION 3.20 (Transitivity of different and discriminant). *Let L/K and M/L be finite separable extensions, $A \subseteq K$ a Dedekind domain with* $\mathrm{Frac}(A) = K$, *and B, C the integral closures of A in L, resp. M. Then*

$$\mathfrak{D}_{C/A} = \mathfrak{D}_{C/B}\mathfrak{D}_{B/A}$$

and

$$\mathfrak{d}_{C/A} = N_{L/K}(\mathfrak{d}_{C/B})\mathfrak{d}_{B/A}^{[M:L]}.$$

PROOF. The formula for the discriminant follows from the formula for the different by applying $N_{M/K}$. The formula for the different follows from the transitivity of the trace, i.e. from the fact that $\mathrm{Tr}_{M/K} = \mathrm{Tr}_{L/K} \circ \mathrm{Tr}_{M/L}$ (exercise). □

The different and discriminant indicate whether an extension is ramified or not. To see this, we now put ourselves in the complete local situation.

LEMMA 3.21. *Let L/K be a finite separable extension of complete discretely valued fields, A the valuation ring of K, B its integral closure in L. Let $x_1, \dots, x_n$ be a basis of B as an A-module. Let K^{sep} be a separable closure of K. Then*

$$\mathfrak{d}_{B/A} = (\det((\mathrm{Tr}_{L/K}(x_i x_j))_{ij})) = (\det((\sigma_i x_j)_{ij}))^2,$$

where σ_i runs through the finite number of K-linear embeddings of L in K^{sep}.

PROOF. First note that for any $a_1, \dots, a_n, b_1, \dots, b_n \in L$, we have

$$\mathrm{Tr}_{L/K}(a_i b_j) = \sum_{k=1}^{n} \sigma_k(a_i)\sigma_k(b_j) = (\sigma_1(a_i), \dots, \sigma_n(a_i)) \cdot \begin{pmatrix} \sigma_1(b_j) \\ \vdots \\ \sigma_n(b_j) \end{pmatrix},$$

so $\det(\mathrm{Tr}_{L/K}(a_i b_j)_{ij}) = \det((\sigma_j(a_i))_{ij}) \det((\sigma_i(b_j))_{ij})$.

The A-module $B^\vee = \mathrm{Hom}_A(B, A)$ is spanned by the basis $x_1', \dots, x_n'$, where $x_i'(x_j) = \mathrm{Tr}_{L/K}(x_i' x_j) = \delta_{ij}$. But $B^\vee$ is a fractional ideal, so $B^\vee = (\beta)$ for some $\beta \in L^\times$, and $\beta x_1, \dots, \beta x_n$ is also an A-basis for $B^\vee$.

Let $M = (m_{ij}) \in \mathrm{GL}_n(A)$ be the base change matrix such that $\beta x_i = \sum_{j=1}^n m_{ij} x_j'$. Then $M \cdot (\sigma_j x_i')_{ij} = (\sigma_j(\sum_{r=1}^n m_{ir} x_r'))_{ij} = (\sigma_j(\beta x_i))_{ij}$. It follows that

$$\begin{aligned}\det(\mathrm{Tr}_{L/K}(\beta x_i x_j)) &= \det(M \cdot (\sigma_j(x_i'))_{ij} \cdot (\sigma_i(x_j))_{ij}) \\ &= \det(M \cdot (\mathrm{Tr}_{L/K}(x_i' x_j))_{ij}) \\ &= \det(M) \in A^\times.\end{aligned}$$

On the other hand, $\det(\mathrm{Tr}_{L/K}(\beta x_i x_j))_{ij} = N_{L/K}(\beta) \det(\mathrm{Tr}_{L/K}(x_i x_j)_{ij})$. It follows that $\mathfrak{d}_{B/A} = (N_{L/K}(\beta^{-1})) = (\det(\mathrm{Tr}_{L/K}(x_i x_j)_{ij})) = (\det((\sigma_i(x_j))_{ij}))^2$. □

PROPOSITION 3.22. *Let L/K be a finite separable extension of complete discretely valued fields. Then L/K is unramified if and only if $\mathfrak{D}_{B/A} = B$, if and only if $\mathfrak{d}_{B/A} = A$.*

PROOF. As $\mathfrak{d}_{B/A} = N_{L/K}(\mathfrak{D}_{B/A})$, we see that $\mathfrak{D}_{B/A} = B$ if and only if $\mathfrak{d}_{B/A} = A$.

By definition, the extension L/K is unramified if and only if $\mathfrak{m}_K B = \mathfrak{m}_L$, and if the residue extension is separable. Let $x_1, \dots, x_r$ be a basis of B over A. Then the residue classes $\bar{x}_i$ form a basis of $B/\mathfrak{m}_K B$ over $A/\mathfrak{m}_K$. Since $B/\mathfrak{m}_K B$ is a local ring, B is unramified over A, if and only if $B/\mathfrak{m}_K B$ is a separable $A/\mathfrak{m}_K$-algebra. By Proposition 3.14 this is equivalent to $\det(\mathrm{Tr}_{L/K}(\bar{x}_i \bar{x}_j)) \not\equiv 0 \mod \mathfrak{m}_K$. According to the lemma this is equivalent to $\mathfrak{d}_{B/A} = A$. □

We need two more facts about the different.

THEOREM 3.23. *Let L/K be a finite separable extension of complete discretely valued fields. Assume that there exists $\alpha \in B$ which generates B as an A-algebra. If $f \in K[X]$ is the monic minimal polynomial of α, then $f \in A[X]$, and $\mathfrak{D}_{B/A} = (f'(\alpha))$.*

PROOF. Let $f(X) \in K[X]$ be the monic minimal polynomial of α. The coefficients of $f(X)$ are symmetric functions in the roots of $f(X)$. Since α is integral over A, the same is true for all roots of $f(X)$ (in a sufficiently large extension of K), so $f(X) \in A[X]$. Let $b_0, \dots, b_{n-1} \in L$ such that

$$\frac{f(X)}{X - \alpha} := b_0 + b_1 X + \ldots + b_{n-1} X^{n-1}. \tag{3.3}$$

The same argument as above shows that $b_0, \dots, b_{n-1}$ are integral over B, hence elements of B.

Note that $1, \alpha, \dots, \alpha^{n-1}$ is a basis of B over A. We compute its dual basis of $B^\vee$.

Let $\alpha = \alpha_1, \dots, \alpha_n$ be the distinct roots of $f(X)$ (in a sufficiently large extension of K). Then $f'(\alpha_i) = \prod_{j\neq i}(\alpha_i - \alpha_j)$. Hence, for $r \geq 0$ we compute

$$\left(\frac{f(X)}{X - \alpha_i}\right)(\alpha_j) \cdot \frac{\alpha_i^r}{f'(\alpha_i)} = \delta_{ij}\alpha_i^r.$$

Consequently, if $r = 0, \dots, n-1$, then

$$\mathrm{Tr}_{L[X]/K[X]}\left(\frac{f(X)}{X - \alpha}\frac{\alpha^r}{f'(\alpha)}\right) = \sum_{i=1}^{n} \frac{f(X)}{X - \alpha_i}\frac{\alpha_i^r}{f'(\alpha_i)} = X^r,$$

as the difference of both sides is a polynomial of degree $< n$ with zeroes $\alpha_1, \dots, \alpha_n$. In turn, we get

$$X^r = \mathrm{Tr}_{L[X]/K[X]}\left(\frac{f(X)}{X - \alpha}\frac{\alpha^r}{f'(\alpha)}\right) = \sum_{j=0}^{n-1} \mathrm{Tr}_{L/K}\left(\alpha^r \frac{b_j}{f'(\alpha)}\right) X^j,$$

which shows that $\frac{b_0}{f'(\alpha)}, \dots, \frac{b_{n-1}}{f'(\alpha)} \in B^\vee$ is the dual basis of $1, \alpha, \dots, \alpha^{n-1}$, and hence

$$B^\vee = \frac{1}{f'(\alpha)} \cdot (b_0 A + \dots + b_{n-1}A) \subseteq \frac{1}{f'(\alpha)} B.$$

Finally, if we write $f(X) = X^n + \sum_{i=0}^{n-1} a_i X^i$ with $a_0, \dots, a_{n-1} \in K$, then from (3.3) we know that

$$\begin{aligned} b_{n-1} &= 1 \\ b_{n-i} - \alpha b_{n-i+1} &= a_{n-i+1} \text{ for } 1 < i < n \\ -b_0\alpha &= a_0. \end{aligned}$$

Inductively it follows that $\alpha^j \in \bigoplus_{i=0}^{n-1} b_i A$ for $j = 0, \dots, n-1$, so $B = \bigoplus_{i=0}^{n-1} b_i A$. It follows that

$$B^\vee = (f'(\alpha))^{-1},$$

which proves the claim. □

The hypothesis of the theorem is always satisified when the residue extension of L/K is separable.

THEOREM 3.24 ([**50**, Ch. III, Prop. 12]). *Let L/K be a finite extension of complete discretely valued fields with separable residue extension. Then there exists an element $x \in B$, generating B as an A-algebra. Moreover, x can be chosen such that there exists a monic polynomial $R \in A[X]$ of degree f such that $R(x)$ is a uniformizer of B. If L/K is totally ramified, any uniformizer x generates B as an A-algebra.*

For future reference, we also prove:

PROPOSITION 3.25. *Let L/K be a finite separable extension of complete discretely valued fields with separable residue extension. Write A for the valuation ring of K and $B = A[\alpha]$ for its integral closure in L.*

(a) *If the ramification index of L/K is e, then*

$$v_L(\mathfrak{D}_{B/A}) \geq e - 1$$

with equality if and only if L/K is tamely ramified (Definition 3.39).

(b) *If L/K is Galois with group G, then*

$$v_L(\mathfrak{D}_{B/A}) = \sum_{\sigma \in G \setminus \{1\}} i_G(\sigma)$$

where $i_G(\sigma) := v_L(\alpha - \sigma(\alpha))$, see Section 3.6.

PROOF. After reducing to the totally ramified case, there exists a uniformizer $\alpha \in B$ such that $B = A[\alpha]$. Let f be the minimal polynomial of α. For both formulas we have to compute $v_L(f'(\alpha))$.

(a) We only sketch the proof, as we will not use this result in the sequel. We already saw that L/K is unramified if and only if $\mathfrak{D}_{B/A} = B$ if and only if $v_L(\mathfrak{D}_{B/A}) = 0$.

The minimal polynomial of α can be seen to be an Eisenstein polynomial $f(X) = X^e + a_{e-1}X^{e-1} + \ldots + a_0$ with $a_i \in A$. We want to compute the valuation of

$$f'(\alpha) = e\alpha^{e-1} + (e-1)a_{e-1}\alpha^{e-2} + \ldots + a_1.$$

The valuation of each summand is

$$v_i := v_L((e-i)a_{e-i}\alpha^{e-i-1}) = ev_K(e-i) + ev_K(a_{e-i}) + e - i - 1$$

for $i = 0, \ldots, e-1$. Note that the v_i are pairwise distinct, because they are modulo e. Hence

$$v_L(f'(\alpha)) = \min_{i=0,\ldots,e-1} \{v_i\}$$

Note that for $i > 0$, $v_i = ev_K(e-i) + ev_K(a_{e-i}) + e - i - 1 \geq e$, as f is an Eisenstein polynomial, and hence $v_K(a_{e-i}) > 0$. It follows that $v_L(f'(\alpha)) = e - 1$ if and only if $v_0 = e - 1$. But $v_0 = ev_K(e) + e - 1$, so $v_0 = e - 1$ if and only if $v_K(e) = 0$ if and only if L/K is tamely ramified.

If L/K is not tamely ramified, then $v_K(e) > 0$, so $v_0 = v_L(e) + e - 1 \geq e$, so $e \leq v_L(f'(\alpha)) \leq v_L(e) + e - 1$. This completes the proof.

(b) This is much easier: $f'(\alpha) = \prod_{\sigma \in G \setminus \{\mathrm{id}\}}(\alpha - \sigma\alpha)$; the formula for $v_L(f'(\alpha)) = v_L(\mathfrak{D}_{B/A})$ follows.

□

REMARK 3.26. Theorem 3.23 allows us to characterize the different $\mathfrak{D}_{B/A}$ geometrically as follows: Let $\mu : B \otimes_A B \to B$ be the multiplication map $\mu(b_1 \otimes b_2) = b_1 b_2$ and I its kernel. Then I/I^2 is a $B \otimes_A B$-module, so it also carries two B-module structures, obtained via the two maps $B \to B \otimes_A B$, $b \mapsto b \otimes 1$ and $b \mapsto 1 \otimes b$. It is not difficult to check that these two B-module structures on I/I^2 coincide. The B-module I/I^2 is denoted by $\Omega^1_{B/A}$, and it is called the module of *Kähler differentials*. The map $d : B \to \Omega^1_{B/A}$, $d(b) = b \otimes 1 - 1 \otimes b$ is an A-derivation.

If $B = A[X]/(f(X))$, write x for the image of X in B. Then $\Omega^1_{B/A}$ is generated by dx, and the annihilator of $\Omega^1_{B/A}$ is precisely the ideal $(f'(x))$. Indeed, there is an exact sequence of B-modules

$$(f(X))/(f(X))^2 \xrightarrow{d \otimes 1} \Omega^1_{A[X]/A} \otimes_A B \to \Omega^1_{B/A} \to 0,$$

and $df(X) = f'(X)dX$.

3.5. Galois theory of discretely valued fields. We are now going to study the Galois theory of finite extensions L/K, where K is equipped with a discrete valuation v_K. We start with the noncomplete setting.

Let (K, v_K) be a discretely valued field, L/K a finite separable extension. Let A be the valuation ring of K and B its integral closure in L. The separability of L/K implies that B is finitely generated over A (Lemma 3.9), so we can apply the results from Section 3.3.

Write $w_1, \dots, w_r$ for the discrete valuations on L extending v_K. We saw in Proposition 3.10 that we have a canonical isomorphism of K_{v_K}-algebras $L \otimes_K K_{v_K} \xrightarrow{\cong} \prod_{i=1}^r L_{w_i}$. From this it follows that the extensions L_{w_i}/K_{v_K} are separable as well. Indeed, the extension L/K is defined by separable polynomials with coefficients in K, hence the finite K_{v_K}-algebra $L \otimes_K K_{v_K}$ is defined by separable polynomials with coefficients in K_{v_K}.

If L/K is in addition a Galois extension with group $G := \mathrm{Gal}(L/K)$, then G acts on the set of valuations $\{w_1, \dots, w_r\}$ over v_K: If $\sigma \in G$, then $w_i \circ \sigma$ is a discrete valuation lying over v_K. Moreover this action is transitive ([**50**, Ch. I, Prop. 19], or [**41**, Ch. II, Prop. 9.1]), so the numbers e_i and f_i are independent of $i \in \{1, \dots, r\}$; we just write e and f. The formula (3.1) becomes $n = efr$, where $n = [L : K]$.

Next, we define subgroups $D_{w_i} := \{\sigma \in G | w_i \circ \sigma = w_i\}$. By construction they are all conjugate subgroups of order ef of G,

DEFINITION 3.27. The subgroup D_{w_i} is called *decomposition group of* w_i.

In ideal theoretic language, if $\mathfrak{P}_i$ is the prime ideal of B corresponding to w_i, then G acts transitively on the set $\{\mathfrak{P}_1, \dots, \mathfrak{P}_r\}$, and $D_{w_i} = \{\sigma \in G | \sigma(\mathfrak{P}_i) = \mathfrak{P}_i\} \subseteq G$.

PROPOSITION 3.28. *If L/K is Galois, then L_{w_i}/K_{v_K} is Galois with group D_{w_i}.*

PROOF. Indeed, we have seen that $[L_{w_i} : K_{v_K}] = ef = |D_{w_i}|$, and $D_{w_i} \subseteq \mathrm{Aut}_{K_{v_K}}(L_{w_i})$. □

Since we are mainly interested in the decomposition group, we now put ourselves in a complete local situation. Let L/K be a finite Galois extension of complete discretely valued fields. Write v for the valuation of K, v_L for the valuation of L, A for the valuation ring of K, and B for its integral closure in L. The Galois group $G = \mathrm{Gal}(L/K)$ is equal to the decomposition group of v_L.

Next, consider the residue extension $k(v_L)/k(v)$. Since v_L is the unique valuation of L lying over v, we get a homomorphism of groups $G \to \mathrm{Aut}_{k(v)}(k(v_L))$.

PROPOSITION 3.29 ([**50**, Ch. I, Prop. 20]). *If the extension $k(v_L)/k(v)$ is separable, then it is Galois, and*

$$G \to \mathrm{Gal}(k(v_L)/k(v))$$

is a surjective homomorphism of groups.

Its kernel is denoted by $I := I_L$.

DEFINITION 3.30. The normal subgroup I of G is called *inertia group of* v_L.

The Galois correspondence shows that I corresponds to a field L^I lying between L and K; the fixed field of I. It is Galois over K, and by construction this is the maximal subextension of L/K which is unramified. Indeed, $|G/I| = f = [L^I : K]$, so the ramification index of L^I/K is 1. On the other hand, the residue extension of L/L^I is trivial, so L/L^I is totally ramified.

The inclusion $I \subseteq G$ is the first step in the so called *ramification filtration.*

3.6. The ramification filtration — lower numbering. We continue with the previous setup: L/K is a finite Galois extension of complete discretely valued fields with separable residue extension. Let $G = \mathrm{Gal}(L/K)$. We also write $p = \mathrm{char}(k(v)) \geq 0$, A, B for the valuation rings of K, L, respectively, and $\mathfrak{m}_K, \mathfrak{m}_L$ for their maximal ideals.

DEFINITION 3.31. Let $i \geq -1$ be an integer. Note that G acts on $B/\mathfrak{m}_L^{i+1}$. Define

$$G_i := \{\sigma \in G | \sigma \text{ acts trivially on } B/\mathfrak{m}_L^{i+1}\}.$$

The G_i form a descending chain of subgroups of G, and G_i is called the *i-th ramification subgroup of G*. This filtration is called the *ramification filtration of G in the lower numbering.*

We make the definition of the G_i a little bit more concrete:

$$G_i = \{\sigma \in G | \forall b \in B : v_L(\sigma(b) - b) \geq i + 1\}.$$

By Theorem 3.24, there exists $x \in B$ such that $B = A[x]$. Then we can also write

$$G_i = \{\sigma \in G | v_L(\sigma(x) - x) \geq i + 1\} \tag{3.4}$$

Indeed, x also generates $B/\mathfrak{m}_L^{i+1}$ as an A-algebra, so σ acting trivially on $B/\mathfrak{m}_L^{i+1}$ is equivalent to $v_L(\sigma(x) - x) \geq i + 1$.

PROPOSITION 3.32 ([**50**, Ch. IV, Prop. 1]). *We have $G_{-1} = G$ and $G_0 = I$ the inertia group. The G_i are normal subgroups of G, and $G_i = \{1\}$ for $i \gg 0$*

PROOF. The identifications of G_0 and G_{-1} follow directly from the definition, and so does the claim that G_i are normal. Finally, since G is a finite group, the description (3.4) shows that $G_i = \{1\}$ for i large enough. □

The description (3.4) shows that if x is a generator of B as an A-algebra, then for $\sigma \in G \setminus \{\mathrm{id}\}$, the integer

$$i_G(\sigma) := v_L(\sigma(x) - x) \tag{3.5}$$

is independent of the choice of x. We obtain a map $i_G : G \to \mathbb{Z}_{\geq 0} \cup \{\infty\}$, which has the property that $G_i = i_G^{-1}([i+1, \infty])$. As the G_i are normal subgroups, we also see that i_G is a class function, i.e. that it is constant on conjugacy classes.

EXAMPLE 3.33. We begin to study an important example: Let k be an algebraically closed field of characteristic $p > 0$, and $k[\![x]\!]$ the ring of formal power series in one indeterminate. Write $K := \mathrm{Frac}(k[\![x]\!]) = k((x))$, and consider the polynomial $t^{p^n} - t - \frac{1}{x} \in K[t]$. It is not difficult to see that it is irreducible. Indeed, if we make the change of coordinates $u = \frac{1}{t}$, we obtain an isomorphism of fields

$$L := K[u]/(u^{p^n} + xu^{p^n-1} - x) \cong K[t]/(t^{p^n} - t - \frac{1}{x}).$$

Note that $u^{p^n} + xu^{p^n-1} - x$ is a separable *Eisenstein polynomial* ([**50**, Ch. I, Prop. 17]). The extension L/K is a Galois extension with group $\mathbb{F}_{p^n}$, where $a \in \mathbb{F}_{p^n}$ acts via $t \mapsto t + a$, i.e. $u \mapsto \frac{u}{1+au}$. This extension is called *Artin-Schreier extension.*

Moreover, according to [**50**, Ch. I, Prop. 17], the integral closure of $k[\![x]\!]$ in L is $k[\![x]\!][u]/(u^{p^n}+xu^{p^n-1}-x)$, which is a discrete valuation ring with uniformizer $u=\frac{1}{t}$. The ramification index of this extension is p^n: $u^{p^n}=x(1-u^{p^n-1})$, and $1-u^{p^n-1}$ is a unit.

To determine the ramification groups of $\mathrm{Gal}(L/K)=\mathbb{F}_{p^n}$, we compute for every $a\in\mathbb{F}_{p^n}\setminus\{0\}$:

$$v_L\left(\frac{u}{1+au}-u\right)=v_L\left(u\left(\frac{1-(1+au)}{1+au}\right)\right)=2.$$

It follows that $\mathbb{F}_{p^n}=G=G_0=G_1\supsetneq G_2=0$.

Following [**33**], one similarly proves that for the $\mathbb{F}_p$-Galois extension $L=K[t]/(t^p-t-x^{-m})$, one has $G_0=G_1=\ldots=G_m\supsetneq G_{m+1}=0$, if $(m,p)=1$.

PROPOSITION 3.34 (Compatibility with taking subgroups). *Let H be a subgroup of G and L^H its fixed field. Then L/L^H is Galois with group H, and $H_i=G_i\cap H$ for all i.*

PROOF. This is clear, since if x generates B over A, then x also generates B over the valuation ring of L^H. □

REMARK 3.35.

(a) If we apply the proposition to the group G_0, then L^{G_0} is the maximal unramified subextension of L/K, and L/L^{G_0} is totally ramified. To study the higher ramification groups G_i, $i>0$, we may assume that L/K is totally ramified. In this case the degree of L/K is equal to the ramification index of L/K.

(b) If H is a normal subgroup of G, then G/H is the Galois group of the extension L^H/K. One might hope that the ramification filtration is compatible with quotients, i.e. that for $i\geq 0$, the image of G_i in G/H is the i-th ramification subgroup of G/H. This is not the case. The image of G_i in G/H is one of the ramification subgroups of G/H, but in general not $(G/H)_i$. The indexing has to be adjusted. To make this precise one introduces the *upper numbering filtration*, see Section 3.7.

Now let π be a uniformizer for L and assume that L/K is totally ramified. In this case $G=G_0$, and π generates B as A-algebra, see Theorem 3.24. For $\sigma\in G_0$, we see that $v_L(\sigma(\pi))=1$, so $\sigma(\pi)/\pi\in B^\times$. Moreover, by (3.4), G_i is characterized as the set of $\sigma\in G_0$, such that $\sigma(\pi)-\pi\in(\pi)^{i+1}$. Consequently,

$$G_i=\left\{\sigma\in G\;\middle|\;\frac{\sigma(\pi)}{\pi}\equiv 1 \mod \mathfrak{m}_L^i\right\}.$$

DEFINITION 3.36. For $i>0$, the set $U_L^i:=1+\mathfrak{m}_L^i\subseteq B^\times$ is easily seen to be a subgroup, the group of *i-th units*. We define $U_L:=U_L^0:=B^\times$.

This defines a descending filtration of the group of units U_L, and the observation that $\sigma(\pi)/\pi\in B^\times$ can be made more precise:

PROPOSITION 3.37. *For $i\geq 0$, the assignment $\sigma\mapsto\sigma(\pi)/\pi$ induces an injective homomorphism of groups*

$$G_i/G_{i+1}\hookrightarrow U_L^i/U_L^{i+1},$$

which is independent of the choice of the uniformizer π.

PROOF. We have seen that $\sigma(\pi)/\pi \in U_L^i$ if $\sigma \in G_i$. If τ is a second uniformizer, then $\tau = u\pi$ with $u \in B^\times$, and

$$\frac{\sigma(\tau)}{\tau} = \frac{\sigma(u)}{u} \cdot \frac{\sigma(\pi)}{\pi}.$$

If $\sigma \in G_i$, then by definition $\sigma(u) \equiv u \mod \mathfrak{m}_L^{i+1}$, and hence $\sigma(u)/u \in U_L^{i+1}$, as $u \in B^\times$. We obtain a map $G_i \to U_L^i/U_L^{i+1}$, independent of the choice of π. This is a group homomorphism: If $\sigma, \sigma' \in G_i$, then

$$\frac{\sigma\sigma'(\pi)}{\pi} = \frac{\sigma'(\pi)}{\sigma'(\pi)} \cdot \frac{\sigma(\sigma'(\pi))}{\pi} = \frac{\sigma'(\pi)}{\pi} \cdot \frac{\sigma(\sigma'(\pi))}{\sigma'(\pi)},$$

and $\sigma'(\pi)$ also is a uniformizer of B.

Finally, $\sigma(\pi)/\pi \in U_L^{i+1}$ if and only if $\sigma \in G_{i+1}$, so the kernel of the map $G_i \to U_L^i/U_L^{i+1}$ is G_{i+1}. This completes the proof. □

We can now use the fairly concrete structure of the groups U_L^i to get some information about the groups G_i:

COROLLARY 3.38. *(a) G_0/G_1 is cyclic, and if $p = \mathrm{char}(k(v_K)) > 0$, then the order of G_0/G_1 is prime to p.*

(b) If $p = 0$, then $G_i = 0$ for $i > 0$.

(c) If $p > 0$, then for $i > 0$, the groups G_i are p-groups, and the quotients G_i/G_{i+1} are abelian p-groups.

(d) G_0 is a semi-direct product of a cyclic group of order prime to p with a p-group. In particular, G_0 is solvable and G_1 is its unique p-Sylow subgroup.

PROOF. (a) We know that the quotients G_i/G_{i+1} are isomorphic to subgroups of U_L^i/U_L^{i+1}. If $i = 0$, then $U_L^0/U_L^1 = k(v_L)^\times$. As every finite subgroup of the multiplicative group of a field is cyclic, we see that G_0/G_1 is cyclic. If $p > 0$, there are no elements of order p in $k(v_L)^\times$, so G_0/G_1 is cyclic of order prime to p.

(b) Note that for $i > 0$, U_L^i/U_L^{i+1} is isomorphic to $\mathfrak{m}_L^i/\mathfrak{m}_L^{i+1}$, which is isomorphic to the additive group underlying $k(v_L)$. If $p = 0$, then the additive group $k(v_L)$ has no nontrivial elements of finite order, so $G_i/G_{i+1} = 0$ for all $i > 0$. As $G_i = 0$ for $i \gg 0$, it follows that $G_1 = 0$.

(c) If $p > 0$ and $i > 0$, then again G_i/G_{i+1} is isomorphic to a finite subgroup of the additive group of the residue field $k(v_L)$, hence an abelian p-group. This implies that the order of G_1 is a p-power, so G_1 is a p-group.

(d) We just have to show that the sequence

$$0 \longrightarrow G_1 \longrightarrow G_0 \longrightarrow G_0/G_1 \longrightarrow 0$$

admits a splitting. But this follows from the fact that the orders of G_0/G_1 and G_1 are coprime.

□

The ramification filtration of the Galois group $G = \mathrm{Gal}(L/K)$ allows us to "measure" the quality of ramification. If L/K is totally ramified, then its ramification index equals the degree $[L : K]$, the more G_i are nonzero, the "worse" is the ramification. We will make this more precise later on. For now, we content ourselves with the following definition:

DEFINITION 3.39. A finite separable extension L/K of complete discretely valued fields is called *tamely ramified* if its ramification index is prime to the characteristic p of the residue field $k(v)$ of K, and if the extension of the residue fields is separable. Otherwise L/K is called *wildly ramified*.

These definitions are easily extended to the case of general separable extensions of discretely valued fields which are not necessarily complete: If (K,v) is a discretely valued field, L/K a finite separable extension and w a discrete valuation on L extending v (Definition 3.8), then $(L,w)/(K,v)$ is *tamely ramified* if the extension L_w/K_v of the completions is tamely ramified, i.e. if the ramification index of w over v is prime to p and if the residue extension is separable. Otherwise, $(L,w)/(K,v)$ is *wildly ramified.*

EXERCISE 3.40. Let L/K be a Galois extension of discretely valued fields, with separable residue extension and Galois group G. Show that L/K is tamely ramified if and only if $G_1 = 0$ if and only if $(|G_0|, p) = 1$.

3.7. The ramification filtration — upper numbering. We continue with the previous setup: Let L/K be a finite Galois extension of complete discretely valued fields with Galois group G and separable residue extension. Let $H \subseteq G$ be a normal subgroup, such that $G/H = \mathrm{Gal}(L^H/K)$. As indicated in Remark 3.35, the image of the ramification filtration of G in G/H is the ramification filtration of $\mathrm{Gal}(L^H/K)$, up to changing the numbering. We start this section by making this remark precise, before we then fix this numbering defect by introducing the *upper numbering for the ramification filtration.* Among other things, this will have the advantage that we can pass to the inverse limit to obtain a ramification filtration by closed subgroups of the absolute Galois group.

Recall that we defined a class function i_G on G, by setting $i_G(\sigma) = v_L(\sigma(x) - x)$ for $\sigma \neq \mathrm{id}$, x a generator of B as A-algebra, and $i_G(\mathrm{id}) = \infty$.

PROPOSITION 3.41. *Let H be a normal subgroup of G, and e' the ramification index of L/L^H. We can then compute $i_{G/H}$ in terms of i_G:*

$$i_{G/H}(s) = \frac{1}{e'} \sum_{\sigma \mapsto s} i_G(\sigma).$$

PROOF. Let A and B be the valuation rings of K, L and B' the valuation ring of L^H. By Theorem 3.24 we find $x \in B$, $y \in B'$ such that $B = A[x]$, $B' = A[y]$. By definition we have for $s \in G/H$:

$$e' i_{G/H}(s) = v_L(s(y) - y).$$

Let $\sigma \in G$ be a fixed element of the preimage of s. Then $\sigma|_{L^H} = s$ and every element of the preimage of s is of the form $\sigma\tau$ for $\tau \in H$. We show that the elements $\sigma y - y$ and $\prod_{\tau \in H}(\sigma\tau x - x)$ of B are associated, i.e. generate the same ideals.

Let $f(X) \in B'[X]$ be the minimal polynomial of x over L^H. Write $f(X) = X^n + b_{n-1}X^{n-1} + \ldots + b_0$ with $b_i \in B'$. The coefficients of the polynomial $(\sigma f)(X) - f(X) \in B'[X]$ are $\sigma(b_i) - b_i$, $i = 0, \ldots, n-1$. Let $B_i(Y) \in A[Y]$ be polynomials such that $B_i(y) = b_i$. Then $\sigma(b_i) - B_i(\sigma(y)) = 0$, so $Y - \sigma(y)$ divides the polynomial $\sigma(b_i) - B_i(Y) \in B'[Y]$, whence $y - \sigma(y)$ divides $\sigma(b_i) - b_i$. This shows that $\sigma y - y$ divides $(\sigma f)(x) - f(x) = \prod_{\tau \in H}(\sigma\tau x - x)$.

Conversely, let $G \in A[X]$ be a polynomial such that $G(x) = y$. Then x is a zero of $G(X) - y \in B'[X]$, so $G(X) - y = f(X)h(X)$ for some polynomial $h(X)$.

Applying σ gives

$$G(X) - \sigma y = (\sigma f)(X)(\sigma h)(X)$$

and finally

$$y - \sigma y = G(x) - \sigma y = (\sigma h)(x) \prod_{\tau \in H} (\sigma\tau x - x).$$

□

For notational reasons, the following definition is convenient:

DEFINITION 3.42. If $u \in [-1, \infty) \subseteq \mathbb{R}$, then $G_u := G_{\lceil u \rceil}$, where $\lceil u \rceil$ denotes the smallest integer $\geq u$.

We will use this notation in the proof of the following corollary.

COROLLARY 3.43 (Herbrand). *In the situation of the proposition, the ramification groups of G/H are the images of the ramification groups of G.*

PROOF. Let $i \in \mathbb{Z}_{\geq -1}$, $\sigma_0 \in G_i$ and let s_0 be its image in G/H. The proposition allows us to compute $i_{G/H}(s_0)$:

$$i_{G/H}(s_0) = \frac{1}{e'} \sum_{\sigma \mapsto s_0} i_G(\sigma).$$

Write $H_j := G_j \cap H$ and $h_j := |H_j|$. We can rewrite the above formula:

$$i_{G/H}(s_0) = \frac{1}{e'} \left(\sum_{j=-1}^{\infty} \sum_{\sigma \in H_j \smallsetminus H_{j+1}} i_G(\sigma_0\sigma) + i_G(\sigma_0) \right).$$

Note that $i_G(\sigma_0\sigma) \geq \min\{i_G(\sigma), i_G(\sigma_0)\}$ with equality if $i_G(\sigma) \neq i_G(\sigma_0)$ (Exercise 3.3), and that $i_G(\sigma) = j+1$ if $\sigma \in H_j \smallsetminus H_{j+1}$. Using that $\sigma_0 \in G_i$, we compute

$$\begin{aligned} i_{G/H}(s_0) &= \frac{1}{e'} \left(\sum_{j=-1}^{i-1} (h_j - h_{j+1})(j+1) + i_G(\sigma_0) + \sum_{j \geq i} \sum_{\sigma \in H_j \smallsetminus H_{j+1}} i_G(\sigma\sigma_0) \right) \\ &\geq \frac{1}{e'} \left(\sum_{j=-1}^{i-1} (h_j - h_{j+1})(j+1) + (i+1) \cdot \left(1 + \sum_{j \geq i, h_j > 1} (h_j - h_{j+1}) \right) \right) \\ &= \frac{1}{e'} \sum_{j=0}^{i} h_j \end{aligned}$$

This last number is important and we give it a name:

$$\varphi_H(i) + 1 := \frac{1}{|G_0 \cap H|} \sum_{j=0}^{i} |G_j \cap H| \in \mathbb{Q}.$$

Our computation shows that the image $G_iH/H \subseteq G/H$ of G_i is a subgroup of $(G/H)_{\varphi_H(i)}$ (see Definition 3.42). Let us show that this is in fact an equality. Let $s \in (G/H)_{\varphi_H(i)} \smallsetminus \{1\}$, and let i' be maximal with the property that there exists a preimage σ_0 of s in $G_{i'}$. Let $\sigma \in H_j \smallsetminus H_{j+1}$. If $j < i'$, then $i_G(\sigma_0\sigma) = i_G(\sigma) = j+1$. If $j \geq i'$, then $i_G(\sigma_0\sigma) \geq i'+1$, and hence by the maximality of i', $i_G(\sigma_0\sigma) = i'+1$.

Our computation from above then shows

$$\begin{aligned} i_{G/H}(s) &= \frac{1}{e'}\left(\sum_{j=-1}^{i'-1}(h_j - h_{j+1})(j+1) + (i'+1)\cdot\left(1 + \sum_{j\geq i', h_j>1}(h_j - h_{j+1})\right)\right) \\ &= \frac{1}{e'}\sum_{j=0}^{i'} h_j \\ &= \varphi_H(i') + 1. \end{aligned}$$

We immediately see that $i_{G/H}(s) \geq \varphi_H(i) + 1$ implies that $i' \geq i$, so $s \in G_i H/H$. This completes the proof that $G_i H/H = (G/H)_{\varphi_H(i)}$. □

To define the upper numbering filtration of G, we extend the definition of the function φ that appeared in the proof of Corollary 3.43.

DEFINITION 3.44. As before, let L/K be a finite Galois extension of complete discretely valued fields with group G, and separable residue extension of degree f.

(a) For $u \in [-1, 0)$, define $(G_0 : G_u) := (G_u : G_0)^{-1}$, i.e. $(G_0 : G_{-1}) = (G_{-1} : G_0)^{-1} = \frac{1}{f}$ and $(G_0 : G_u) = 1$ if $u \in (-1, 0)$.

(b) Define the function $\varphi_{L/K} : [-1, \infty) \to [-1, \infty)$ by

$$\varphi_{L/K}(u) = \int_0^u \frac{1}{(G_0 : G_t)} \mathrm{d}t$$

If $u \geq 0$, then

$$\varphi_{L/K}(u) + 1 = \frac{1}{|G_0|}\left(|G_0| + |G_1| + \ldots + |G_{\lfloor u \rfloor}| + (u - \lfloor u \rfloor)|G_{\lfloor u \rfloor + 1}|\right), \tag{3.6}$$

where $\lfloor u \rfloor$ is the largest integer $\leq u$. In particular, if $u = m \in \mathbb{Z}_{\geq 0}$, then

$$\varphi_{L/K}(m) + 1 = \frac{1}{|G_0|}\sum_{i=0}^{m} |G_i|.$$

This formula shows how the function $\varphi_{L/K}$ is related to the function φ_H from the proof of Corollary 3.43: $\varphi_{L/L^H} = \varphi_H$.

Using this notation, we restate the result of our calculation from the proof of Corollary 3.43 for future reference.

COROLLARY 3.45. *If $H \subseteq G$ is a normal subgroup, then for $u \in [-1, \infty) \subseteq \mathbb{R}$ we have*

$$G_u H/H = (G/H)_{\varphi_{L/L^H}(u)}.$$

The basic properties of $\varphi_{L/K}$ are easy to verify:

PROPOSITION 3.46. *(a) $\varphi_{L/K}$ is continuous, piecewise linear, increasing, and concave.*

(b) $\varphi_{L/K}(0) = 0$.

(c) $\varphi_{L/K}$ is a homeomorphism of $[-1, \infty)$ onto itself.

DEFINITION 3.47. The inverse of $\varphi_{L/K}$ is denoted $\psi_{L/K}$.

PROPOSITION 3.48. *(a) $\psi_{L/K}$ is continuous, piecewise linear, increasing, and convex.*

(b) $\psi_{L/K}(0) = 0$.

(c) $\psi_{L/K}([-1, \infty) \cap \mathbb{Z}) \subseteq \mathbb{Z}$.

EXERCISE 3.49. Prove that $\psi_{L/K}(\mathbb{Z}\cap[-1,\infty))\subseteq\mathbb{Z}$ (Use formula (3.6)).

EXAMPLE 3.50. Lets compute $\varphi_{L/K}$ and $\psi_{L/K}$ in two simple special cases.

(a) If L/K is unramified, then $\varphi_{L/K}=\psi_{L/K}=\mathrm{id}$. Indeed, $(\varphi_{L/K})|_{[-1,0)}=\mathrm{id}_{[-1,0)}$ is always true. If $u\geq 0$, then

$$\varphi_{L/K}(u)=\frac{1}{|G_0|}\left(|G_1|+\ldots+|G_{\lfloor u\rfloor}|+(u-\lfloor u\rfloor)|G_{\lfloor u\rfloor+1}|\right)=u$$

as $|G_i|=1$ for $i\geq 0$.

(b) Now assume that L/K is Galois with Galois group $\mathbb{Z}/\ell\mathbb{Z}$, where ℓ is a prime number. Moreover, assume that $G=G_0=G_1=\ldots=G_t$, and that $G_i=\{1\}$ for $i>t$. For $0\leq u\leq t$, we compute

$$\varphi_{L/K}(u)=\frac{u\cdot\ell}{\ell}=u.$$

For $u>t$, we get

$$\varphi_{L/K}(u)=t+\frac{u-t}{\ell}$$

so

$$\psi_{L/K}(u)=\ell(u-t)+t,$$

for $u>t$.

This computation applies to Example 3.33.

The next proposition shows that the functions φ and ψ are transitive with respect to Galois subextensions.

PROPOSITION 3.51. *With the notations from above, let $H\subseteq G$ be a normal subgroup. Then*

$$\varphi_{L/K}=\varphi_{L^H/K}\circ\varphi_{L/L^H}\ \textit{and}\ \psi_{L/K}=\psi_{L/L^H}\circ\psi_{L^H/K}.$$

PROOF. Clearly it is enough to prove the claim about the function $\varphi_{L/K}$. Note that φ_{L/L^H} is piecewise linear and hence almost everywhere continuously differentiable. More precisely, it is differentiable on $[-1,\infty)\smallsetminus\mathbb{Z}$, because if $u\notin\mathbb{Z}$, then the left and right derivatives of φ_{L/L^H} at u are equal to $(H_0:H_u)^{-1}$. Similarly, ψ_{L/L^H} is continuously differentiable at all u, for which $\psi_{L/L^H}(u)$ is not an integer, and its derivative at such a u is

$$\psi'_{L/L^H}(u)=(\varphi'_{L/L^H}(\psi_{L/L^H}(u)))^{-1}=\left(H_0:H_{\psi_{L/L^H}(u)}\right). \tag{3.7}$$

Next, recall Proposition 3.34, which tells us that $H_u=H\cap G_u$, for all $u\in[-1,\infty)$.

Using Corollary 3.45 we compute

$$\begin{aligned}\varphi_{L^H/K}\circ\varphi_{L/L^H}(u)&=\int_0^{\varphi_{L/L^H}(u)}\frac{dt}{((G/H)_0:(G/H)_t)}\\&=\int_0^{\varphi_{L/L^H}(u)}\frac{\left(H_0:H_{\psi_{L/L^H}(t)}\right)}{\left(G_0:G_{\psi_{L/L^H}(t)}\right)}dt.\end{aligned}$$

Now, with a certain degree of sloppiness, we use (3.7) to write

$$\int_0^{\varphi_{L/L^H}(u)}\frac{\left(H_0:H_{\psi_{L/L^H}(t)}\right)}{\left(G_0:G_{\psi_{L/L^H}(t)}\right)}dt=\int_0^{\varphi_{L/L^H}(u)}\frac{\psi'_{L/L^H}(t)}{\left(G_0:G_{\psi_{L/L^H}(t)}\right)}dt.$$

Of course ψ_{L/L^H} is not everywhere differentiable, so to be precise, one should do this calculation on every piece where it is differentiable and then sum up.

In the end, we obtain

$$\begin{aligned}\varphi_{L^H/K}\circ\varphi_{L/L^H} &= \int_0^{\varphi_{L/L^H}(u)} \frac{\psi'_{L/L^H}(t)}{\left(G_0 : G_{\psi_{L/L^H}(t)}\right)}\mathrm{d}t \\ &= \int_0^u \frac{\mathrm{d}t}{(G_0 : G_t)} \\ &= \varphi_{L/K}(u)\end{aligned}$$

as claimed. □

Now we have gathered all the facts to define the upper numbering for the ramification filtration.

DEFINITION 3.52. We continue to use the notations from above. In particular $G = \mathrm{Gal}(L/K)$ is the Galois group of a finite Galois extension of complete discretely valued fields with separable residue extension. For a real number $v \in [-1,\infty)$, define $G^v := G_{\psi_{L/K}(v)}$ or equivalently $G^{\varphi_{L/K}(u)} = G_u$.

Note that $G = G^{-1}$, $G^0 = G_0$.

As promised, the upper numbering filtration is compatible with passing to quotients:

PROPOSITION 3.53 (Upper numbering filtration and quotients, Herbrand's Theorem). *Let $H \subseteq G$ be a normal subgroup. For every real number $v \in [-1,\infty)$, we have*

$$G^v H/H = (G/H)^v.$$

PROOF. Most of the work has already been done in the proof of Corollary 3.43 and Proposition 3.51:

$$\begin{array}{ccccc} G^v H/H & & (G/H)^{\varphi_{L^H/K}(\varphi_{L/L^H}(\psi_{L/K}(v)))} & \overset{\text{Prop. 3.51}}{=\!=\!=} & (G/H)^v \\ \Big\| \text{Def.} & & \Big\| \text{Def.} & & \\ G_{\psi_{L/K}(v)}H/H & \overset{\text{Prop. 3.45}}{=\!=\!=} & (G/H)_{\varphi_{L/L^H}(\psi_{L/K}(v))} & & \end{array}$$

□

Proposition 3.53 allows us to define a ramification filtration of the absolute Galois group of K.

DEFINITION 3.54. Let L/K be a (possibly infinite) Galois extension, such that the extension of the residue fields is separable. For a real number $u \in [-1,\infty)$, we define $\mathrm{Gal}(L/K)^u := \varprojlim_{L'} \mathrm{Gal}(L'/K)^u$, where L' runs through the set of finite Galois subextensions of L/K.

REMARK 3.55. If the residue field of K is perfect this applies to a separable closure K^{sep} of K. We obtain a descending filtration of $\mathrm{Gal}(K^{\mathrm{sep}}/K)$.

3.8. The norm. Let L/K be an extension of discretely valued fields, and recall the definition of the norm map $N_{L/K}$ (Definition 3.12). Note that if L/K is Galois with Galois group G, then

$$N_{L/K} : L^\times \to K^\times,\ x \mapsto \prod_{\sigma \in G} \sigma(x).$$

PROPOSITION 3.56. *The norm $N_{L/K}$ is a homomorphism of abelian groups. If L/K is a Galois extension of complete discretely valued fields with separable residue extension of degree f, then*

(a) $v_K(N_{L/K}(x)) = f v_L(x)$
(b) $N_{L/K}(B^\times) \subseteq A^\times$, *where A, B are the valuation rings of K, L.*

PROOF. Exercise. □

PROPOSITION 3.57 ([**50**, Ch. V]). *Let L/K be a totally ramified Galois extension of complete discretely valued fields and write $N := N_{L/K}$.*

(a) *For $m \geq 0$, N induces a homomorphism of groups*
$$N : U_L^{\psi_{L/K}(m)} \to U_K^m$$
which is surjective if $G_{\psi_{L/K}(m)} = 0$ or if the residue field of K is algebraically closed.

(b) *For $m \geq 0$, N induces a homomorphism of groups*
$$N : U_L^{\psi_{L/K}(m)+1} \to U_K^{m+1},$$
which is surjective if $G_{\psi_{L/K}(m+1)} = 0$ or if the residue field of K is algebraically closed.

(c) *From the above we get homomorphisms*
$$N_m : U_L^{\psi_{L/K}(m)} / U_L^{\psi_{L/K}(m)+1} \to U_K^m / U_K^{m+1},$$
such that the sequence
$$0 \to G_{\psi_{L/K}(m)} / G_{\psi_{L/K}(m)+1} \xrightarrow{\theta_m} U_L^{\psi_{L/K}(m)} / U_L^{\psi_{L/K}(m)+1} \xrightarrow{N_m} U_K^m / U_K^{m+1}$$
is exact, where the map θ_m is the map defined in Proposition 3.37.

(d) *N_m is surjective if one of the following holds:*
- *The residue field of K is algebraically closed.*
- *The residue field of K is perfect, and $G_{\psi_{L/K}(m)} = G_{\psi_{L/K}(m)+1}$.*
- $G_{\psi_{L/K}(m)} = 0$.

PROOF. This is rather lengthy. We will prove a more explicit version of (d) by induction:

CLAIM. Let k be the residue field of L and K. Recall from the proof of Corollary 3.38 that we have group isomorphisms $U_L^0/U_L^1 \cong U_K^0/U_K^1 \cong k^\times$, and that after choosing uniformizers $\tau \in \mathcal{O}_K$, $\pi \in \mathcal{O}_L$, we have group isomorphisms $U_L^i/U_L^{i+1} \cong U_K^i/U_K^{i+1} \cong (k, +)$ for $i > 0$. We claim:

- Identifying U_L/U_L^1 and U_K/U_K^1 with $k^\times$, $N_0 : k^\times \to k^\times$ is given by the nonconstant monomial $X^{|G_0|}$.
- For $m \geq 1$, if we identify N_m with a map $k \to k$, then N_m is given by a nonconstant polynomial of the form $P_m(X)$ which is a linear combination of monomials of the form X^{p^h}, where $\mathrm{char}(k) = p \geq 0$. Moreover $\deg P_m = |G_{\psi_{L/K}(m)}|$.

- Keeping the notations, if k is perfect, then there exists a polynomial Q_m with $P_m = (Q_m)^{p^h}$, where $p^h = |G_{\psi_{L/K}(m)+1}|$.

Part (d) follows from the claim. Indeed, if the residue field k is algebraically closed, this is obvious. If k is perfect and $m \geq 0$ such that $G_{\psi_{L/K}(m)} = G_{\psi_{L/K}(m)+1}$, we distinguish the cases $m = 0$ and $m > 0$. If $m = 0$, then $G_0 = G_1$ is a nontrivial p-group and the claim states that $N_0 : k^\times \to k^\times$ can be identified with the polynomial map $X^{|G_0|}$, which is surjective, as k is assumed to be perfect. If $m > 0$, then the claim shows that N_m can be identified with $P_m = (Q_m)^{|G_{\psi_{L/K}(m)+1}|}$ and that $\deg P_m = |G_{\psi_{L/K}(m)}|$ so $\deg Q_m = 1$. As k is assumed to be perfect, it follows that N_m is surjective. Finally, if $G_{\psi_{L/K}(m)} = 0$ then $m > 0$, as otherwise L/K would be trivial, and the claim implies that N_m can be identified with a nonconstant polynomial of degree 1, so N_m is surjective. This shows that the claim implies (d).

The surjectivity statements from (a) and (b) are implied by (d) using a completeness argument ([**50**, Ch. V, Lemma 2]).

It remains to prove (c) and the claim. Since the Galois group of L/K is solvable, it has a quotient which is cyclic of prime order. Using induction and the transitivity of the norm and ψ (Proposition 3.51), one reduces the proof of (c) and the claim to the case where the Galois group of L/K is $G = G_0 = \mathbb{Z}/\ell\mathbb{Z}$ with ℓ a prime (see [**50**, Ch. V, §6]). We assume that $G_0 = G_1 = \ldots = G_t$, and $G_i = \{1\}$ for $i > t$.

In this case we already computed the functions $\varphi_{L/K}$ and $\psi_{L/K}$ in Example 3.50. We will use the following lemma:

LEMMA 3.58. *Let B/A be the extension of discrete valuation rings attached to L/K. For brevity, we write $N := N_{L/K}$ and $\mathrm{Tr} := \mathrm{Tr}_{L/K}$.*

(i) $\mathfrak{D}_{B/A} = \mathfrak{m}_L^{(t+1)(\ell-1)}$.

(ii) If $n \geq 0$, then

$$\mathrm{Tr}(\mathfrak{m}_L^n) = \mathfrak{m}_K^{\left\lfloor \frac{(t+1)(\ell-1)+n}{\ell} \right\rfloor}.$$

(iii) If $x \in \mathfrak{m}_L^n$, then

$$N(1+x) \equiv 1 + \mathrm{Tr}(x) + N(x) \mod \mathrm{Tr}(\mathfrak{m}_L^{2n}).$$

PROOF. (i) Proposition 3.25, (b).

(ii) Recall that if $\mathfrak{b} \subseteq L$ and $\mathfrak{a} \subseteq K$ are fractional ideals, then $\mathrm{Tr}(\mathfrak{b}) \subseteq \mathfrak{a}$ if and only if $\mathfrak{b} \subseteq \mathfrak{a}\mathfrak{D}_{B/A}^{-1}$ (Lemma 3.17). It follows that $\mathrm{Tr}(\mathfrak{m}_L^n) \subseteq \mathfrak{m}_K^r = \mathfrak{m}_L^{\ell r}$ if and only if $r \leq \frac{(t+1)(\ell-1)+n}{\ell}$.

(iii) Exercise (see [**50**, Ch. V, §3., Lemma 5]).

□

We can proceed with the proof of (c) and the claim.

$m = 0$: Clearly $N(U_L) \subseteq U_K$ and $N(U_L^1) \subseteq U_K^1$. The map $N_0 : k^\times \to k^\times$ can be identified with the polynomial X^ℓ. Let us determine the kernel of N_0: Note that $\ell = p := \mathrm{char}(k)$ if and only if $t > 0$, as G_1 is the unique p-Sylow subgroup of G_0 (Corollary 3.38). Thus, if $t = 0$, then the kernel of N_0 is of order ℓ and contains $\theta_0(G_0)$ which is also of order ℓ. If $t > 0$, then $\ell = p$ and hence N_0 is injective. Thus for $m = 0$ we have checked that (c) holds.

$1 \leq m < t$: In this case $\psi_{L/K}(m) = m$ (Example 3.50), so we first show that $N(U_L^m) \subseteq U_K^m$. If $x \in \mathfrak{m}_L^m$, then $\mathrm{Tr}(x) \in \mathfrak{m}_K^{m+1}$ according to Lemma 3.58, as

$$\left\lfloor \frac{(t+1)(\ell-1)+m}{\ell} \right\rfloor > \left\lfloor \frac{\ell m + (\ell-1)}{\ell} \right\rfloor \geq m.$$

This also shows that $\mathrm{Tr}(\mathfrak{m}_L^{2m}) \subseteq \mathfrak{m}_K^{m+1}$, so Lemma 3.58 yields

$$N(1+x) \equiv 1 + N(x) \mod \mathfrak{m}_K^{m+1}.$$

But $N(x) \in \mathfrak{m}_K^m$, so we see that $N(U_L^m) \subseteq U_K^m$, and also that $N(U_L^{m+1}) \subseteq U_K^{m+1}$.

To make N_m explicit, note that since $t > 0$ we have $\ell = p$, and if $x = u\pi^m$ with $u \in U_L$ and π a uniformizer, then $N(x) \equiv u^p N(\pi)^m \mod \mathfrak{m}_K^{m+1}$ (recall that $\sigma u - u \in \mathfrak{m}_L^{t+1}$ for all u). If τ is a uniformizer for K, then $N(\pi^m) = a_m \tau^m$, with $a_m \in U_K$. Then $N(1+u\pi^m) \equiv 1 + a_m u^p \tau^m \mod \mathfrak{m}_K^{m+1}$. Thus N_m can be identified with the polynomial $\alpha_m X^p$, where α_m is the image of a_m in $k^\times$. From this we also see that N_m is injective, and that it is surjective if k is perfect.

$m = t$: Let $x \in \mathfrak{m}_L^t$. Lemma 3.58 shows that

$$N(1+x) \equiv 1 + \mathrm{Tr}(x) + N(x) \mod \mathfrak{m}_K^{t+1}.$$

Moreover,

$$\left\lfloor \frac{(t+1)(\ell-1)+t}{\ell} \right\rfloor = t$$

and

$$\left\lceil \frac{(t+1)(\ell-1)+t}{\ell} \right\rceil = t+1$$

so $N(U_L^t) \subseteq U_K^t$ and $N(U_L^{t+1}) \subseteq U_K^{t+1}$. As $\psi_{L/K}(t) = t$, this is what we wanted. Let us make it more explicit: Again let τ be a uniformizer of K, and write $\mathrm{Tr}(\pi^t) = b\tau^t$, $N(\pi^t) = a\tau^t$, with $a, b \in A$. Every element $x \in U_L^t / U_L^{t+1}$ is the image of an element of the form $1 + u\pi^t$ with $u \in A$. We compute:

$$\begin{aligned} N(1+u\pi^t) &\equiv 1 + \mathrm{Tr}(u\pi^t) + N(u\pi^t) \mod \mathfrak{m}_K^{t+1} \\ &\equiv 1 + (bu + au^p)\tau^t \mod \mathfrak{m}_K^{t+1}. \end{aligned}$$

Writing α, β for the images of a, b in k, we see that N_t can be written as the polynomial $\beta X + \alpha X^p$. As $\alpha \neq 0$, we see that the kernel of N_m has order at most $p = \ell$, but it contains $\theta_t(G_t)$ (Proposition 3.37), so it has to be equal to $\theta_t(G_t)$. Hence $\beta \neq 0$, and (c) is proved.

$m > t$: We computed in Example 3.50 that $\psi_{L/K}(m) = t + \ell(m-t)$. We have

$$\frac{(t+1)(\ell-1) + \psi_{L/K}(m)}{\ell} = \frac{\ell-1}{\ell} + m, \tag{3.8}$$

so $\mathrm{Tr}(\mathfrak{m}_L^{\psi_{L/K}(m)}) \subseteq \mathfrak{m}_K^m$. Moreover, $\psi_{L/K}(m) \geq m+1$ and $N(\mathfrak{m}_L^{\psi_{L/K}(m)}) \subseteq \mathfrak{m}_K^{\psi_{L/K}(m)}$, so if $x \in \mathfrak{m}_L^{\psi_{L/K}(m)}$, then

$$N(1+x) \equiv 1 + \mathrm{Tr}(x) \mod \mathfrak{m}_K^{m+1}.$$

This shows that $N(U_L^{\psi_{L/K}(m)}) \subseteq U_K^m$ and similarly also $N(U_L^{\psi_{L/K}(m)+1}) \subseteq U_K^{m+1}$. Again, if $x = u\pi^{\psi_{L/K}(m)}$ with $u \in A$, and $\mathrm{Tr}(\pi^{\psi_{L/K}(m)}) = a_m \tau^m$, then $N(1 + u\pi^{\psi_{L/K}(m)}) \equiv 1 + a_m u \tau^m \mod \mathfrak{m}_K^{m+1}$, so N_m can be written as the polynomial $\alpha_m X$, if α_m is the image of a_m in k. Finally, $\alpha_m \neq 0$, as otherwise $\mathrm{Tr}(\mathfrak{m}_L^{\psi_{L/K}(m)}) \subseteq \mathfrak{m}_K^{m+1}$, which is impossible according to (3.8) and Lemma 3.58.

□

For future reference, we prove a fact about the norm of a cyclic extension.

THEOREM 3.59 (Hilbert's Theorem 90). *Let L/K be a Galois extension of fields with cyclic Galois group G and generator g (no other assumptions are necessary). If $v \in L^\times$ has norm 1, then there exists $x \in L^\times$, such that $v = \frac{g(x)}{x}$. In other words, the quotient group*

$$\ker(N_{L/K}) \Big/ \left\{ \frac{g(x)}{x} \middle| x \in L^\times \right\}$$

is trivial.

PROOF. Write $V = \ker(N_{L/K})$ and $\overline{V} := V / \left\{ \frac{g(x)}{x} \middle| x \in L^\times \right\}$. The group $\overline{V}$ can be identified with the group cohomology $H^1(G, L^\times)$. A brief reminder: A crossed map $\varphi : G \to L^\times$, is a map φ, such that $\varphi(g_1 g_2) = g_1(\varphi(g_2)) \cdot \varphi(g_1)$, $g_1, g_2 \in G$. A particular example of a crossed map is $g \mapsto \frac{gx}{x}$ for $x \in L^\times$; such a crossed map is called principal. The set of crossed maps inherits a group structure, and

$$H^1(G, L^\times) = \{\text{crossed maps}\}/\{\text{principal crossed maps}\}.$$

If G is a cyclic group with generator g and if $\varphi : G \to L^\times$ is a crossed map, then $\varphi(g) \in L^\times$ has norm 1. It is not difficult to check that this defines a homomorphism $u : \{\text{crossed maps}\} \to \ker(N_{L/K})$. Moreover, the map u is surjective: If $x \in L^\times$ has norm 1, then $\psi(1) := 1$, $\psi(g^r) := \prod_{i=0}^{r-1} g^i x$, $1 \le r < |G|$, defines a crossed map with $\psi(g) = x$. Finally, one checks that

$$u^{-1}\left(\left\{ \frac{g(x)}{x} \middle| x \in L^\times \right\} \right) = \{\text{principal crossed maps}\},$$

from which it follows that u induces an isomorphism $H^1(G, L^\times) \xrightarrow{\cong} \overline{V}$.

To prove the theorem, it hence remains to show that every crossed map $G \to L^\times$ is principal, and this is true even if $G = \mathrm{Gal}(L/K)$ is not cyclic.

Let φ be a crossed map and for $y \in L^\times$ write

$$x := \sum_{\sigma \in G} \varphi(\sigma) \cdot \sigma(y).$$

If there exists y such that $x \neq 0$, then φ is principal. Indeed, in this case, for any $\tau \in G$ we can write

$$\tau(x) = \sum_{\sigma \in G} \tau(\varphi(\sigma)) \cdot \tau\sigma(y).$$

Since $\varphi(\tau\sigma) = \tau\varphi(\sigma) \cdot \varphi(\tau)$, we see that

$$\tau(x) = \varphi(\tau)^{-1} \sum_{\sigma \in G} \varphi(\tau\sigma)\tau\sigma(y) = \varphi(\tau)^{-1} \cdot x$$

which proves that φ is principal, if $x \neq 0$.

To see that we can find $y \in L^\times$ such that $x \neq 0$, one observes (this is not entirely trivial) that the family of maps $\sigma : L^\times \to L^\times$, $\sigma \in G$ is linearly independent over L. In particular: $\sum_{\sigma \in G} \varphi(\sigma) \cdot \sigma : L \to L$ is not the zero-map. □

EXERCISE 3.60. (a) Complete the proof that $H^1(G, L^\times) \cong \overline{V}$.

(b) If T is a group and $f_1, \ldots, f_r : T \to L^\times$ pairwise distinct homomorphisms, prove that the f_i are linearly independent over K, i.e. if there are $a_1, \ldots, a_r \in L$, such that $\sum_{i=1}^r a_i f_i = 0$ as map $T \to L$, then $a_1 = a_2 = \ldots = a_r = 0$.

3.9. Jumps and the theorem of Hasse-Arf. We denote by L/K a finite Galois extension of complete discretely valued fields with separable residue extension. Let G be its Galois group. We know that there are only finitely many subgroups which appear in the ramification filtration, regardless of whether we use the upper or lower numbering. The real numbers $u \in [-1, \infty)$, where the upper numbering filtration $\{G^u\}$ transitions from one group to the next, are of special interest.

DEFINITION 3.61. The numbers $u \in [-1, \infty)$, such that $G^u \supsetneq G^{u+\varepsilon}$ for all $\varepsilon > 0$ are called *jumps* or *breaks* of this filtration.

The jumps are not necessarily integers, if L/K is not abelian.

EXAMPLE 3.62. In [**47**, Sec. 4], an example of a totally ramified, finite Galois extension $L/\mathbb{Q}_2$ is constructed, for which the jumps of the upper numbering filtration of its Galois group are not integers. More precisely, the Galois group $G := \mathrm{Gal}(L/\mathbb{Q}_2)$ is the group of quaternions $\{\pm 1, \pm i, \pm j, \pm k\}$ with $(-1)^2 = 1$, $i^2 = j^2 = k^2 = ijk = -1$. Its center is $Z(G) = \{\pm 1\}$, and $L/\mathbb{Q}_2$ has the ramification filtration $G_0 = G_1 = G$, $G_2 = G_3 = Z(G)$, $G_n = \{1\}$, $n > 3$. It follows that the two jumps are 1 and 3/2.

THEOREM 3.63 (Hasse-Arf). *If the group G is* abelian, *then the jumps are integers.*

REMARK 3.64. We already know a few jumps of the ramification filtration: Since $\varphi_{L/K}(u) = u$ for $u \in [-1, 0]$, we see that the jumps in the filtration which happen in $[-1, 0]$ can only happen at -1 and 0. Thus, in the proof of the Hasse-Arf theorem, it suffices to treat the jumps > 0.

This deep theorem is fundamental for everything which we want to cover later on, so we will sketch a proof. For those who know local class field theory, the theorem of Hasse-Arf is easy to derive from it, at least in the case that the residue field of K is a finite field.

EXAMPLE 3.65. Fix a separable closure K^{sep} of K, and let K^{ab} be the compositum of all finite abelian extensions of K contained in K^{sep}. If the residue field of K is finite, local class field theory states among other things that there exists a continuous morphism $\theta : K^\times \to \mathrm{Gal}(K^{\mathrm{ab}}/K)$, the *local Artin map* or *local reciprocity map*, which becomes an isomorphism after profinite completion. Here, the profinite completion of the (non-discrete) topological group $K^\times$ is to be understood as $\varprojlim K^\times/U$, where U runs through the finite index, normal, open subgroups.

More relevantly for us, if L/K is a finite abelian extension, then the composition

$$K^\times \xrightarrow{\theta} \mathrm{Gal}(K^{\mathrm{ab}}/K) \twoheadrightarrow \mathrm{Gal}(L/K)$$

maps U_K^v onto $\mathrm{Gal}(L/K)^v$ for all real $v \in [0, \infty)$, see [**10**, Thm. 1, Ch. VI.4]. Here we use the definition $U_K^v := U_K^{\lceil v \rceil}$. The jumps of the filtration $\{U_K^v\}$ are obviously integers, hence the same is true for the upper numbering filtration of $\mathrm{Gal}(L/K)$, whence the Hasse-Arf theorem follows from local class field theory, if the residue field of K is finite.

Of course proving the existence of the local reciprocity map and establishing its properties is difficult. We will follow [**50**] and give a more elementary proof of the Hasse-Arf theorem for arbitrary residue fields.

The proof of the Hasse-Arf Theorem 3.63 is fairly long and intricate, so we first make our life easier by simplifying the situation.

PROPOSITION 3.66. *Recall that L/K is a finite Galois extension of complete discretely valued fields with separable residue extension. Additionally, assume that L/K is totally ramified and that the group $G = \mathrm{Gal}(L/K)$ is* cyclic.

If μ is the largest integer such that $G_\mu \neq 1$, then $\varphi_{L/K}(\mu)$ is an integer.

This special case of Theorem 3.63 actually implies the theorem.

LEMMA 3.67. *If Proposition* 3.66 *is true, then so is the Theorem* 3.63 *of Hasse-Arf.*

PROOF. Let L/K be a finite abelian extension with Galois group G and let $v \in [0,\infty)$ be a jump, i.e. $G^v \supsetneqq G^{v+\varepsilon}$ for all $\varepsilon > 0$. By Remark 3.64 we may assume that $v > 0$, hence $G^v \subsetneqq G^0 = G_0$. Replacing K by L^{G_0}, we may assume that L/K is totally ramified.

Let $\varepsilon_0 > 0$ be sufficiently small, so that $v + \varepsilon_0$ is smaller than the next jump in the filtration. Since G is a finite abelian group, so is $G/G^{v+\varepsilon_0}$, and the structure theorem for such groups tells us that there exists a cyclic group H, and a surjective map $\gamma : G \twoheadrightarrow H$, such that $\gamma(G^{v+\varepsilon_0}) = 1$, and $\gamma(G^v) \neq 1$.

The cyclic group H is the Galois group of a subextension L'/K of L/K, and by Proposition 3.53, we know that $H^u = \gamma(G^u)$ for all $u \in [-1,\infty)$. It follows that $H^{v+\varepsilon_0} = 1$ and $H^v \neq 1$, and that there are no jumps between v and $v+\varepsilon_0$. Finally, write $v = \varphi_{L'/K}(\mu)$. Since $H_\mu = H^{\varphi_{L'/K}(\mu)}$, by continuity of φ, it follows that $H_{\mu+\varepsilon} = 1$ for all $\varepsilon > 0$. This means that $\lceil \mu \rceil < \lceil \mu + \varepsilon \rceil$ for all $\varepsilon > 0$, and hence μ is an integer; in fact it is the largest integer such that $H_\mu \neq 1$. Now we are in the situation of Proposition 3.66, which implies, if true, that $v = \varphi_{L'/K}(\mu)$ is an integer. □

To prove the Hasse-Arf Theorem 3.63, we can thus concentrate on totally ramified, cyclic extensions L/K, which we will do in the following section.

Before we proceed with the proof of the Hasse-Arf Theorem, we continue Example 3.33.

EXAMPLE 3.68. Let $K = k((x))$ with k an algebraically closed field of characteristic $p > 0$. Let L be the Galois extension given by $L := K[t]/(t^p - t - x^{-m})$, where $(m,p) = 1$; its Galois group is $G = \mathbb{F}_p$. We saw in Example 3.33 that $G_0 = G_1 = \ldots = G_m \supsetneqq G_{m+1} = 0$. Let us find the jumps of L/K. For this we just have to compute $\varphi_{L/K}(m)$, which is easy because of (3.6):

$$\varphi_{L/K}(m) = \sum_{i=0}^{m} \frac{|G_i|}{|G_0|} - 1 = m.$$

We see that in this simple example, the conclusion of the Hasse-Arf theorem holds true.

3.10. The ramification filtration of a cyclic extension. In this section, let L/K be a totally ramified, finite cyclic Galois extension of complete discretely valued fields, G its Galois group and g a generator. Let V be the kernel of the norm $N : L^\times \to K^\times$. By Hilbert's Theorem 90 (Theorem 3.59) $V = \{\frac{gy}{y} | y \in L^\times\}$. Define

$$W := \left\{ x \in V | x = \frac{gy}{y} \text{ for some } y \in U_L \right\} \subseteq V,$$

where B is the valuation ring of L and $U_L = B^\times$. Clearly, W is a subgroup of V.

For $i \geq 0$, define $V_i := V \cap U_L^i$ and $W_i := W \cap U_L^i$. The quotients V_i/W_i define a filtration of the finite cyclic group V/W.

PROPOSITION 3.69. *Let $\pi \in B$ be a uniformizer and consider G to be filtered by the ramification filtration $\{G_i\}$ in the lower numbering.*

(a) *The assignment $\theta(\sigma) := \frac{\sigma\pi}{\pi}$ defines an isomorphism of groups $\theta : G \to V/W$.*
(b) *θ respects the filtrations, i.e. $\theta|_{G_i} : G_i \hookrightarrow V_i/W_i$ for all $i \geq 0$.*

PROOF. The following argument partly follows lecture notes by I. Fesenko. First, note that for any $\sigma \in G$, $\theta(\sigma) = \frac{\sigma(\pi)}{\pi} \in V$. Indeed, $v_L(\theta(\sigma)) = v_L(\sigma(\pi)) - v_L(\pi) = 0$, so $\theta(\sigma) \in U_L$. Also,

$$N_{L/K}(\theta(\sigma)) = \prod_{\tau \in G} \frac{\tau\sigma(\pi)}{\tau(\pi)} = \frac{\prod_{\tau \in G} \tau\sigma(\pi)}{\prod_{\tau \in G} \tau(\pi)} = 1.$$

Let us see that θ is a homomorphism of groups: For any $\tau, \sigma \in G$ we have

$$\frac{\tau\sigma(\pi)}{\pi} \cdot \frac{\pi^2}{\sigma(\pi)\tau(\pi)} = \frac{\sigma\left(\frac{\tau(\pi)}{\pi}\right)}{\frac{\tau(\pi)}{\pi}},$$

and $\frac{\tau(\pi)}{\pi} \in U_L$. Hence $\theta(\sigma\tau) \equiv \theta(\sigma)\theta(\tau) \mod W$.

To see that θ is injective, assume that $\frac{g^r(\pi)}{\pi} \in W$. As we just saw that θ is a homomorphism, this means that there exists $u \in U_L$ such that $\theta(g^r) = \theta(g)^r = \left(\frac{g(\pi)}{\pi}\right)^r = \frac{g(u)}{u}$. But this means $g\left(\frac{\pi^r}{u}\right) = \frac{\pi^r}{u}$, so $\pi^r/u \in K$. Hence the degree $[L:K]$ divides r, so $g^r = 1$, and θ is injective.

For the surjectivity, note that every element in V has the shape $\frac{g(u\pi^i)}{u\pi^i}$, for $u \in U_L$ and $i \in \mathbb{Z}$. But in V/W we have $\frac{g(u\pi^i)}{u\pi^i} = \frac{g(\pi^i)}{\pi^i} = \theta(g^i)$.

Finally, if $\sigma \in G_i$, then $\frac{\sigma(\pi)}{\pi} \in U_L^i$ by definition, so θ indeed induces injective morphisms $\theta_i : G_i \to V_i/W_i$. □

PROPOSITION 3.70. *In addition to the assumptions and notations of Proposition 3.69, assume that the residue field of K is not the prime field $\mathbb{F}_p$. Then $G_m = 0$ if and only if $V_m/W_m = 0$.*

PROOF. We have seen that θ restricts to an injective map $G_m \hookrightarrow V_m/W_m$, so it just remains to show that $V_m/W_m = 0$ if $G_m = 0$. We proceed by decending induction on m. We know that $G_m = 0$ for $m \gg 0$, and the same is true for V_m/W_m. Indeed, as L/K is separable, there exists $t \in L$ such that $\mathrm{Tr}_{L/K}(t) = 1$ (Proposition 3.14). Write $M := \max\{0, -v_L(t)\}$. We claim that $V_m = W_m$ for all $m > M$. Let

$x \in V_m$, and write

$$y := t + \sum_{i=1}^{|G|-1} xg(x)g^2(x)\cdot\ldots\cdot g^{i-1}(x)g^i(t).$$

Subtracting $\mathrm{Tr}_{L/K}(t) = 1$ from both sides we obtain

$$y - 1 = \sum_{i=1}^{|G|-1} (xg(x)g^2(x)\cdot\ldots\cdot g^{i-1}(x) - 1)g^i(t),$$

and hence $v_L(y-1) > 0$. It follows that $y \in U_L^1$, and that $\frac{y}{g(y)} = x$, since $N_{L/K}(x) = 1$:

$$\begin{aligned}\frac{y}{g(y)} &= \frac{xy}{xg(y)}\\ &= \frac{xy}{\sum_{i=1}^{|G|-1} xg(x)\cdot\ldots\cdot g^{i-1}(x)g^i(t) + N_{L/K}(x)t}\\ &= x.\end{aligned}$$

Thus $x = \frac{y}{g(y)} \in W_m$, so $W_m = V_m$ for $m > M$.

Now let $G_m = 0$, and assume that we know that $V_{m+1}/W_{m+1} = 0$. Our goal is to show that $V_m/W_m = 0$.

We will actually prove something stronger:

LEMMA 3.71. *We continue to use the notations and assumptions from Proposition* 3.70. *In particular, the residue field of K is not the prime field. Write $n + 1 := \lceil \varphi_{L/K}(m)\rceil$. If $V_{m+1}/W_{m+1} = 0$ and $G_{\psi_{L/K}(n+1)} = 0$, then $V_m/W_m = 0$.*

PROOF. Note that $\psi_{L/K}(n+1)$ is an integer $\geq m$ (Exercise 3.49), with equality if and only if $\varphi_{L/K}(m)$ is an integer.

We treat these two cases separately: $\varphi_{L/K}(m) \in \mathbb{Z}$ and $\varphi_{L/K}(m) \notin \mathbb{Z}$.

First, let us assume that $\varphi_{L/K}(m) \in \mathbb{Z}$, i.e. $\varphi_{L/K}(m) = n + 1$. Proposition 3.57 then produces a commutative diagram

$$\begin{array}{ccccccccc}
 & & 0 & & 0 & & 0 & & \\
 & & \downarrow & & \downarrow & & \downarrow & & \\
0 & \longrightarrow & V_{m+1} & \longrightarrow & V_m & \longrightarrow & G_m/G_{m+1} & & \\
 & & \downarrow & & \downarrow & & \downarrow{\scriptstyle \theta_m} & & \\
0 & \longrightarrow & U_L^{m+1} & \longrightarrow & U_L^m & \longrightarrow & U_L^m/U_L^{m+1} & \longrightarrow & 0\\
 & & \downarrow{\scriptstyle N} & & \downarrow{\scriptstyle N} & & \downarrow{\scriptstyle N_m} & & \\
0 & \longrightarrow & U_K^{n+2} & \longrightarrow & U_K^{n+1} & \longrightarrow & U_K^{n+1}/U_K^{n+2} & \longrightarrow & 0.
\end{array}$$

As $V_{m+1}/W_{m+1} = 0$ by assumption, we see that the composition

$$V_m \to G_m/G_{m+1} \to (V_m/W_m)/(V_{m+1}/W_{m+1}) = V_m/W_m$$

is the canonical quotient map. But $G_m = G_{\psi_{L/K}(n+1)} = 0$, so $V_m/W_m = 0$.

Now assume that $\varphi_{L/K}(m) \notin \mathbb{Z}$, i.e.

$$n < \varphi_{L/K}(m) < n+1 \text{ and } \psi_{L/K}(n) + 1 \leq m < \psi_{L/K}(n+1).$$

We have $V_m/V_{m+1} \subseteq U_L^m/U_L^{m+1}$, and by assumption $V_{m+1} = W_{m+1}$. Thus, to show that $V_m/W_m = 0$, it suffices to show that V_m and W_m have the same image in U_L^m/U_L^{m+1}.

Let $x \in U_L^m$ be arbitrary. We have the inclusion $U_L^{\psi_{L/K}(n+1)} \subseteq U_L^m$, and the norm induces a surjective morphism $N : U_L^{\psi_{L/K}(n+1)} \twoheadrightarrow U_K^{n+1}$ by Proposition 3.57. It follows that there exists $y \in U_L^{\psi_{L/K}(n+1)}$ with $N(y) = N(x)$. Then $N(xy^{-1}) = 1$ and hence $xy^{-1} \in V_m$. As $\psi_{L/K}(n+1) \geq m+1$, we see that $y \in U_L^{m+1}$, so $x \equiv xy^{-1}$ mod U_L^{m+1}. It follows that V_m maps surjectively to U_L^m/U_L^{m+1}.

As for the image of W_m, from what we showed in the previous paragraph, it follows that V_m/W_m maps surjectively to $(U_L^m/U_L^{m+1})/W_m$, so this group is also cyclic. But for $m > 0$, the group U_L^m/U_L^{m+1} is isomorphic to the additive group underlying the residue field of K, which is not cyclic, as the residue field of K is assumed not to be the prime field. Thus the image of W_m in U_L^m/U_L^{m+1} is nontrivial.

Let $x \in W_m \smallsetminus U_L^{m+1}$. Then there exists $y \in U_L$ such that $x = \frac{g(y)}{y}$. By modifying y with an element from U_K, we may assume that $y \in U_L^1$, i.e. $y = 1 + z$, $v_L(z) > 0$. For $a \in U_K$ define $y_a := 1 + az$ and $x_a := \frac{g(y_a)}{y_a}$.

Denote by $\gamma : U_L^{m+1}/U_L^m \to \mathfrak{m}_L^m/\mathfrak{m}_L^{m+1}$ the isomorphism given by $\gamma(1 + u\pi^m) = u\pi^m \mod \mathfrak{m}_L^{m+1}$, where π is a uniformizer of L and $\mathfrak{m}_L$ the maximal ideal of the discrete valuation ring of L. We claim that $x_a \in W_m$, and that $\gamma(x_a) = a\gamma(x)$ mod $\mathfrak{m}_L^{m+1}$. Since $\mathfrak{m}_L^m/\mathfrak{m}_L^{m+1}$ is a 1-dimensional vector space over the residue field of K, this would show that W_m maps surjectively to U_L^m/U_L^{m+1}, which is what we wanted to show.

As for the claim, we compute

$$\begin{aligned} x_a - 1 &= \frac{g(y_a) - y_a}{y_a} \\ &= \frac{1 + ag(z) - (1 + az)}{y_a} \\ &= \frac{a(gz - z)}{y_a} \\ &= a\frac{g(y) - y}{y_a} \\ &= a\frac{y}{y_a}(x - 1). \end{aligned}$$

It follows that $v_L(x_a - 1) = v_L(a) + m \geq m$, so $x_a \in W_m$. Finally, to see that $\gamma(x_a) = a\gamma(x)$, it suffices to remark that $\frac{y}{y_a} \in U_L^1$. □

□

REMARK 3.72. If the residue field of K is algebraically closed, almost the same proof shows that the isomorphism $\theta : G \xrightarrow{\cong} V/W$ induces isomorphisms $G_i \xrightarrow{\cong} V_i/W_i$ for all i.

Finally we can prove Proposition 3.66 and hence the Hasse-Arf theorem 3.63.

PROOF OF PROPOSITION 3.66. Recall that L/K is a totally ramified, cyclic extension of complete discretely valued fields with separable residue extension. If the residue field of K is the prime field, we can replace K by an unramified extension with larger residue field ([**50**, V, §4]). This does not change G_i for $i \geq 0$. If μ is the largest positive integer such that $G_\mu \neq 0$, then we want to show that $\varphi_{L/K}(\mu)$ is an integer. If it is not an integer, then there exists an integer ν, such that $\nu < \varphi_{L/K}(\mu) < \nu + 1$, and hence $\mu < \psi_{L/K}(\nu + 1)$. It follows that $G_{\psi_{L/K}(\nu+1)} = 0$. Now Lemma 3.71 applies, and thus $V_\mu/W_\mu = 0$. But we have seen in Proposition 3.69 that $\theta : G \xrightarrow{\cong} V/W$ respects the filtrations, so $G_\mu \hookrightarrow V_\mu/W_\mu = 0$, which is a contradiction. Thus $\varphi_{L/K}(\mu) \in \mathbb{Z}$, and the proof is complete. □

4. The Swan representation

Let K be a complete discretely valued field with perfect residue field and K^{sep} a separable closure. If ℓ is a prime number different from the residue characteristic of K and E a finite extension of $\mathbb{Q}_\ell$, then we want to study continuous morphisms $\rho : \mathrm{Gal}(K^{\mathrm{sep}}/K) \to \mathrm{GL}_r(E)$, where $\mathrm{GL}_r(E)$ is considered as a topological group with the topology induced by E (see Definition 4.66).

In Section 4.4 will define the *Swan conductor* $\mathrm{sw}(\rho) \in \mathbb{Z}$ of such a representation, which is additive in short exact sequences and zero if and only if ρ is *tame*, i.e. if ρ restricts to the trivial representation on the wild ramification group. We will see in Section 10 that this invariant has a cohomological interpretation. Before we come to the Swan conductor, we will use the theorem of Hasse-Arf to attach to a finite Galois extension L/K with group G, first a complex representation, then a projective $\mathbb{Z}_\ell[G]$-module Sw_G, the *Swan representation.*

4.1. A minimal amount of representation theory of finite groups. We roughly follow [**49**], but [**32**] and [**53**] are also helpful.

4.1.1. *Basics.* Let G be a finite group, E a field and V a finite dimensional E-vector space. A morphism $\rho : G \to \mathrm{GL}(V)$ is called a *representation of* E. If $\rho' : G \to \mathrm{GL}(V')$ is a second representation, then a *morphism of representations* $\alpha : \rho \to \rho'$ is an E-linear map $\varphi : V \to V'$, such that for every $g \in G$, $v \in V$, we have $\varphi(\rho(g)(v)) = \rho'(g)(\varphi(v))$. We write $\mathrm{Repf}_E\, G$ for the category of representations of G on finite dimensional E-vector spaces. This category is equivalent to the category of finitely generated left modules over the group ring $E[G]$. Recall that $E[G] = E \otimes_{\mathbb{Z}} \mathbb{Z}[G]$, where $\mathbb{Z}[G]$ is the free $\mathbb{Z}$-module with basis G, and multiplication induced by the rule $ag \cdot bg' = (ab)gg' \in \mathbb{Z}[G]$, for $a, b \in \mathbb{Z}$. From this perspective it is clear that we have the usual notions of subrepresentations, quotients and constructions like direct sums, products, etc. In the sequel, we will use the notions of "representation" and "finitely generated $E[G]$-module" interchangeably. Unless explicitly stated otherwise, the word "$E[G]$-module" means "finitely generated left-$E[G]$-module".

Let $\varphi : H \to G$ be a homomorphism of finite groups, and $\psi : E[H] \to E[G]$ the induced morphism of E-algebras. If V is a left-$E[G]$-module, then using ψ we can also consider it as an left-$E[H]$-module. The corresponding representation of H is called the *restriction of* V *to* H, *denoted* $\mathrm{Res}_\varphi V$. Similarly, if W is a left-$E[H]$-module, and if we consider $E[G]$ as a right-$E[H]$-module via φ, then the left-action of $E[G]$ on itself makes $E[G] \otimes_{E[H]} W$ into a left-$E[G]$-module. The corresponding representation of G is said to be *the representation induced by* φ *and* W, *denoted* $\mathrm{Ind}_\varphi W$. These constructions define functors $\mathrm{Ind}_\varphi : \mathrm{Repf}_E\, H \to \mathrm{Repf}_E\, G$, and $\mathrm{Res}_\varphi : \mathrm{Repf}_E\, G \to \mathrm{Repf}_E\, H$. Considering the module theoretic definitions of these constructions, we see that these functors are actually adjoint:

$$\mathrm{Hom}_{\mathrm{Repf}_E H}(-, \mathrm{Res}_\varphi(-)) = \mathrm{Hom}_{\mathrm{Repf}_E G}(\mathrm{Ind}_\varphi(-), -). \tag{4.1}$$

Here are some important examples of representations.

EXAMPLE 4.1.

(a) The 1-dimensional vector space E with G-action given by $ge = e$ for all $g \in G$, $e \in E$. This is the *trivial representation of rank* 1, which we denote by $\mathbf{1}_G$ or just E. If $\rho : G \to \mathrm{GL}(V)$ is a finite dimensional E-representation of G, such that $\rho(g) = \mathrm{id}$ for all $g \in G$, then ρ is isomorphic to a direct sum of finitely many copies of $\mathbf{1}_G$. Such a representation ρ is called *trivial*. (The trivial $E[G]$-module is the trivial G-representation of rank 0.)

If V is an $E[G]$-module, we write $V^G := \{v \in V | \forall g \in G : gv = v\}$. This is the maximal trivial subrepresentation of V.

(b) The ring $E[G]$ is a left-$E[G]$-module. The corresponding representation is called the *regular representation* of G over E.

(c) We have a surjective morphism of $E[G]$-modules $E[G] \to E$, $g \mapsto 1$. In other words: The trivial representation of rank 1 is a quotient of the regular representation. Its kernel is the ideal $\mathrm{Aug}_E(G) := (g - 1 | g \in G)$, and it is called the *augmentation representation*.

(d) If the order of G is invertible in E, then the short exact sequence

$$0 \to \mathrm{Aug}_E(G) \to E[G] \to E \to 0$$

is split by the map $\theta : E \to E[G]$, $\theta(e) = \frac{e}{|G|} \sum_{g \in G} g$.

(e) If V, W are two finite dimensional representations of G over E, then $\mathrm{Hom}_E(V, W)$ has a canonical structure of a representation of G: If $\varphi \in \mathrm{Hom}_E(V, W)$ and $g \in G$, then define $g \bullet \varphi$ by $(g \bullet \varphi)(v) = g\varphi(g^{-1}v)$. Clearly $\mathrm{Hom}_E(V, W)^G = \mathrm{Hom}_{E[G]}(V, W)$.

We write $\underline{\mathrm{Hom}}(V, W)$ for the representation on $\mathrm{Hom}_E(V, W)$ just defined.

(f) If V is a finite dimensional representation of G, then $\underline{\mathrm{Hom}}(V, \mathbf{1}_G)$ is a finite dimensional representation of G, the *dual representation* $V^\vee$ *of* ρ. The underlying vector space is the dual vector space $\mathrm{Hom}_E(V, E)$ of V and G acts via $(g \bullet \varphi)(v) = \varphi(g^{-1}v)$.

(g) If V and W are finite dimensional representations of G, then the tensor product $V \otimes_E W$ carries a canonical G-action, defined via $g(v \otimes w) = gv \otimes gw$ for $g \in G$, $v \in V$, $w \in W$. Using this definition we see that the natural isomorphism of E-vector spaces $V^\vee \otimes_E W \to \mathrm{Hom}_E(V, W)$ is G-invariant, and hence induces a natural isomorphism of representations

$$V^\vee \otimes W \cong \underline{\mathrm{Hom}}(V, W).$$

Similarly we see that if U is a third representation of G, then there is a natural isomorphism

$$\underline{\mathrm{Hom}}(U \otimes V, W) \cong \underline{\mathrm{Hom}}(U, \underline{\mathrm{Hom}}(V, W)).$$

DEFINITION 4.2. A representation V of G over E is called

- *irreducible* or *simple* if it is irreducible as $E[G]$-module, i.e., if it is nonzero and does not have nontrivial sub-$E[G]$-modules.
- *semi-simple* if V is a direct sum of irreducible $E[G]$-modules.

- *decomposable* if $V \cong V_1 \oplus V_2$ with $V_1 \ncong V$ and $V_2 \ncong V$.
- *indecomposable* if V is not decomposable.

THEOREM 4.3 (Maschke). *If G is a finite group with order coprime to* $\mathrm{char}(E)$, *then every representation of G on a finite dimensional E-vector space is semi-simple.*

PROOF. Let V be a finite dimensional E-representation of G, and $W \subseteq V$ a subrepresentation. Pick a projection $P : V \twoheadrightarrow W$ of E-vector spaces, which splits the inclusion $W \subseteq V$, i.e. such that $P|_W = \mathrm{id}_W$, and define

$$P_0 := \frac{1}{|G|} \sum_{g \in G} gPg^{-1}.$$

It is easy to check that P_0 is a morphism of $E[G]$-modules, and we still have $P_0|_W = \mathrm{id}_W$. It follows that $\ker(P_0)$ is a subrepresentation of V, and $V = W \oplus \ker(P_0)$ as $E[G]$-modules. □

To see the contrast to the situation where $|G|$ and $\mathrm{char}(E)$ are not coprime, do the following exercise.

EXERCISE 4.4. Let $p = \mathrm{char}(E)$. If G is a finite group, then the two cases $(|G|, p) = 1$ and $(p, |G|) > 1$ are very different.

As a "worst case" scenario, prove that if $p > 0$ and if G is a p-group acting on a finite dimensional E-vector space $V \neq 0$, then $V^G \neq 0$. Here $V^G = \{v \in V | \forall g \in G : gv = v\}$.

Deduce that V is a successive extension of the trivial rank 1 representation of G by itself.

PROPOSITION 4.5. *Let E be a field containing all roots of unity of order dividing $|G|$. If G is abelian, then every irreducible representation of G over E has rank* 1. *If $|G|$ is invertible in E, then the converse is also true.*

PROOF. Let $\rho : G \to \mathrm{GL}(V)$ be a nonzero irreducible representation of G. For $g \in G$, the order of $\rho(g)$ divides $|G|$. Let $v \in V$ be an eigenvector of $\rho(g)$ and λ its eigenvalue. Then λ is a root of unity of order dividing $|G|$, thus $\lambda \in E$ by assumption. For any other $h \in G$ we have $\rho(gh)(v) = \rho(hg)(v) = \lambda\rho(h)(v)$. Thus the λ-eigenspace of $\rho(g)$ is a nonzero subrepresentation of ρ, hence equal to ρ as ρ is irreducible. This shows that every element $g \in G$ acts via scalar multiplication on V. As ρ is irreducible, $\dim V = 1$.

If $|G|$ is invertible in E, let ρ be a faithful representation of G, i.e. an injective map $\rho : G \to \mathrm{GL}(V)$ (such a representation always exists). If the irreducible representations of G have rank 1, Maschke's Theorem shows that the matrices $\{\rho(g) | g \in G\}$ are simultaneously diagonizable. It follows that G is abelian. □

It is convenient to introduce the following notation.

DEFINITION 4.6. If G is a finite group and E a field, we write $R_E(G)$ for the *Grothendieck ring* of the category $\mathrm{Repf}_E\, G$. Recall that $R_E(G)$ is the abelian group generated by isomorphism classes $[V]$ of finite dimensional E-representations of G, subject to the relations $[V] = [V_1] + [V_2]$ if there exists a short exact sequence

$$0 \to V_1 \to V \to V_2 \to 0.$$

The tensor product over E makes $R_E(G)$ into a commutative ring (*Exercise*: Check this!). It is easy to see that $R_E(G)$ is the free $\mathbb{Z}$-module generated by the isomorphism classes of irreducible representations of G. We will see in Proposition 4.7 that there are only finitely many of those.

There is a natural map of monoids

$$(\{\text{isomorphism classes of objects in } \mathrm{Repf}_E G\}, \oplus) \to R_E(G). \tag{4.2}$$

Its image is the subset of all linear combinations $\sum_{V_i \text{irreducible}} a_i[V_i]$, $a_i \geq 0$, and is denoted by $R_E^+(G)$.

Maschke's Theorem 4.3 can be phrased as follows: If the order G is prime to the characteristic of E, then the map (4.2) is injective.

Next, we see that there are only finitely many irreducible representations of a finite group.

PROPOSITION 4.7. *If G is a finite group and E a field, then any irreducible finite dimensional representation of G on E is a composition factor of the regular representation $E[G]$. In particular, up to isomorphism there are only finitely many irreducible representations of G on E.*

PROOF. As $E[G]$ and all its subrepresentations are finite dimensional E-vector spaces, $E[G]$ admits a finite composition series and thus, up to isomorphism, $E[G]$ has only finitely many composition factors due to the Jordan-Hölder Theorem (see e.g. [**8**, I, §4.7, Thm. 6]).

If V is an irreducible $E[G]$-module, pick any $v \in V \setminus \{0\}$. The map $x \mapsto xv$, is an $E[G]$-linear morphism $E[G] \to V$, which is surjective as V is irreducible. Thus V is a composition factor of $E[G]$. □

We close this section with a fundamental but easy lemma.

LEMMA 4.8 (Schur's Lemma). *Let V_1, V_2 be two finite dimensional* irreducible *$E[G]$-modules. If $f : V_1 \to V_2$ is a morphism of representations, then precisely one of the following statements is true:*

(a) $f = 0$.

(b) f is an isomorphism and if E is algebraically closed, then for any other isomorphism $g : V_1 \xrightarrow{\cong} V_2$, there exists $a \in E$, such that $f = ag$.

In particular, if V is an irreducible $E[G]$-module, then $\mathrm{End}_{E[G]}(V)$ *is a division algebra, and* $\mathrm{End}_{E[G]}(V) = E$ *if E is algebraically closed.*

PROOF. If f is not an isomorphism, then $f = 0$, due to the irreducibility of V_1, V_2. Now assume that f and g are isomorphisms of $E[G]$-modules and that E is algebraically closed. Let a be an eigenvalue of $g^{-1}f : V_1 \to V_1$. Then $g^{-1}f - a\,\mathrm{id}_{V_1}$ has a nonzero kernel, so $g^{-1}f = a \cdot \mathrm{id}_{V_1}$ as claimed. □

We give an exemplary application:

PROPOSITION 4.9. *Let G be a group and E a field such that $|G|$ is invertible in E. If V is a representation of G such that* $\mathrm{End}_{E[G]}(V) = E$, *then V is irreducible. If E is algebraically closed, the converse is true as well.*

PROOF. Since $|G|$ is invertible V can be written as $V = V_1 \oplus V_2$ with V_1 irreducible. We compute

$$\dim_E \mathrm{End}_{E[G]}(V) \geq \dim_E \mathrm{End}_{E[G]}(V_1) + \dim_E \mathrm{End}_{E[G]}(V_2) \geq 2$$

if $V_2 \neq 0$. It follows that $V = V_1$ if $\mathrm{End}_{E[G]}(V) = E$. The converse is just Schur's Lemma. □

4.1.2. *Extension of scalars.* If $E \subseteq F$ is an extension of fields, then $- \otimes_E F$ gives rise to a functor $\mathrm{Repf}_E(G) \to \mathrm{Repf}_F(G)$. In this section, we analyze this functor.

LEMMA 4.10. *Let $E \subseteq F$ be an extension of fields and G a finite group. If V is an $E[G]$-module, then $V \otimes_E F = F[G] \otimes_{E[G]} V$ is an $F[G]$-module. The natural map*

$$V^G \otimes_E F \to (V \otimes_E F)^G$$

is an isomorphism. Consequently, if W is a second $E[G]$-module, the natural map

$$\mathrm{Hom}_{E[G]}(V, W) \otimes_E F \cong \mathrm{Hom}_{F[G]}(V \otimes_E F, W \otimes_E F)$$

is an isomorphism.

PROOF. Since F is flat over E, the natural map of F-vector spaces

$$\mathrm{Hom}_E(V, W) \otimes_E F \to \mathrm{Hom}_F(V \otimes_E F, W \otimes_E F),$$

is an isomorphism. It is easily seen to be G-equivariant, so $\underline{\mathrm{Hom}}(V, W) \otimes_E F \cong \underline{\mathrm{Hom}}(V \otimes_E F, W \otimes_E F)$ as $F[G]$-modules, and the second statement of the lemma follows from the first as $\mathrm{Hom}_{E[G]}(V, W) = \underline{\mathrm{Hom}}(V, W)^G$.

For the first statement, consider the E-linear map $\varphi : V \to \bigoplus_{g \in G} V$, $v \mapsto (gv - v)_{g \in G}$. Its kernel is the E-vector space V^G. Tensoring with F gives the map

$$\varphi \otimes \mathrm{id}_F : V \otimes_E F \to (V \otimes_E F)^{|G|}, \quad v \otimes \lambda \mapsto (g(v \otimes \lambda) - v \otimes \lambda)_{g \in G}$$

Its kernel is $(V \otimes_E F)^G$, but it is also $V^G \otimes_E F$, as F is flat over E. □

PROPOSITION 4.11. *Let $E \subseteq F$ be an arbitrary field and G a finite group. The ring homomorphism $\varphi : R_E(G) \to R_F(G)$ induced by $- \otimes_E F$ is injective.*

PROOF. Recall that $R_E(G)$ (resp. $R_F(G)$) is the free abelian group generated by the classes of irreducible $E[G]$-modules (resp. classes of irreducible $F[G]$-modules). If $c \in R_E(G)$ we can write $c = c_+ - c_-$ with $c_+, c_- \in R_E^+(G)$, where $R_E^+(G)$ is the submonoid of $\mathbb{Z}_{\geq 0}$-linear combinations of classes of irreducible $E[G]$-modules. We have $\varphi(c) = 0$ if and only if $\varphi(c_+) = \varphi(c_-)$. Thus it suffices to show that the map $R_E^+(G) \to R_F^+(G)$ is injective.

Let V_1, V_2 be irreducible finitely generated $E[G]$-modules. If $\mathrm{char}(E) = 0$, then by Maschke's Theorem 4.3 the $F[G]$-modules $V_1 \otimes_E F$ and $V_2 \otimes_E F$ are semi-simple, i.e. direct sums of irreducible $F[G]$-modules.

If $V_1 \not\cong V_2$, then the decompositions over F have no common factors. Indeed, otherwise Lemma 4.10 would imply that

$$0 = \dim_E \mathrm{Hom}_{E[G]}(V_1, V_2) = \dim_F \mathrm{Hom}_{F[G]}(V_1 \otimes_E F, V_2 \otimes_E F) > 0.$$

This shows that $R_E^+(G) \to R_F^+(G)$ is injective, if $\mathrm{char}(E) = 0$.

If $\mathrm{char}(E) > 0$, we have to generalize the argument a little ([**32**, (7.13)]): If $V_1 \not\cong V_2$ are irreducible, then $V_1 \otimes_E F$ and $V_2 \otimes_E F$ are perhaps not semi-simple, but we show that their composition series have no common factor. To this end, we show that there is an element $e \in E[G]$ acting as the identity on V_1 and as 0 on V_2. Then $e \otimes 1$ acts as identity on all composition factors of $V_1 \otimes_E F$ and as 0 on all composition factors of $V_2 \otimes_E F$, which would complete the proof of the proposition.

Every simple $E[G]$-module is of the shape $E[G]/M$, where $M \subseteq E[G]$ is a maximal left-ideal. Indeed, if V is a simple $E[G]$-module, pick $v \in V \setminus \{0\}$. The map $\varphi_v : E[G] \to V$, $x \mapsto xv$ is surjective and its kernel is maximal, as V is simple. Let $\mathrm{rad}(E[G])$ denote the intersection of all maximal left-ideals. If $x \in \mathrm{rad}(E[G])$, and if V is a simple $E[G]$-module, then for any $v \in V$, $x \in \ker(\varphi_v)$, so x acts trivially on every simple $E[G]$-module. Conversely, if $x \in E[G]$ acts trivially on every simple $E[G]$-module, then $x \in \mathrm{rad}(E[G])$, as for every maximal left-ideal M, x must map to 0 in the simple module $E[G]/M$.

Thus $\mathrm{rad}(E[G])$ is the set of elements acting trivially on every simple $E[G]$-module. This shows that $\mathrm{rad}(E[G])$ is a two-sided ideal, so that $E[G]/\mathrm{rad}(E[G])$ is a ring. As a left-$E[G]$-module, $E[G]/\mathrm{rad}(E[G])$ is semi-simple; in fact it is the direct sum of all simple $E[G]$-modules, with multiplicities (see Lemma 4.56). If, say,

$$E[G]/\mathrm{rad}(E[G]) = V_1 \oplus \ldots \oplus V_n$$

with $V_1, \ldots, V_n$ simple $E[G]$-modules (hence left-ideals of $E[G]/\mathrm{rad}(E[G])$), then there are orthogonal idempotents $e_1, \ldots, e_n \in E[G]/\mathrm{rad}(E[G])$, such that $1 = e_1 + \ldots + e_n$, and such that for every $j = 1, \ldots, n$, e_j acts as the identity on V_j, but as 0 on all V_i with $i \neq j$. If $f_j \in E[G]$ is a lift of e_j, then f_j acts as identity on V_j and as 0 on V_i, $i \neq j$. The proof is complete. □

DEFINITION 4.12. Let G be a finite group and E a field. We say that a representation V of G on E is *absolutely irreducible* if $V \otimes_E F$ is irreducible for all fields $F \supseteq E$.

If $|G|$ is invertible in E, then we can easily give a criterion for absolute irreducibility:

PROPOSITION 4.13. *Let G be a group and E a field such that $|G|$ is invertible in E. The following statements for a representation V of G on E are equivalent:*

(a) $\mathrm{End}_{E[G]}(V) = E$
(b) *V is absolutely irreducible.*
(c) *There exists an algebraically closed field $F \supseteq E$ such that $V \otimes_E F$ is irreducible.*

PROOF. Assume (a) and let F be an extension of E. Then $\mathrm{End}_{F[G]}(V \otimes_E F) = F$ by Lemma 4.10, so $V \otimes_E F$ is irreducible according to Proposition 4.9, so V is absolutely irreducible.

(b) trivially implies (c). Assume (c) holds. Then $\dim_F \mathrm{End}_{F[G]}(V \otimes_E F) = 1$ according to Proposition 4.9, so again we see that $\mathrm{End}_{E[G]}(V) = E$. This completes the proof. □

REMARK 4.14. An analogous criterion is true even if $|G|$ is not invertible in E, but the proof is more complicated, see e.g. [**32**, Thm. 7.5].

DEFINITION 4.15. We say that E is a *splitting field* for G, if every irreducible representation of G over E is absolutely irreducible.

PROPOSITION 4.16. *If G is a finite group and E a splitting field, then any field $F \supseteq E$ is a splitting field. In this case the map $R_E(G) \to R_F(G)$ is an isomorphism.*

PROOF. Assume that E is a splitting field. If $0 \subsetneq C_1 \subsetneq \ldots \subsetneq C_n = E[G]$ is a composition series, then

$$0 \subsetneq C_1 \otimes_E F \subsetneq \ldots \subsetneq C_n \otimes_E F = F[G]$$

is a composition series of $F[G]$, as $(C_i \otimes_E F)/(C_{i-1} \otimes_E F) = (C_i/C_{i-1}) \otimes_E F$ is (absolutely) irreducible for $i = 1, \ldots, n$ by assumption. If W is an irreducible $F[G]$-module, then $W \cong (C_i/C_{i-1}) \otimes_E F$ for some i according to Proposition 4.7, so W is absolutely irreducible and F a splitting field. Moreover, the class of W in $R_F(G)$ lies in the image of the map $R_E(G) \hookrightarrow R_F(G)$. □

Proposition 4.13 translates to:

COROLLARY 4.17. *Let E be a field and G a group such that $|G|$ is invertible in E. If E is algebraically closed, then E is a splitting field.*

REMARK 4.18. The same statement is true if $|G|$ is not necessarily invertible, but the proof is a little bit more complicated, see [**32**, Thm. 8.3].

PROPOSITION 4.19. *Let G be a finite group and E a field. If E is a splitting field (e.g. algebraically closed), then every irreducible representation of G on E comes via base change from a finite extension E_0 of the prime field.*

PROOF. Write $\mathbb{F}$ for the prime field of E. Let $\overline{E}$ be an algebraic closure of E and V an irreducible representation of G over E. If E is a splitting field, then $V \otimes \overline{E}$ is also irreducible, so it comes via base change from $\overline{\mathbb{F}} \subseteq \overline{E}$ according to Proposition 4.16, and hence from a finite extension $\mathbb{F} \subseteq E_V$ contained in $\overline{\mathbb{F}}$. Let E_0 be the compositum of the fields E_V in $\overline{\mathbb{F}}$. As there are only finitely many isomorphism classes of irreducible representations of G (Proposition 4.7), E_0 is finitely generated over $\mathbb{F}$. □

REMARK 4.20. If $\mathrm{char}(E) = 0$, we will see in Corollary 4.43 that one can always take $E_0 = \mathbb{Q}(\zeta_m)$ as a splitting field, where m is the exponent of G and ζ_m an m-th root of unity. Moreover, the functor $\mathrm{Repf}_{\mathbb{Q}(\zeta_m)}(G) \to \mathrm{Repf}_{\overline{E}}(G)$ is an equivalence because of Maschke's Theorem and Lemma 4.10.

If $\mathrm{char}(E) = p > 0$, then one can show that the field $\mathbb{Z}[\zeta_m]/\mathfrak{p}$ is a splitting field for G, where $\mathfrak{p}$ is any prime ideal of $\mathbb{Z}[\zeta_m]$ lying over p.

4.1.3. *Characters.* We continue to denote by E a field and by G a finite group.

DEFINITION 4.21. Let G be a finite group. A map $\varphi : G \to E$ satisfying the condition

$$\varphi(hgh^{-1}) = \varphi(g)$$

for all $g, h \in G$, is called *class function.*

If $\rho : G \to \mathrm{GL}(V)$ is a representation of G on a finite dimensional E-vector space, then $\chi_\rho : G \to E$, $\chi_\rho(g) := \chi_V(g) := \mathrm{Tr}(\rho(g))$ is a class function; it is called the *character of* ρ.

If ρ is irreducible, then its character is also called *irreducible.*

LEMMA 4.22. *If $\rho : G \to \mathrm{GL}(V)$ and $\rho' : G \to \mathrm{GL}(V')$ are isomorphic representations over E of the finite group G, then $\chi_\rho = \chi_{\rho'}$.*

PROOF. If $\gamma : V \to V'$ is a G-invariant E-isomorphism, then $\rho(g) = \gamma\rho'(g)\gamma^{-1}$, and $\mathrm{Tr}(\rho'(g)) = \mathrm{Tr}(\gamma\rho'(g)\gamma^{-1})$. □

EXAMPLE 4.23. (a) Let V_1, V_2 be two $E[G]$-modules with characters χ_1, χ_2.

- If V is an $E[G]$-module with character χ, which is an extension of V_1 by V_2, then $\chi = \chi_1 + \chi_2$.

- The character of $V_1 \otimes V_2$ is $\chi_1 \cdot \chi_2$.
- If $\chi_1^\vee$ is the character of the dual representation $V_1^\vee$, then we see directly from the definition that $\chi_1^\vee(g) = \chi_1(g^{-1})$ for all $g \in G$.
  If E is a subfield of $\mathbb{C}$, then $\chi_1^\vee(g) = \overline{\chi_1(g)}$, where $\overline{(-)}$ denotes complex conjugation. Indeed, the eigenvalues of each g acting on V_1 are roots of unity, and if $\xi \in \mathbb{C}$ is a root of unity, then $\xi^{-1} = \overline{\xi}$.
- The representation $\underline{\mathrm{Hom}}(V_1, V_2)$ has the character $\chi_2 \cdot \chi_1^\vee$, so if $E \subseteq \mathbb{C}$, then we can write $\chi_2 \cdot \overline{\chi_1}$.

(b) Let E be the trivial rank 1 representation. Its character is the constant map $g \mapsto 1 \in E$. We denote it by 1_G.

(c) Let r_G denote the character of the regular representation $E[G]$ of G. Then $r_G(1) = |G|$, and $r_G(g) = 0$ for $g \neq 1$. Indeed, if $g \neq 1$, then $gh \neq h$ for all h, so thc $|G| \times |G|$-matrix of g acting on $E[G]$ (with respect to standard basis) has only zeroes on the diagonal.

(d) From the previous two examples we can compute the character of the augmentation representation (Example 4.1), which we denote by u_G: $u_G = r_G - 1_G$, so $u_G(1) = |G| - 1$ and $u_G(g) = -1$ for $g \neq 1$.

Definition 4.24. Assume that $|G|$ is invertible in E. Let G be a finite group and write $\mathbf{C}_{E,G}$ for the E-vector space of all class functions $G \to E$. For arbitrary maps $\varphi, \psi : G \to E$ we define

$$\langle \varphi, \psi \rangle_G := \frac{1}{|G|} \sum_{g \in G} \varphi(g)\psi(g^{-1}).$$

This is a symmetric bilinear form on the space of maps $G \to E$, and also on $\mathbf{C}_{E,G}$. If confusion is unlikely, we drop the subscript G from $\langle - | - \rangle_G$.

Proposition 4.25 (Orthogonality of characters). *As before, assume that $|G|$ is invertible in E.*

(a) *If V_1, V_2 are arbitrary representations with characters χ_1, χ_2, then*

$$\langle \chi_1, \chi_2 \rangle = \dim_E(\mathrm{Hom}_{E[G]}(V_1, V_2)).$$

(b) *Let χ_1, χ_2 characters of irreducible representations V_1, V_2 of G. Then*

$$\langle \chi_1, \chi_2 \rangle = \begin{cases} a \in \mathbb{Z}_{\geq 1} & \text{if } V_1 \cong V_2 \\ 0 & \text{otherwise.} \end{cases}$$

(c) *If E is algebraically closed then $a = 1$.*

Proof. Taking Schur's Lemma 4.8 into account, it follows that we just have to prove (a). First let V be any representation and write $V = V/V^G \oplus V^G$. The element $\frac{1}{|G|}\sum_{g \in G} g \in E[G]$ induces an $E[G]$-linear map $V \to V$ whose image is V^G, and which is the identity restricted to V^G. It follows that

$$\mathrm{Tr}\left(\frac{1}{|G|}\sum_{g \in G} g\right) = \dim_E V^G.$$

We apply this observation to $\underline{\mathrm{Hom}}(V_1, V_2)$. The character of $\underline{\mathrm{Hom}}(V_1, V_2)$ is $\chi_{V_1}^\vee \cdot \chi_{V_2}$, so

$$\begin{aligned}\dim_E \mathrm{Hom}_{E[G]}(V_1, V_2) &= \dim_E \left(\underline{\mathrm{Hom}}(V_1, V_2)\right)^G \\ &= \frac{1}{|G|} \sum_{g\in G} \chi_{V_1}(g^{-1})\chi_{V_2}(g) \\ &= \langle \chi_{V_1}, \chi_{V_2} \rangle .\end{aligned}$$

□

COROLLARY 4.26. *Assume that $|G|$ is invertible in E. The isomorphism class of a representation over E is determined by its character.*

More precisely, if V is an E-representation of G, then

$$V \cong \bigoplus_{W \text{ irreducible}} W^{\frac{\langle \chi_V, \chi_W \rangle}{\langle \chi_W, \chi_W \rangle}}$$

where the direct sum is over the distinct irreducible representations of G on E.

Proposition 4.13 translates to:

COROLLARY 4.27. *As before, assume that $|G|$ is invertible in E. If V is a representation of G over E, then V is absolutely irreducible if and only if $\langle \chi_V, \chi_V \rangle = 1$.*

DEFINITION 4.28. We say that an irreducible representation W is *contained in V with multiplicity m*, if $\frac{\langle \chi_V, \chi_W \rangle}{\langle \chi_W, \chi_W \rangle} = m$, i.e. if m is maximal such that $W^{\oplus m} \subseteq V$.

COROLLARY 4.29. *If G is a finite group and $|G|$ invertible in E, then an irreducible representation W of G with $\dim_E W = d$ is contained in the regular representation $E[G]$ with multiplicity $d/\langle \chi_W, \chi_W \rangle$.*

If $W_1, \dots, W_r$ are the distinct irreducible representations of G, then

$$|G| = \sum_{i=1}^{r} \frac{(\dim W_i)^2}{\langle \chi_{W_i}, \chi_{W_i} \rangle}.$$

PROOF. If W is an irreducible representation of G, then by Example 4.23 we can compute

$$\langle r_G, \chi_W \rangle = \frac{1}{|G|} \sum_{g\in G} r_G(g) \chi_W\left(g^{-1}\right) = \chi_W(1) = \dim W.$$

It follows that

$$E[G] = \sum_{i=1}^{r} W_i^{\frac{\dim W_i}{\langle \chi_{W_i}, \chi_{W_i} \rangle}},$$

so

$$|G| = \dim_E E[G] = \sum_{i=1}^{r} \frac{(\dim W_i)^2}{\langle \chi_{W_i}, \chi_{W_i} \rangle}$$

as claimed. □

REMARK 4.30. (a) If G is a finite group and $|G|$ invertible in E, then we will identify the group $R_E(G)$ (Definition 4.6) with the free abelian group on the finite set of irreducible characters of G on E.

(b) If $E \subseteq F$ is an extension of fields and χ a character of a representation V of G over E, then the character of $V \otimes_E F$ is the composition $G \xrightarrow{\chi} E \hookrightarrow F$, and abusing notation we will also write χ for the character of $V \otimes_E F$.

PROPOSITION 4.31. *We continue to assume that $|G|$ is invertible in E. Let $E \subseteq F$ be an extension of fields. A representation of G over F comes from a representation of G over E if and only if its class lies in $R_E(G) \subseteq R_F(G)$ (Proposition 4.11).*

PROOF. Clearly, if a representation comes from E, then its class lies in $R_E(G)$. Conversely, let χ be the character of a representation of G over F, and assume that $\chi \in R_E(G)$. Let let $\psi_1, \dots, \psi_r$ be the irreducible characters of G over E. By assumption we can write $\chi = \sum_{i=1}^r a_i \psi_i$, with $a_i \in \mathbb{Z}$. Since χ and ψ_i are characters of representations over F (Remark 4.30), we have $\langle \psi_i, \psi_i \rangle > 0$, and

$$a_i \langle \psi_i, \psi_i \rangle = \langle \chi, \psi_i \rangle \geq 0,$$

so $a_i \geq 0$. □

With the preceding corollary in mind, we make a detour to complex representations.

4.1.4. *Complex representations.* In this section let $E = \mathbb{C}$.

DEFINITION 4.32. Recall that $\mathbf{C}_{\mathbb{C},G}$ denotes the space of class functions $G \to \mathbb{C}$. For $\varphi, \psi \in \mathbf{C}_{\mathbb{C},G}$ we define

$$(\varphi|\psi)_G = \frac{1}{|G|} \sum_{g \in G} \varphi(g)\overline{\psi(g)}.$$

If no confusion is possible we write $(-,-) = (-,-)_G$. This is a scalar product on $\mathbf{C}_{E,G}$: $(\varphi|\psi) = \overline{(\psi|\varphi)}$, $(\varphi|\varphi) > 0$ for $\varphi \neq 0$, and $(\varphi|\psi)$ is linear (resp. semi-linear) in φ (resp. ψ).

LEMMA 4.33. *If χ is the character of a complex representation ρ of G, and $\varphi \in \mathbf{C}_{\mathbb{C},G}$, then*

$$(\varphi|\chi) = \langle \varphi, \chi \rangle.$$

PROOF. This is clear as $\chi(g)$ is a sum of roots of unity, and if $\xi \in \mathbb{C}$ is a root of unity then $\xi^{-1} = \overline{\xi}$. □

We have seen that the irreducible characters of G form an orthonormal system in $\mathbf{C}_{\mathbb{C},G}$ with respect to $\langle -,- \rangle_G$. In fact, they also span $\mathbf{C}_{\mathbb{C},G}$.

THEOREM 4.34. *Let G be a finite group. Its irreducible characters $\chi_1, \dots, \chi_r$ form an orthonormal basis of the space of class functions $\mathbf{C}_{\mathbb{C},G}$.*

PROOF. Let $\varphi \in \mathbf{C}_{\mathbb{C},G}$ be a class function such that $\langle \varphi, \chi \rangle = 0$ for all irreducible characters χ of G. We want to show that $\varphi = 0$.

If $\rho : G \to \mathrm{GL}(V)$ is any representation of G, define $\rho_\varphi := \sum_{g \in G} \varphi(g)\rho(g^{-1})$. Then ρ_φ is an endomorphism of V, and it is easily checked that ρ_φ is actually an endomorphism of the representation ρ. If ρ is irreducible, then Schur's Lemma 4.8 shows that $\rho_\varphi = \lambda \cdot \mathrm{id}_V$ for some $\lambda \in \mathbb{C}$. We have

$$\lambda \dim V = \mathrm{Tr}(\rho_\varphi) = |G| \langle \varphi, \chi_\rho \rangle = 0.$$

It follows that $\rho_\varphi = 0$ even if ρ is not irreducible, since ρ is the direct sum of irreducible subrepresentations. Now let ρ be the regular representation of G, i.e. G

acts on the $|G|$-dimensional E-vector space with basis $\{e_g | g \in G\}$ via $\rho(g)(e_h) = e_{gh}$. Then

$$0 = \rho_\varphi(e_1) = \sum_{g\in G} \varphi(g)\rho(g^{-1})(e_1) = \sum_{g\in G} \varphi(g)e_{g^{-1}}.$$

It follows that $\varphi = 0$, which is what we wanted to show. □

COROLLARY 4.35. *If G is a finite group, then the number of nonisomorphic irreducible complex representations is equal to the number of conjugacy classes in G. This is also true over any splitting field of characteristic* 0.

PROOF. It is clear that the dimension of $\mathbf{C}_{\mathbb{C},G}$ is the number of conjugacy classes in G. By the theorem, this dimension is precisely the number of irreducible characters. If E is a splitting field of characteristic 0, then there exists a splitting field $E' \subseteq E$ which is finitely generated over $\mathbb{Q}$, such that every representation of G over E comes from E'. The field E' can be embedded in $\mathbb{C}$, so the claim follows from Proposition 4.16. □

The following corollary will be crucial for Section 4.2.

COROLLARY 4.36. *A class function φ is the character of a complex representation of G if and only if it is a linear combination*

$$\varphi = a_1\chi_1 + \ldots + a_r\chi_r$$

with $a_i \in \mathbb{Z}_{\geq 0}$ and χ_i characters of irreducible representations of G.

4.1.5. *Brauer's Theorem.* We will need one more tool in our bag before we can return to extensions of local fields. We continue to work over the field of complex numbers.

DEFINITION 4.37. Let $\alpha : H \to G$ be a morphism of groups. If $\varphi \in \mathbf{C}_{\mathbb{C},G}$, then the composition $\varphi \circ \alpha$ is a class function on H, which we call *the restriction of* φ, and denote by $\alpha^*\varphi$.

Clearly, if φ is the character of a representation ρ of G, i.e. if $\varphi \in R^+_{\mathbb{C}}(G)$, then $\alpha^*\varphi$ is the character of the representation $\operatorname{Res}_\varphi \rho$.

If $\varphi \in \mathbf{C}_{\mathbb{C},H}$, there is an *induced class function* on G denoted by $\alpha_*\varphi$, which can be computed using the following two special cases:

(a) If α is injective, then

$$\alpha_*\varphi(g) := \frac{1}{|H|} \sum_{\substack{x\in G\\ xgx^{-1}\in H}} \varphi(xgx^{-1}).$$

(b) If α is surjective, then

$$\alpha_*\varphi(g) := \frac{1}{|\ker(\alpha)|} \sum_{h\mapsto g} \varphi(h).$$

In general we define $\alpha_*\varphi$ by factoring α into a surjective map followed by an injective map.

LEMMA 4.38. *If $\alpha : H \to G$ is a morphism of finite groups and $\rho : H \to \mathrm{GL}(V)$ a complex representation of H with character χ_ρ, then $\alpha_*\chi_\rho$ is the character of $\operatorname{Ind}_\alpha \rho$.*

PROOF. Since α is the composition of an injection and a surjection, it suffices to check that the character of $\operatorname{Ind}_\alpha \rho$ satisfies the two formulas (a) and (b) if α is either injective, or surjective.

In the first case this is a simple calculation, which we leave as an exercise. In the second case, α induces an isomomorphism $H/\ker(\alpha) \to G$. Since $\ker(\alpha)$ is a normal subgroup of H, the subspace $V' := \{v - hv \in V | h \in \ker(\alpha), v \in V\}$ is a subrepresentation, and $\ker(\alpha)$ acts trivially on V/V'. It is not difficult to show that the map $\operatorname{Ind}_\alpha \rho = E[H/\ker(\alpha)] \otimes_{E[H]} V \to V/V'$, defined by $h\ker(\alpha) \otimes v \mapsto hv + V'$ is an isomorphism of H (or $H/\ker(\alpha)$)-representations. Without loss of generality we may assume that ρ is irreducible. Hence $V' = V$ or $V' = 0$. If $V' = 0$, then $\ker(\alpha)$ acts trivially on V, so $\chi_\rho(hh') = \chi_\rho(h)$ if $h' \in \ker(\alpha)$. Hence the formula.

Otherwise, $V' = V$, $\operatorname{Ind}_\alpha \rho = 0$, and we check that $\alpha_* \chi_\rho(g) = 0$. Indeed, if $h_0 \in \ker(\alpha)$ then for all $v \in V'$ we see

$$\sum_{h \in \ker(\alpha)} h(v - h_0 v) = 0.$$

As $V = V'$, $\sum_{h \in \ker(\alpha)} hv = 0$ for all $v \in V$. This means that

$$\sum_{h \mapsto g} \operatorname{Tr}_V(h) = \operatorname{Tr}_V\left(\tilde{g} \sum_{h \in \ker(\alpha)} h\right) = 0$$

for every $g \in H/\ker(\alpha)$ with lift $\tilde{g} \in H$. Hence $\alpha_* \chi_\rho = 0$. □

The adjunction relation between induction and restriction of representations induces a relation for induction and restriction of class functions:

PROPOSITION 4.39 (Frobenius reciprocity). *If $\alpha : H \to G$ is a homomorphism of groups, ψ a class function on H, φ a class function on G, then*

$$(\psi | \alpha^* \varphi) = (\alpha_* \psi | \varphi).$$

PROOF. By Theorem 4.34, we know that ψ is a $\mathbb{C}$-linear combination of characters of H, and similarly for φ. We may thus assume that $\psi = \chi_W$, $\varphi = \chi_V$. In this case we have

$$\begin{aligned}(\chi_W | \alpha^* \chi_V) &= \dim \operatorname{Hom}_{E[H]}(W, \operatorname{Res}_\alpha V)\\ &= \dim \operatorname{Hom}_{E[G]}(\operatorname{Ind}_\alpha W, V)\\ &= (\alpha_* \chi_W | \chi_V).\end{aligned}$$

□

In practice, a question about a general representation can often be reduced to simpler representations on a different group.

THEOREM 4.40 (Brauer). *If G is a finite group and χ a character of a finite dimensional complex representation of G, then χ is a $\mathbb{Z}$-linear combination of characters of the form $\alpha_{i,*}\chi_i$, where $\alpha_i : H_i \hookrightarrow G$ is the inclusion of a subgroup, and χ_i is the character of a 1-dimensional representation of H.*

Theorem 4.40 follows from a slight variant, also due to Brauer. Recall that if p is a prime number, a group G is called *p-elementary* if G is the direct product of a cyclic group of order prime to p with a p-group.

THEOREM 4.41. *Let G be a finite group and $V_p \subseteq R(G)$ the subgroup generated by those characters, which are induced by p-elementary subgroups of G. The index of V_p in $R(G)$ is finite and prime-to-p.*

We do not give a proof of Theorem 4.41 (see [**49**, Ch. 10, Thm. 18]), but we show how to deduce Theorem 4.40 from this.

PROOF OF THEOREM 4.40. If G is abelian then we know from Proposition 4.5 that every irreducible representation has dimension 1, so we assume that G is not abelian.

By Theorem 4.41, $R(G) = \sum_{p \text{ prime}} V_p$, and thus every character of G is a $\mathbb{Z}$-linear combination of characters induced by a character of a p-elementary subgroup, for some p. Hence we may assume that G is a p-elementary group. Let $\rho : G \to \mathrm{GL}(V)$ be an irreducible representation and χ its character. By induction on $|G|$ we may assume that ρ is injective. Write $G = C \times P$, where C is cyclic of order prime to p, and P a p-group. Clearly C is contained in the center of G, and by assumption $G/Z(G)$ is a nonzero p-group. Thus $Z(G/Z(G)) \neq 0$, and $G/Z(G)$ contains a cyclic normal p-group; let $H \subseteq G$ be its preimage in G. Then H is abelian and is not contained in $Z(G)$. Since ρ is injective by assumption, $\rho(H)$ is abelian and not contained in the center of $\rho(G)$, i.e. there exists $h \in H$, such that $\rho(h)$ is not given by multiplication with a scalar. This means that the irreducible decomposition of $\rho|_H : H \to \mathrm{GL}(V)$ contains nonisomorphic representations, since H is abelian. Let $V = V_1 \oplus \ldots \oplus V_n$ be the isotypical decomposition of V as an $E[H]$-module, i.e. the irreducible representations contained in V_i are all isomorphic, and V_i, V_j contain no common isomorphic irreducible subrepresentations, unless $i = j$. The elements of G permute the V_i transitively, as V is an irreducible $E[G]$-module.

Define $H' := \{g \in G | gV_1 = V_1\}$. This is a subgroup of G, containing H, and as V is irreducible $H' \neq G$. We claim that ρ is induced by a representation of H'. Indeed, H' acts on V_1, and $E[G] \otimes_{E[H']} V_1 \xrightarrow{\simeq} V$, as $V = \sum_{g \in G} gV_1$.

We have seen that V is induced by a representation of $H' \subsetneq G$, so the claim follows by induction. □

COROLLARY 4.42. *Let G be a finite group of exponent m and E an algebraically closed field of characteristic 0. Let $\zeta_m \in E$ be a primitive m-th root of unity. The injection $R_{\mathbb{Q}(\zeta)}(G) \to R_E(G)$ is an isomorphism. In other words, $\mathbb{Q}(\zeta)$ is a splitting field for G.*

PROOF. Clearly E contains an algebraic closure of $\mathbb{Q}$, which we can embed into $\mathbb{C}$. Proposition 4.16 hence reduces the proof to the case $E = \mathbb{C}$. Let χ be the character of a complex representation of G. By Brauer's theorem, χ can be written

$$\chi = \sum_{i=1}^{r} a_i \alpha_{i,*} \varphi_i$$

with $a_i \in \mathbb{Z}$, $\alpha_i : H_i \hookrightarrow G$ a subgroup and φ_i a 1-dimensional character of H_i. In particular: $\varphi_i : H_i \to \mathbb{C}^\times$ is a homomorphism, so φ_i has image contained in $\mathbb{Q}(\zeta_m)$ (even in its ring of integers), so φ_i comes from $\mathbb{Q}(\zeta_m)$, and hence so does $\alpha_{i,*}\varphi_i$. □

COROLLARY 4.43. *With the notation of the previous corollary, scalar extension induces an equivalence* $\mathrm{Repf}_{\mathbb{Q}(\zeta)} G \to \mathrm{Repf}_E G$.

PROOF. This is just Proposition 4.31 together with Lemma 4.10. □

4.2. Existence of the complex Artin and Swan representations. We now return to the study of a finite Galois extension L/K of complete discretely valued fields L/K with separable residue extension of degree f. We fix $G := \mathrm{Gal}(L/K)$. Also recall the notations v, v_L for the valuations of K and L, and A, B for the discrete valuation rings of K, L.

Recall that we defined the function $i_G : G \to \mathbb{Z}_{\geq 0} \cup \{\infty\}$ as $i_G(g) = v_L(gx - x)$. Here $x \in B$ is an element generating B as an A-algebra. Such an element exists by Theorem 3.24. We also saw that $g \in G_n$ if and only if $i_G(g) \geq n+1$, which implies that $i_G(hgh^{-1}) = i_G(g)$, for any $h \in G$, as G_n is normal. Thus i_G is a class function.

THEOREM 4.44 (Artin). *In the situation above, the function $a_G : G \to \mathbb{Z}$,*

$$a_G(g) := \begin{cases} -f i_G(g) & \text{if } g \neq 1 \\ f \sum_{g' \neq 1} i_G(g') & \text{if } g = 1 \end{cases}$$

is the character of a representation of G over the complex numbers.

The proof of this theorem is the object of this section, and it justifies the following definition.

DEFINITION 4.45. The function a_G attached to the extension L/K is called the *Artin character of G* (or *of L/K*). Once we know that a_G is the character of a representation, we will see in Corollary 4.49 that the function $\mathrm{sw}_G := a_G - (r_G - r_{G/G_0})$ is also a character of G. It is called the *Swan character of L/K*. Here we write r_{G/G_0} for the composition $G \to G/G_0 \xrightarrow{r_{G/G_0}} \mathbb{C}$. If L/K is totally ramified, i.e. if $f = 1$, then $\mathrm{sw}_G = a_G - u_G$, where $u_G = r_G - 1_G$ is the augmentation character of G (see Example 4.23).

Since i_G is a class function, so is a_G, and by Corollary 4.36, Theorem 4.44 is equivalent to the following theorem.

THEOREM 4.46. *For any character χ of a complex representation of G, we have $f(\chi) := (a_G|\chi) \in \mathbb{Z}_{\geq 0}$.*

Note that $(a_G|\chi) = (\chi|a_G)$, as a_G takes values in $\mathbb{Z}$.

The proof of the theorem consists of two steps. First, we show that $(a_G|\chi)$ is always a non-negative *rational* number. Second, to show that $(a_G|\chi)$ is an integer, we use Brauer's theorem 4.40 to reduce to the case where χ is induced by a degree 1 character of a subgroup. If χ is itself a character of a representation ρ of dimension 1, then $G/\ker(\rho)$ is an abelian group corresponding to a Galois subextension K'/K of L/K. In this case, we can relate the number $(a_G|\chi)$ to the jumps in the ramification filtration of $G/\ker(\rho)$, which we know to be integers by the Theorem of Hasse-Arf 3.63.

Let us first see that without loss of generality we may assume that L/K is totally ramified, i.e. that the residue extension is trivial.

LEMMA 4.47. *Write $\alpha : G_0 = G^0 \hookrightarrow G$ for the inclusion of the inertia group into G. Then $\alpha_* a_{G_0} = a_G$ and $\mathrm{sw}_G = \alpha_* \mathrm{sw}_{G_0}$.*

PROOF. Recall that for $g \in G$, we have

$$\alpha_* a_{G_0}(g) = \frac{1}{|G_0|} \sum_{\substack{x \in G \\ xgx^{-1} \in G_0}} a_{G_0}(xgx^{-1}),$$

according to Definition 4.37. As G_0 is normal in G, $xgx^{-1} \in G_0$ if and only if $g \in G_0$. It follows that for $g \in G \setminus G_0$,

$$\alpha_* a_{G_0}(g) = 0 = -f i_G(g) = a_G(g).$$

Note that if $g \in G_0$, then $i_{G_0}(xgx^{-1}) = i_G(xgx^{-1}) = i_G(g)$. Thus, if $g \neq 1$, we get

$$\alpha_* a_{G_0}(g) = -\frac{|G|}{|G_0|} i_G(g) = a_G(g).$$

Finally, if $g = 1$,

$$\alpha_* a_{G_0}(1) = \frac{|G|}{|G_0|} \sum_{g \in G_0 \setminus \{1\}} i_{G_0}(g) = f \sum_{g \in G \setminus \{1\}} i_G(g) = a_G(1),$$

as $i_G(g) = 0$ for $g \notin G_0$. This proves the statement about a_G.

For sw_G, we just have to note that $\alpha_* r_{G_0} = r_G$ and $\alpha_* 1_{G_0} = r_{G/G_0}$. □

Hence, for the rest of the section, we assume that L/K is totally ramified, i.e. that $G = G_0$ and $f = 1$.

LEMMA 4.48. *For every complex character χ of G, we have $f(\chi) = (a_G|\chi) \in \mathbb{Q}_{\geq 0}$. More precisely, if χ is the character of a representation V, then*

$$f(\chi) = \sum_{i \geq 0} \frac{1}{|G : G_i|} \left(\dim V - \dim V^{G_i} \right). \tag{4.3}$$

PROOF. First recall that $u_{G_i} = r_{G_i} - 1_{G_i}$ is the augmentation character of the i-th ramification group in the lower numbering. If $\alpha_i : G_i \hookrightarrow G = G_0$ is the inclusion map, then $\alpha_{i,*} u_{G_i}(g) = 0$ if $g \notin G_i$, and if $g \in G_i \setminus \{1\}$ we get

$$\alpha_{i,*} u_{G_i}(g) = -\frac{|G|}{|G_i|} = -|G : G_i|.$$

Hence, if $g \in G_r \setminus G_{r+1}$, then

$$a_G(g) = -(r+1) = \sum_{i=0}^{\infty} \frac{1}{|G : G_i|} \alpha_{i,*} u_{G_i}(g).$$

For $1 \in G$, we compute

$$\sum_{i=0}^{\infty} \frac{\alpha_{i,*} u_{G_i}(1)}{|G : G_i|} = \sum_{i=0}^{\infty} (|G_i| - 1) = \sum_{g \neq 1} i_G(g) = a_G(1),$$

so we have proved that

$$a_G = \sum_{i=0}^{\infty} \frac{1}{|G : G_i|} \alpha_{i,*} u_{G_i}.$$

If φ is a class function on G, then we use Frobenius reciprocity (Proposition 4.39) to compute

$$\begin{aligned} f(\varphi) &= (a_G|\varphi) \\ &= \sum_{i=0}^{\infty} \frac{1}{|G : G_i|} (\alpha_{i,*} u_{G_i}|\varphi) \\ &= \sum_{i=0}^{\infty} \frac{1}{|G : G_i|} (u_{G_i}|\alpha_i^* \varphi) \end{aligned}$$

Recall that $u_{G_i} = r_{G_i} - 1_{G_i}$, and $(r_{G_i}|\alpha_i^*\varphi) = \overline{\varphi(1)}$. We obtain

$$f(\varphi) := \sum_{i=0}^{\infty} \frac{1}{|G:G_i|}\left(\overline{\varphi(1)} - (1_{G_i}|\alpha^*\varphi)\right)$$

Finally, if $\varphi = \chi_V$ for V a $E[G]$-module, then $\overline{\varphi(1)} = \varphi(1) = \dim V$ and

$$(1_{G_i}|\alpha_i^*\varphi) = \dim_{E[G]}(\mathbf{1}_{G_i}, V) = \dim V^{G_i},$$

so we obtain (4.3). □

COROLLARY 4.49. *If a_G is the character of a complex representation of G, then so is sw_G.*

PROOF. We again assume that L/K is totally ramified, see Lemma 4.47. If χ is the character of the representation V, then $\langle r_G - 1_G, \chi\rangle = \dim V/V^G$. As $G = G_0$, the preceeding lemma shows that

$$\langle \mathrm{sw}_G, \chi\rangle = f(\chi) - \dim V/V^G = \sum_{i\geq 1} \frac{1}{|G:G_i|} \dim V/V^{G_i} \geq 0.$$

Thus, if a_G is a character of G, then $f(\chi)$ is an integer and hence $\langle \mathrm{sw}_G, \chi\rangle \in \mathbb{Z}_{\geq 0}$, which implies that sw_G is a character of G. □

Now we concentrate on characters of 1-dimensional representations.

LEMMA 4.50. *Let χ be the character of a 1-dimensional complex representation V of G. Then $\chi(G) \subseteq \mathbb{C}^\times$, and $\chi : G \to \mathbb{C}^\times$ is a homomorphism of groups. Let $H := \ker(\chi)$ and K'/K the associated subextension of L/K. Write c'_χ for the largest integer for which $(G/H)_{c'_\chi} \neq \{1\}$. Then*

$$f(\chi) = \varphi_{K'/K}(c'_\chi) + 1 \in \mathbb{Z},$$

where $\varphi_{K'/K}$ is the function from Defintion 3.44.

PROOF. Write $c_\chi := \psi_{L/K'}(c'_\chi)$. From Corollary 3.45 we know that $(G/H)_{c'_\chi}$ is the image of G_{c_χ} in G/H. Then c_χ is the largest integer, such that G_{c_χ} does not act trivially on V. Since V is 1-dimensional, this means that

$$V^{G_i} = \begin{cases} V & \text{if } i > c_\chi \\ 0 & \text{else,} \end{cases}$$

so by (4.3), we obtain

$$f(\chi) = \sum_{i=0}^{c_\chi} \frac{1}{|G:G_i|} = \varphi_{L/K}(c_\chi) + 1,$$

according to (3.6) below Definition 3.44. Finally, we have seen in Proposition 3.51 that

$$\varphi_{L/K}(c_\chi) = \varphi_{K'/K}\varphi_{L/K'}(\psi_{L/K'}(c'_\chi)) = \varphi_{K'/K}(c'_\chi),$$

so we get $f(\chi) = \varphi_{K'/K}(c'_\chi) + 1$ as claimed. Moreover, as $G/H \subseteq \mathbb{C}^\times$, the extension K'/K is abelian, so $\varphi_{K'/K}(c'_\chi) \in \mathbb{Z}$ according to the Hasse-Arf Theorem 3.63. □

Now we can complete the proof of Theorem 4.46 and hence of Theorem 4.44.

PROOF OF THEOREM 4.46. Let χ be the character of a $\mathbb{C}[G]$-module V. By Lemma 4.48 we know that $f(\chi) \in \mathbb{Q}_{\geq 0}$. It thus remains to show that $f(\chi)$ is an integer. Brauer's Theorem 4.40 tells us that χ can be written as

$$\chi = \sum_{i=1}^{r} a_i \chi_i'$$

with $a_i \in \mathbb{Z}$ and $\chi_i' = \gamma_{i,*}\chi_i$, where $\gamma_i : H_i \hookrightarrow G$ is a subgroup and χ_i the character of a 1-*dimensional* representation of H_i. Thus it remains to show that $f(\chi_i') \in \mathbb{Z}$.

Note that $f(\chi_i') = (a_G|\chi_i') = (\gamma_i^* a_G|\chi_i)$ by Frobenius reciprocity, so we have to understand the class function $\gamma_i^* a_G$.

LEMMA 4.51. *If $\gamma : H \hookrightarrow G$ is a subgroup, then there exists a non-negative integer λ, such that*

$$\gamma^* a_G = \lambda r_H + a_H.$$

PROOF. First let $h \in H \smallsetminus \{1\}$. Then $r_H(h) = 0$, and $i_G(h) = i_H(h)$, so $\gamma^* a_G(h) = \lambda r_H(h) + a_H(h)$ for any λ.

The interesting case is $h = 1$. Recall that A, B denote the valuation rings of K, L. Let x be a generator of B as an A-algebra. Then Theorem 3.23 shows that $\mathfrak{D}_{B/A} = (f'(x))$, if f is the monic minimal polynomial of x. Note that $f'(x) = \prod_{g \neq 1}(x - gx)$, so $v_L(\mathfrak{D}_{B/A}) = a_G(1)$, see Proposition 3.25. Similary, if $K' = L^H$ and if B' denotes the valuation ring of K', then $a_H(1) = v_L(\mathfrak{D}_{B/B'})$. Finally, Proposition 3.20 shows that

$$\mathfrak{D}_{B/A} = \mathfrak{D}_{B/B'}\mathfrak{D}_{B'/A},$$

and applying v_L gives

$$a_G(1) = a_H(1) + v_L(\mathfrak{D}_{B'/A}) = a_H(1) + |H| v_{K'}(\mathfrak{D}_{B'/A}).$$

Since $r_H(1) = |H|$, the proof is complete, taking $\lambda = v_{K'}(\mathfrak{D}_{B'/A})$. □

Returning to the proof of Theorem 4.46, we use the lemma to compute

$$\begin{aligned} f(\chi_i') &= (\gamma_i^* a_G|\chi_i) \\ &= \lambda(r_{H_i}|\chi_i) + (a_{H_i}|\chi_i) \\ &= \lambda\chi_i(1) + (a_{H_i}|\chi_i) \end{aligned}$$

Since χ_i is the character of a 1-dimensional representation of H, we have $\chi_i(1) = 1$, and we know that $(a_H|\chi_i) \in \mathbb{Z}$, by Lemma 4.50. □

REMARK 4.52. Note that the existence of the Artin representation implies the Hasse-Arf theorem: Let L/K be an abelian Galois extension with group G. As before, it suffices to prove Proposition 3.66. We may hence assume that G is cyclic of order n. Fixing a primitive n-th root of unit in E, we get an injective homomorphism $\chi : G \to E^\times$. Let μ be the largest integer such that $G_\mu \neq 0$. The existence of the Artin representation shows that $f(\chi) = (a_G|\chi) \in \mathbb{Z}$. But as in Lemma 4.50 we see that $f(\chi) = \varphi_{L/K}(\mu) + 1$, so $\varphi_{L/K}(\mu) \in \mathbb{Z}$, which implies Proposition 3.66.

4.3. Integrality over $\mathbb{Z}_\ell$. As before, let L/K be a Galois extension of complete discretely valued fields with group G, and assume that the extension of the residue fields is separable of degree f. We have seen in the previous section that the Artin function a_G and the Swan function $\mathrm{sw}_G := a_G - (r_G - r_{G/G_0})$ are characters of complex representations. For this reason, we call these $\mathbb{Z}$-valued functions the *Artin character* and the *Swan character*. By Proposition 4.19, we know that the attached representations are realizable over $\overline{\mathbb{Q}}$, and hence also over $\overline{\mathbb{Q}_\ell}$, for any prime ℓ. It is therefore meaningful to ask, whether the Artin and Swan representations are defined over a "nice" subfield (or ring) of $\overline{\mathbb{Q}_\ell}$. Answering this question is the main purpose of this section.

THEOREM 4.53. *Let ℓ be a prime different from the residue characteristic of K. Then*

(a) The Artin and Swan representations are realizable over $\mathbb{Q}_\ell$.
(b) There exists a finitely generated projective left-$\mathbb{Z}_\ell[G]$-module Sw_G, which is unique up to isomorphism, such that $\mathrm{Sw}_G \otimes_{\mathbb{Z}_\ell} \mathbb{Q}_\ell$ is isomorphic to the Swan representation, i.e. has character sw_G.

REMARK 4.54. (a) The analogous question about the realizability of sw_G over $\mathbb{Q}$ or even $\mathbb{Z}$ has a negative answer. Indeed, there are even examples of extensions where sw_G is not realizable over $\mathbb{R}$, [**47**].
(b) There is no direct construction of the $\mathbb{Z}_\ell[G]$-module Sw_G known, but there is a cohomological description of Sw_G, see section 10.2. Note however, that for this construction, one needs (amongst other ingredients), Theorem 4.53.

Clearly (b) of Theorem 4.53 implies (a). In the proof of Theorem 4.53, (b), one needs a good understanding of the representation theory of G over three different bases: $\mathbb{Q}_\ell$, $\mathbb{Z}_\ell$ and $\mathbb{F}_\ell$. Until now, we have mainly discussed representations over a field of characteristic prime to the order of the group (recall that even if L/K is totally ramified, G is an extension of a cyclic group of order prime to p and the p-group G_1, so $|G|$ will not necessarily be prime to ℓ). A complete presentation of the necessary representation theory is beyond the scope of these notes, but we will nonetheless try to give an overview.

4.3.1. *Projective representations.* If E is a field and G a group such that $|G|$ is invertible in E, then it is a consequence of Maschke's Theorem 4.3 that two finitely generated $E[G]$-modules are isomorphic if and only if their classes in $R_E(G)$ agree. This is no longer the case if $|G|$ is not invertible in E (Example 4.4). We can recover this nice property, once we concentrate on *projective* $E[G]$-modules. We write $P_E(G)$ for the Grothendieck group of projective finitely generated $E[G]$-modules.

Recall that if R is a (not necessarily commutative) ring, and M a left-R-module, then M is called *projective*, if there exists a free R-module F, such that M is a direct summand of F.

In the proof of Proposition 4.7 we saw that an irreducible representation of a finite group G on a field E is a quotient of $E[G]$. Thus an irreducible representation is projective if and only if it is a direct summand of $E[G]$.

DEFINITION 4.55. Let R be a ring.

- A surjective morphism $f : V \to W$ of left R-modules is an *essential epimorphism* if for every proper submodule $V' \subsetneq V$ we have $f(V') \subsetneq W$.

- An essential epimorphism $f : P \to V$ with P projective is called *projective cover* or *projective envelope* of V.

LEMMA 4.56. *Let E be a field, G a finite group and V a finitely generated $E[G]$-module. Up to unique isomorphism there exists a unique quotient $q_{\max} : V \twoheadrightarrow S_{\max}$ with $S_{\max}$ semi-simple and $\dim_E S_{\max}$ maximal among all semi-simple quotients of V.*

If $f : V \twoheadrightarrow S'$ is any quotient with S' semi-simple, then S' is isomorphic to $S_{\max}$ if and only if f is an essential epimorphism.

PROOF. Simple quotients of V correspond to maximal submodules of V. Thus, as $\dim_E V < \infty$, there exists a quotient $q_{\max} : V \twoheadrightarrow S_{\max}$ such that $S_{\max}$ is semi-simple and such that its dimension is maximal among the semi-simple quotients of V. If $q : V \twoheadrightarrow S$ is a simple quotient, then $q_{\max} \oplus q : V \to S_{\max} \oplus S$ is not surjective, and as S is simple, this implies that the image of $q_{\max} \oplus q$ does not contain $0 \oplus S \subseteq S_{\max} \oplus S$, so $\operatorname{im}(q_{\max}) \cap (0 \oplus S) = 0$. It follows that $\ker(q_{\max}) \subseteq \ker(q)$.

Let $\operatorname{rad}(V)$ denote the intersection of all maximal submodules of V. we have proved that $\ker(q_{\max}) \subseteq \operatorname{rad}(V)$. On the other hand, an element $v \in \operatorname{rad}(V)$ maps to 0 in every simple quotient, and hence also in every semi-simple quotient. Thus $\ker(q_{\max}) = \operatorname{rad}(V)$ and up to unique isomorphism, $q_{\max}$ is the quotient morphism $V \twoheadrightarrow V/\operatorname{rad}(V)$.

Note that the maximal semi-simple quotient is an essential epimorphism: Otherwise there would exist a maximal submodule $M \subsetneq V$ such that $q_{\max}(M) = S_{\max}$. This would mean that M maps surjectively onto every simple quotient of V, which is impossible, as $V/M \neq 0$ is simple.

If $q' : V \to S'$ is a semi-simple quotient, then S' is isomorphic to a direct summand of $S_{\max}$. In particular $q'(q_{\max}^{-1}(S')) = S'$, so if q' is an essential epimorphism, then $q_{\max}^{-1}(S') = V$ and hence $S_{\max} = S'$. □

DEFINITION 4.57. Thanks to the lemma, we can talk about the *maximal semi-simple quotient of V*.

PROPOSITION 4.58. *Let G be a group and E a field.*

(a) If $|G|$ is invertible in E, then every finitely generated $E[G]$-module is projective.

(b) For every finitely generated $E[G]$-module V, there exists a projective envelope $f_V : P_V \to V$. The pair (P_V, f_V) is unique up to isomorphism.

(c) Every finitely generated projective $E[G]$-module P is the projective envelope of its maximal semi-simple quotient.

(d) An indecomposable finitely generated projective $E[G]$-module is the projective envelope of a simple $E[G]$-module.

(e) If $P_E(G)$ denotes the Grothendieck ring of finitely generated projective $E[G]$-modules, then $P_E(G)$ is freely generated by the projective envelopes of simple $E[G]$-modules. In particular, two finitely generated projective $E[G]$-modules P, P' are isomorphic if and only if their classes in $P_E(G)$ agree.

PROOF. (a) We have seen that if $|G|$ is invertible, then every finitely generated $E[G]$-module is the direct sum of irreducible $E[G]$-modules and that an irreducible $E[G]$-module is a direct summand of the regular representation, which is a free $E[G]$-module.

(b) We just give a sketch, for full details see [**49**, Prop. 41]. If V is a finitely generated $E[G]$-module, there exists a finitely generated projective $E[G]$-module Q and a surjection $Q \twoheadrightarrow V$. Let $P \subseteq Q$ be a submodule of minimal length amongst the submodules M for which the composition $M \to Q \to V$ is surjective. Clearly the map $P \to V$ is an essential epimorphism. One proceeds to show that P is a direct summand of Q and hence projective.
The uniqueness of projective envelopes is not difficult to prove.

(c) This follows directly from the lemma.

(d) One easily sees that the projective envelope of a direct sum is the direct sum of the envelopes of the summands. We also know from (c) that every finitely generated projective $E[G]$-module is the projective envelope of a semi-simple $E[G]$-module. Hence the claim.

(e) If P is a finitely generated projective $E[G]$-module, P can be written as a finite direct sum of indecomposable finitely generated projective $E[G]$-modules, each of which is the projective envelope of a simple $E[G]$-module V_i. Thus $P_E(G)$ is generated by the projective envelopes of simple $E[G]$-modules. On the other hand, every extension $0 \to P_1 \to P \to P_2 \to 0$ of projective $E[G]$-modules splits.

□

EXAMPLE 4.59. Assume $\mathrm{char}(E) = p > 0$ and let G be a finite p-group. Then every nonzero $E[G]$-module has nontrivial invariants (Exercise 4.4). This means that every simple $E[G]$-module is the trivial representation E. Given a G-equivariant map $\varphi : E[G] \to E$, we see that $\varphi(g) = g\varphi(1) = \varphi(1)$ for every $g \in G$. This means that every simple quotient of $E[G]$ is isomorphic to $E[G] \twoheadrightarrow E$, $g \mapsto 1$. Consequently, the maximal semi-simple quotient of $E[G]$ is E, as a map $E[G] \to E \oplus E$ cannot be surjective.

Next, we show that $E[G]$ is indecomposable. Indeed, if $E[G] = U \oplus V$ with $U, V \subsetneq E[G]$, then $\dim_E E[G]^G \geq 2$. But $E[G]^G = E \cdot \sum_{g \in G} g$ is 1-dimensional. Proposition 4.58, (c), shows that the morphism $E[G] \to E$ is the projective envelope of E.

Now let E'/E be an extension of fields; we have seen in Proposition 4.11 that $R_E(G) \to R_{E'}(G)$ is injective. The same is true for $P_E(G)$:

PROPOSITION 4.60. *The map $P_E(G) \to P_{E'}(G)$ is injective.*

PROOF. We have seen in Proposition 4.11 that $R_E(G) \to R_{E'}(G)$ is injective. More precisely, we have seen that if V_1, V_2 are nonisomorphic irreducible $E[G]$-modules, then $V_1 \otimes_E E'$ and $V_2 \otimes_E E'$ have no common composition factors.

If $f : W \to V$ is an essential epimorphism of finitely generated $E[G]$-modules, then so is $f' : W \otimes_E E' \to V \otimes_E E'$. To see this, fix a basis $\{e_i\}_{i \in I}$ of E' over E. Then $W \otimes_E E'$ considered as a (not necessarily finitely generated) $E[G]$-module is just $\bigoplus_i W \otimes e_i \cong W^{\oplus |I|}$ and similarly for $V \otimes_E E'$. If $M \subseteq W \otimes_E E'$ is a maximal submodule such that $f'(M) = V \otimes_E E'$, then $M \cap (W \otimes e_i) \twoheadrightarrow V \otimes e_i$. As f is an essential epimorphism it follows that $M = W \otimes_E E'$, and the claim is proved.

Thus, if V_1, V_2 are nonisomorphic irreducible $E[G]$-modules with projective envelopes P_1, P_2, then $P_i \otimes E' \to V_i \otimes E'$, $i = 1, 2$, are projective envelopes. The maximal semi-simple quotient of $P_i \otimes E'$ is the maximal semi-simple quotient of $V_i \otimes E'$ according to Lemma 4.56. It follows that the maximal semi-simple quotients of $P_1 \otimes E'$ and $P_2 \otimes E'$ have no common factors, and thus $P_E(G) \to P_{E'}(G)$ is injective. □

4.3.2. *Representation theory in mixed characteristic.* Now let E be a complete discretely valued field of characteristic 0 with residue field k of characteristic $\ell > 0$ and let G be a finite group. If $A := \mathcal{O}_E$ is the valuation ring of E, we write $P_A(G)$ for the Grothendieck ring of finitely generated projective $A[G]$-modules. Reduction modulo the maximal ideal $\mathfrak{m}$ of A gives us a homomorphism of groups $P_A(G) \to P_k(G)$.

PROPOSITION 4.61. *$P_A(G) \to P_k(G)$ is an isomorphism.*

PROOF. We only give a rough sketch of the proof. For details, see [**49**, §14.4]. If P_1, P_2 are projective $A[G]$-modules, and $\bar{f} : \overline{P}_1 \to \overline{P}_2$ is an isomorphism of $k[G]$-modules, then projectivity implies that $\bar{f}$ lifts to a morphism of $A[G]$-modules $f : P_1 \to P_2$, which is an isomorphism by Nakayama's Lemma. This shows that the map $P_A(G) \to P_k(G)$ is injective.

To see that it is surjective, one uses that A is complete. Let $\mathfrak{m}$ be the maximal ideal of A. Write $A_n := A/\mathfrak{m}^n$. If $\overline{P}$ is a projective $k[G]$-module, let $P_n \to \overline{P}$ denote the projective envelope of $\overline{P}$ considered as an $A_n[G]$-module. It is not difficult to verify that $P_n/\mathfrak{m}P_n \xrightarrow{\cong} \overline{P}$ as $k[G]$-modules. Passing to the limit over n produces an $A[G]$-module $\hat{P}$, which is free as an A-module and whose reduction is isomorphic to the projective $k[G]$-module P. A calculation is required to show that this implies that $\hat{P}$ is actually projective as an $A[G]$-module. □

PROPOSITION 4.62. *There is a commutative diagram of rings:*

$$\begin{array}{ccc} P_A(G) \cong P_k(G) & \xrightarrow{\quad c \quad} & R_k(G) \\ & \searrow^{e} \qquad\qquad {}^{d}\nearrow & \\ & R_E(G) & \end{array}$$

compatible with base change along finite extensions $E \subseteq E'$.

PROOF. The map c is just given by "forgetting" the projectivity of a module, while e is given by $- \otimes_A E$. Only d requires work: Given a $E[G]$-module V, pick a finitely generated free A-submodule W of V with $\operatorname{rank}_A W = \dim_E V$. Such a submodule is called *lattice.* Define $W' := \sum_{g \in G} gW$. This is also a lattice, and an $A[G]$-module. We claim that the class of $W'/\mathfrak{m}W'$ in $R_k(G)$ is independent of the choice of the lattice W (the isomorphism class of the representation $W'/\mathfrak{m}W'$ is not necessarily independent of the choice of W!). Let W_1, W_2 be G-stable lattices in V, without loss of generality we may assume $W_2 \subseteq W_1$: There exists some $r \geq 0$, such that $\mathfrak{m}^r W_2 \subseteq W_1$.

There exists a minimal n such that $\mathfrak{m}^n W_1 \subseteq W_2$. We proceed by induction on n. If $n = 0$, there is nothing to show. If $n = 1$, the module W_1/W_2 is killed by multiplication with a uniformizer π of A. This means we have an short exact sequence of $k[G]$-modules

$$0 \to W_1/W_2 \xrightarrow{\pi} W_2/\mathfrak{m}W_2 \to W_1/\mathfrak{m}W_1 \to W_1/W_2 \to 0$$

This means that the classes of $W_1/\mathfrak{m}W_1$ and $W_2/\mathfrak{m}W_2$ agree in $R_k(G)$, as claimed.

If $n > 1$, take $W_3 := \mathfrak{m}^{n-1}W_1 + W_2$. Then $\mathfrak{m}W_3 \subseteq W_2 \subseteq W_3$, so by the previous calculation the classes of $W_2/\mathfrak{m}W_2$ and $W_3/\mathfrak{m}W_3$ agree in $R_k(G)$. On the other hand, $\mathfrak{m}^{n-1}W_1 \subseteq W_3 \subseteq W_1$, so by induction the classes of $W_3/\mathfrak{m}W_3$ and $W_1/\mathfrak{m}W_1$ agree. This concludes the construction of the map $d\colon R_E(G) \to R_k(G)$.

It is easy to see that this map is a ring homomorphism, and the compatibility with base change is also clear. □

EXERCISE 4.63. The triangle (**??**) is not so interesting if the order $|G|$ is prime to the residue characteristic $\mathrm{char}(k) = \ell$. Indeed, under that assumption, prove that

(a) Every finitely generated $k[G]$-module is projective.
(b) Every finitely generated $A[G]$-module which is free over A is projective.
(c) A projective $A[G]$-module V is simple if and only if $V \otimes_A k$ is a simple $k[G]$-module.
(d) The maps c, d, e in the triangle (**??**) are isomorphisms.

We will give – without proof – some important properties of the maps (**??**).

THEOREM 4.64.

(a) The map $d\colon R_E(G) \to R_k(G)$ is surjective.
(b) The map $e\colon P_k(G) \to R_E(G)$ is injective.
(c) Let E' be a finite extension of the complete, discretely valued field E, and write k' for its residue field. We then have a commutative diagram

$$\begin{array}{ccc} P_{k'}(G) & \xrightarrow{e'} & R_{E'}(G) \\ \uparrow & & \uparrow \\ P_k(G) & \xrightarrow{e} & R_E(G), \end{array}$$

and a class $x \in R_{E'}(G)$ lies in $\mathrm{im}(e)$ if and only if its character χ_x has values in E and if $\chi_x(g) = 0$ whenever $\ell | \mathrm{ord}(g)$.

Here, the character χ_x is to be understood as follows: E' has characteristic 0, and therefore x can be uniquely written a $\mathbb{Z}$-linear combination of classes $[V_i]$ of irreducible E'-representations; $x = \sum_{i=1}^n a_i[V_i]$, $a_i \in \mathbb{Z}$. We then define $\chi_x := \sum_{i=1}^n a_i \chi_{V_i} : G \to E'$.

The proof of Theorem 4.64 is long and rather complicated, see [**49**, Ch. 17]. We omit it. Note that statement (c) can be understood as a generalization of Proposition 4.31.

Admitting the theorem, we can prove Proposition 4.65, which was the reason for our discussion.

PROPOSITION 4.65. *We continue to denote by E a complete discretely valued field of characteristic 0 with valuation ring A and residue field k of characterisitc $\ell > 0$*

Let E' be a finite extension of E and A' the ring of integers of E'. A class $x \in R_{E'}(G)$ arises from a projective $A[G]$-module if and only if

(a) The character χ_x takes values in E.
(b) There exists $n \in \mathbb{N}$, such that nx is the class of a E'-representation, which comes from a projective $A'[G]$-module.

Moreover, the projective $A[G]$-module realizing x is unique up to isomorphism.

PROOF. Clearly, if the class x comes from a projective $A[G]$-module, then (a) and (b) are satisfied.

Conversely, assume that $x \in R_{E'}(G)$ satisfies (a) and (b). Let V' be the projective $A'[G]$-module corresponding to nx. Write V for the $A[G]$-module obtained from V' via the inclusion $A \subseteq A'$. Then V is a projective $A[G]$-module as A' is free over A.

Since E'/E is finite, $V \otimes_A E$ is an E-vector space of dimension $[E' : E] \cdot \dim_{E'}(V' \otimes_{A'} E')$. We compute the character of $V \otimes_A E$. Let $b_1, \dots, b_r \in E'$ be a basis of E' as a E-vector space such that $b_1 = 1$, and fix $g \in G$. If $v' \in V'$ is an eigenvector of g acting on V', then $v, b_2 v, b_3 v, \dots$ are linearly independent eigenvectors of g acting on V. Thus $\mathrm{Tr}(g|V \otimes_A E) = [E' : E]\,\mathrm{Tr}(g|V' \otimes_{A'} E')$. Using the notation from Proposition ??, we see that $e([V]) = [E' : E]nx \in R_{E'}(G)$. But by construction $e([V]) \in R_E(G) \subseteq R_{E'}(G)$, so by Theorem 4.64 we see that $\chi_{e([V])}(g) = 0$ whenever $\ell|\operatorname{ord}(g)$. This means that $\chi_x(g) = 0$ whenever $\ell|\operatorname{ord}(g)$. Again by Theorem 4.64, it follows that $x \in \operatorname{im}(P_A(G) \xrightarrow{e} R_E(G))$. Let $y \in P_A(G)$ be an element such that $e(y) = x$. Since e is injective, $[V] = [E' : E]ny$. As $V \in P_A^+(G)$, the same is true for y, i.e. there exists a projective $A[G]$-module Y, such that $Y \otimes_A E' \cong V'$, i.e. $[Y \otimes_A E] = x \in R_E(G)$.

The projective $A[G]$-module Y is unique up to isomorphism by Proposition 4.58, (e): The isomorphism class of Y is determined by its class in $P_A(G)$, and by the injectivity of e, this class is uniquely determined. □

4.3.3. *The Swan character comes from $\mathbb{Z}_\ell$.* Using Proposition 4.65, we can now finally prove the ℓ-adic integrality of the Swan conductor.

PROOF OF THEOREM 4.53. Recall that we defined $\mathrm{sw}_G := a_G - (r_G - r_{G/G_0})$, and that we saw that a_G, hence sw_G, can be realized over a finite extension $E/\mathbb{Q}_\ell$. We want to apply Proposition 4.65. We know that sw_G takes values in $\mathbb{Z}$, so to prove that sw_G can be realized by a projective $\mathbb{Z}_\ell[G]$-module, we just have to check that some multiple $n\,\mathrm{sw}_G$ can be realized by a projective $\mathcal{O}_E[G]$-module, where $\mathcal{O}_E$ is the integral closure of $\mathbb{Z}_\ell$ in E. To see this, note that by Lemma 4.47 we know that sw_G is induced by sw_{G_0}, so we may assume that L/K is totally ramified, i.e. that $G_0 = G$. We computed in Lemma 4.48 that

$$a_G = \sum_{i=0}^{\infty} \frac{1}{|G : G_i|} \alpha_{i,*} u_{G_i},$$

where $\alpha_i : G_i \hookrightarrow G$ are the inclusions of the ramification subgroups. It follows that

$$|G|\,\mathrm{sw}_G = \sum_{i=1}^{\infty} |G_i| \alpha_{i,*} u_{G_i}.$$

Since $u_{G_i} = r_{G_i} - 1$, each $\alpha_{i,*} u_{G_i}$ can be realized over $\mathcal{O}_E[G]$ (even over $\mathbb{Z}_\ell[G]$). It remains to see hat the augmentation representations U_{G_i}, $i > 0$, are projective $\mathbb{Z}_\ell[G_i]$-modules.

As the groups G_i are p-groups, for $i > 0$, and since $p \neq \ell$, every finitely generated $\mathbb{Z}_\ell[G_i]$-module V which is free (=torsion-free) over $\mathbb{Z}_\ell$ is projective. Indeed, one finds a surjection $F \twoheadrightarrow V$, with F a free $\mathbb{Z}_\ell[G_i]$-module, and as $(|G_i|, \ell) = 1$ for $i > 0$, this surjection splits, and V is a direct summand of F. In particular the augmentation representation of G_i is projective over $\mathbb{Z}_\ell[G_i]$ for $i > 0$.

It follows that $|G|\,\mathrm{sw}_G$ can be realized by a projective $\mathcal{O}_E[G]$-module, and Proposition 4.65 shows that sw_G can be realized by a finitely generated projective $\mathbb{Z}_\ell[G]$-module, unique up to isomorphism. □

4.4. Swan conductor of an ℓ-adic Galois representation. In this section let K be a complete discretely valued field with perfect residue field k of characteristic $p > 0$. Let K^{sep} be a separable closure of K, and write $G_K := \text{Gal}(K^{\text{sep}}/K)$. Fix a second prime number $\ell \neq p$. We are now interested in *continuous representations* $\rho : G_K \to \text{GL}(V)$, where V is a finite dimensional vector space over a finite extension E of $\mathbb{Q}_\ell$, equipped with its ℓ-adic topology. Such a representation is called *ℓ-adic Galois representation.*

DEFINITION 4.66. Let us recall the definition of the topology on $\text{GL}(V)$: The topology on V makes $\text{End}(V)$ into a topological ring, and we can identify the group $\text{GL}(V)$ with the subspace of $\text{End}(V) \times \text{End}(V)$ defined by $XY = 1$, i.e. with the set of pairs (A, A^{-1}), $A \in \text{GL}(V)$. Using this identification we equip $\text{GL}(V)$ with the subspace topology to make it a topological group. Analogously, if $\mathcal{V}$ is a free $\mathcal{O}_E$-module of finite rank, we can make $\text{Aut}_{\mathcal{O}_E}(\mathcal{V})$ into a topological group.

LEMMA 4.67. *Let E be a finite extension of $\mathbb{Q}_\ell$, V a finite dimensional E-vector space, G a profinite group and $\rho : G \to \text{GL}(V)$ continuous representation. Writing $\mathcal{O}_E$ for the ring of integers of E, there exists a free $\mathcal{O}_E$-submodule $\mathcal{V} \subseteq V$, such that $V = \mathcal{V} \otimes_{\mathcal{O}_E} E$ and such that ρ factors*

$$\rho : G \to \text{GL}(\mathcal{V}) \to \text{GL}(V),$$

where $\text{GL}(\mathcal{V}) := \text{Aut}_{\mathcal{O}_E}(\mathcal{V})$.

PROOF. Let $e_1, \dots, e_r$ be a basis of V and use it to identify $\text{GL}(V)$ with $\text{GL}_r(E)$. The inclusion $\text{GL}_r(\mathcal{O}_E) \subseteq \text{GL}_r(E)$ given by the choice of the basis makes $\text{GL}_r(\mathcal{O}_E)$ into an *open* subgroup of $\text{GL}_r(E)$. Then $\{\rho^{-1}(M\,\text{GL}_r(\mathcal{O}_E)) | M \in \text{GL}_r(E)\}$ is an open covering of G, and as G is compact, it follows that there exists a minimal $n \geq 1$ and matrices $M_1, \dots, M_n \in \text{GL}_r(E)$, such that $\text{im}(\rho) \subseteq M_1\,\text{GL}_r(\mathcal{O}_E) \cup \dots \cup M_n\,\text{GL}_r(\mathcal{O}_E)$.

If $\mathcal{V}' := \sum_{i=1}^r e_i \mathcal{O}_E \subseteq V$, then for every $g \in G$ there exists i such that $\rho(g)(\mathcal{V}') = M_i \mathcal{V}'$, and conversely, for every i there exists $g \in G$ such that $M_i \mathcal{V}' = \rho(g)\mathcal{V}'$.

Defining $\mathcal{V} := \sum_{i=1}^n M_i \mathcal{V}' \subseteq V$ hence gives a G-stable, free sub-$\mathcal{O}_E$-module of V, such that $\mathcal{V} \otimes_{\mathcal{O}_E} E = V$. □

We will be particularly interested in how the *wild ramification subgroup* acts in an ℓ-adic representation. Recall that the upper numbering filtration on finite Galois groups is compatible with taking quotients (Herbrand's Theorem, Proposition 3.53), which allowed us to introduce the *upper numbering filtration* on the infinite, absolute Galois group G_K (Definition 3.54). We saw that $G_K^0 \trianglelefteq G_K$ is the kernel of the projection $G_K \twoheadrightarrow \text{Gal}(\bar{k}/k)$. If L/K is a finite Galois extension, we also showed that $\text{Gal}(L/K)_1 = \text{Gal}(L/K)^{\varphi_{L/K}(1)}$ is the unique p-Sylow subgroup of $\text{Gal}(L/K)$, and that the quotient $\text{Gal}(L/K)/\text{Gal}(L/K)_1$ is cyclic of order prime-to-p (Corollary 3.38). Passing to the inverse limit over all finite Galois extensions we obtain a closed normal subgroup $P_K \trianglelefteq G_K$, such that P_K is a pro-p-group, and such that G_K/P_K is pro-cyclic with every finite quotient of order prime to p. In particular, if $H \trianglelefteq G_K$ is an open (hence finite index) normal subgroup, then the image of P_K in the finite Galois group G_K/H is $(G_K/H)_1$.

DEFINITION 4.68. The pro-p-group P_K is called the *wild inertia group* or *wild ramification group*. Let R be a commutative ring and $\rho : G_K \to \text{GL}_r(R)$ a homomorphism.

(a) ρ is called *unramified* if $G_K^0 \subseteq \ker(\rho)$.
(b) ρ is called *tame* or *tamely ramified* if $P_K \subseteq \ker(\rho)$. Otherwise ρ is called *wild* or *wildly ramified*.

Our main interest will be in the cases $R = \mathcal{O}_E, E$ or $\mathbb{F}_\lambda$, and ρ continuous.

We now reduce a Galois representation modulo ℓ.

DEFINITION 4.69. Let E be a finite extension of $\mathbb{Q}_\ell$ with valuation ring $\mathcal{O}_E$, uniformizer λ, and finite residue field $\mathbb{F}_\lambda$. If $\mathcal{V}$ is a free finitely generated $\mathcal{O}_E$-module, and $\rho : G_K \to \mathrm{GL}(\mathcal{V})$ a continuous representation, then the composition $\bar{\rho} : G_K \to \mathrm{GL}(\mathcal{V}) \to \mathrm{GL}(\overline{\mathcal{V}})$, with $\overline{\mathcal{V}} = \mathcal{V}/\lambda\mathcal{V}$, is called the *reduction modulo* λ of ρ.

The next lemma shows that in the situation of the definition, the pro-p-group $P_K \subseteq G_K$ acts on $\mathcal{V}$ and $\overline{\mathcal{V}}$ through the same *finite* group.

LEMMA 4.70. *Let $\ell \neq p$ be two primes, E a finite extension of $\mathbb{Q}_\ell$, $\mathcal{O}_E$ its valuation ring and $\mathbb{F}_\lambda$ its residue field. If P is a pro-p-group and $\rho : P \to \mathrm{GL}_r(\mathcal{O}_E)$ a continuous representation, then the image of ρ is finite, and $\rho(P) \cap \ker(\mathrm{GL}_r(\mathcal{O}_E) \twoheadrightarrow \mathrm{GL}_r(\mathbb{F}_\lambda)) = \{1\}$.*

If $\rho : P \to \mathrm{GL}_r(E)$ is a continuous representation, then ρ factors through a finite quotient of P.

PROOF. In Lemma 4.67, we saw that any continuous map $P \to \mathrm{GL}_r(E)$ factors through (a conjugate of) the inclusion $\mathrm{GL}_r(\mathcal{O}_E) \subseteq \mathrm{GL}_r(E)$, so the final statement follows from the first.

The kernel of the projection $\mathrm{GL}_r(\mathcal{O}_E) \to \mathrm{GL}_r(\mathbb{F}_\lambda)$ is an open subgroup which can be written as $H := \mathrm{id} + \lambda M_r(\mathcal{O}_E)$, where λ is a uniformizer of $\mathcal{O}_E$ and $M_r(\mathcal{O}_E)$ the ring of $r \times r$-matrices with entries in $\mathcal{O}_E$. By continuity $\rho^{-1}(H)$ is an open subgroup of P containing the kernel of ρ. The group H is a pro-ℓ-group, and it is not difficult to see that there are no non-trivial maps between pro-p and pro-ℓ-groups, if $\ell \neq p$. It follows that $\rho^{-1}(H) \subseteq \ker(\rho)$, and hence $\rho(P) \cap H = \{1\}$. As H has finite index in $\mathrm{GL}_r(\mathcal{O}_E)$, $\ker(\rho)$ has finite index in P, which completes the proof. □

COROLLARY 4.71. *If $\mathcal{V}$ is a free $\mathcal{O}_E$-module and $\rho : G_K \to \mathrm{GL}(\mathcal{V})$ a continuous representation, then ρ is tame if and only if the reduction $\bar{\rho} : G_K \to \mathrm{GL}(\mathcal{V}) \to \mathrm{GL}(\overline{\mathcal{V}})$ is tame.*

PROOF. The lemma shows that $\rho(P_K) \cong \bar{\rho}(P_K)$. □

We now use the Swan representation to define a measure for "how wild" a representation is.

DEFINITION 4.72. Let E be a finite extension of $\mathbb{Q}_\ell$ with valuation ring $\mathcal{O}_E$, uniformizer λ, and residue field $\mathbb{F}_\lambda$. Let $\mathcal{V}$ be a free finitely generated $\mathcal{O}_E$-module, and $\rho : G_K \to \mathrm{GL}(\mathcal{V})$ a continuous representation.

(a) Since $\mathbb{F}_\lambda$ is a finite field, $G := G_K/\ker(\bar{\rho})$ is the Galois group of a finite extension L/K. Following Serre, we define

$$b(\rho) := b(\bar{\rho}) := \dim_{\mathbb{F}_\lambda} \mathrm{Hom}_{\mathbb{F}_\lambda[G]}(\mathrm{Sw}_G \otimes_{\mathbb{Z}_\ell} \mathbb{F}_\lambda, \bar{\rho}).$$

We also write $b(\mathcal{V})$, if the representation ρ is implicitly given. Note that since Sw_G is a projective $\mathbb{Z}_\ell[G]$-module, $\overline{\mathrm{Sw}_G}$ is a projective $\mathbb{F}_\lambda[G]$-module, so the number $b(\mathcal{V})$ only depends on the class of $\overline{\mathcal{V}}$ in $R_{\mathbb{F}_\lambda}(G)$.

(b) Let V be a finite dimensional E-vector space and $\rho : G_K \to \mathrm{GL}(V)$ a continuous representation. If $\mathcal{V} \subseteq V$ is a G_K-stable $\mathcal{O}_E$-lattice for V, then the number $b(\mathcal{V})$ only depends on $\rho : G_K \to \mathrm{GL}(V)$, not on the choice of the lattice. Indeed, since Sw_G is projective, $b(\mathcal{V})$ only depends on the class of $\overline{\mathcal{V}}$ in $R_{\mathbb{F}_\lambda}(G)$, which only depends on ρ (same argument as in proof as Proposition 4.62). We denote this number by $b(V)$ or $b(\rho)$.

REMARK 4.73.

(a) In the definition of $b(\rho)$ we used the group $G = G_K/\ker(\bar{\rho})$. Instead, we could have used G_K/N for any open normal subgroup N of G_K contained in $\ker(\bar{\rho})$ without changing the result. See the final comments in the proof of Theorem 4.86.

(b) Taking this into account, we see that in the situation of Definition 4.72, if ρ itself factors through a finite quotient G of G_K, then

$$\begin{aligned} b(\rho) &= \dim_{\mathbb{F}_\lambda[G]} \mathrm{Hom}_{\mathbb{F}_\lambda[G]}(\mathrm{Sw}_G \otimes_{\mathbb{Z}_\ell} \mathbb{F}_\lambda, \bar{\rho}) \\ &= \mathrm{rank}_{\mathcal{O}_E} \mathrm{Hom}_{\mathcal{O}_E[G]}(\mathrm{Sw}_G \otimes_{\mathbb{Z}_\ell} \mathcal{O}_E, \rho) \\ &= \dim_E \mathrm{Hom}_{E[G]}(\mathrm{Sw}_G \otimes_{\mathbb{Q}_\ell} E, \rho \otimes E). \end{aligned}$$

To see this, apply the exact functor $\mathrm{Hom}_{\mathcal{O}_E[G]}(\mathrm{Sw}_G \otimes_{\mathbb{Z}_\ell} \mathcal{O}_E, -)$ to the short exact sequence of $\mathcal{O}_E[G]$-modules

$$0 \to \mathcal{V} \xrightarrow{\cdot\lambda} \mathcal{V} \to \overline{\mathcal{V}} \to 0.$$

PROPOSITION 4.74. *In the situation of the definition, let $\mathcal{V}$ be a finitely generated, free $\mathcal{O}_E$-module equipped with a continuous representation $\rho : G_K \to \mathrm{GL}(\mathcal{V})$. If $G := G_K/\ker(\bar{\rho})$, then*

$$b(\mathcal{V}) = \sum_{i=1}^{\infty} \frac{|G_i|}{|G_0|} \dim_{\mathbb{F}_\lambda}(\overline{\mathcal{V}}/\overline{\mathcal{V}}^{G_i}) \tag{4.4}$$

where $\overline{\mathcal{V}} = \mathcal{V}/\lambda\mathcal{V}$, with λ a uniformizer of $\mathcal{O}_E$.

PROOF. We have seen that $|G_0| \cdot \mathrm{sw}_G = \sum_{i \geq 1} |G_i| u^*_{G_i}$, where $u^*_{G_i}$ denotes the character of the augmentation representation (Example 4.1, (c)) $U_{G_i, \mathcal{O}_E}$ of G_i, induced along the inclusion $G_i \subseteq G$. It follows that

$$\begin{aligned} |G_0| b(\mathcal{V}) &= \dim_{\mathbb{F}_\lambda}(\mathrm{Hom}_{\mathbb{F}_\lambda[G]}(\overline{\mathrm{Sw}_G^{\oplus |G_0|}}, \overline{\mathcal{V}})) \\ &= \sum_{i \geq 1} \dim_{\mathbb{F}_\lambda}(\mathrm{Hom}_{\mathbb{F}_\lambda[G]}(\overline{\mathrm{Ind}_{G_i}^G U_{G_i, \mathcal{O}_E}}^{\oplus |G_i|}, \overline{\mathcal{V}})) \\ &= \sum_{i \geq 1} |G_i| \dim_{\mathbb{F}_\lambda} \mathrm{Hom}_{\mathbb{F}_\lambda[G]}(\overline{\mathrm{Ind}_{G_i}^G U_{G_i, \mathcal{O}_E}}, \overline{\mathcal{V}}) \end{aligned}$$

As the G_i, $i > 0$, are p-groups, the augmentation representation is a projective $\mathcal{O}_E[G_i]$-module (Exercise 4.63), its reduction is the augmentation representation of G_i over $\mathbb{F}_\lambda$, because we already saw that $P_A(G_i) \to P_k(G_i)$ is an isomorphism. It is easily checked that induction and reduction modulo λ commute, so $\overline{\mathrm{Ind}_{G_i}^G U_{G_i, \mathcal{O}_E}} = \mathrm{Ind}_{G_i}^G U_{G_i, \mathbb{F}_\lambda}$.

Finally, note that by the properties of induction

$$\mathrm{Hom}_{\mathbb{F}_\lambda[G]}(\mathrm{Ind}_{G_i}^G U_{G_i, \mathbb{F}_\lambda}, \overline{\mathcal{V}}) = \mathrm{Hom}_{\mathbb{F}_\lambda[G_i]}(U_{G_i, \mathbb{F}_\lambda}, \mathrm{Res}_{G_i}^G \overline{\mathcal{V}}) = \overline{\mathcal{V}}/\overline{\mathcal{V}}^{G_i}$$

The claim follows. ☐

By definition, a representation $\mathcal{V}$ of G_K is tame, if P_K acts trivially. On the other hand, the invariant $b(\mathcal{V})$ only depends on the reduction of $\mathcal{V}$ modulo the maximal ideal. Nonetheless, as promised, $b(\mathcal{V})$ does measure whether $\mathcal{V}$ is tame or not.

PROPOSITION 4.75. *Let $\rho : G_K \to \mathrm{GL}_r(\mathcal{O}_E)$ be a continuous representation. Then the following are equivalent:*

(a) $\rho \otimes E : G_K \xrightarrow{\rho} \mathrm{GL}_r(\mathcal{O}_E) \hookrightarrow \mathrm{GL}_r(E)$ *is tame.*
(b) ρ *is tame.*
(c) $\bar{\rho} : G_K \xrightarrow{\rho} \mathrm{GL}_r(\mathcal{O}_E) \twoheadrightarrow \mathrm{GL}_r(\mathbb{F}_\lambda)$ *is tame.*
(d) $b(\rho) = 0$.

PROOF. The equivalence of (a) and (b) is clear, while the equivalence of (b) and (c) is Corollary 4.71. The equivalence of (c) and (d) follows from the formula of Proposition 4.74. □

4.4.1. *The break decomposition.* We continue to use the notations from the previous section. The simple observation made in Lemma 4.70 tells us that in any ℓ-adic representation $\rho : G_K \to \mathrm{GL}_r(E)$, where E is a finite extension of $\mathbb{Q}_\ell$, $\ell \neq p$, the wild inertia group P_K always acts through a finite quotient.

We follow Katz [**27**, Ch. 1] in studying fairly general representations of the pro-p-group P_K (Definition 4.68), which factor through a finite quotient. See also [**34**, §2.].

Let us first establish some more facts about the ramification filtration on G_K.

LEMMA 4.76. *For $\lambda \in \mathbb{R}_{\geq 0}$, write*

$$G_K^{\lambda+} := \overline{\bigcup_{\lambda' > \lambda} G_K^{\lambda'}}$$

where $\overline{(-)}$ denotes the closure in the topological group G_K.

The decreasing filtration G_K^λ, $\lambda \in \mathbb{R}_{\geq 0}$, satisfies the following properties.

(a) $\bigcap_{\lambda > 0} G_K^\lambda = \{1\}$
(b) For $\lambda > 0$ we have

$$G_K^\lambda = \bigcap_{0 < \lambda' < \lambda} G_K^{\lambda'}.$$

(c) $P_K = G_K^{0+}$

PROOF. (a) If L/K is a finite Galois extension, the quotient map $G_K \twoheadrightarrow \mathrm{Gal}(L/K)$ induces surjections $G_K^\lambda \to \mathrm{Gal}(L/K)^\lambda$. Thus, if $g \in G_K$ lies G_K^λ for all $\lambda > 0$, then g maps to 1 in $\mathrm{Gal}(L/K)$ for every finite Galois extension L/K. The claim follows.

(b) The profinite group G_K/G_K^λ corresponds to an algebraic extension L/K (not necessarily finite) and the ramification filtration on its Galois group G_K/G_K^λ is given by the image of the filtration of G_K. Thus the claim follows from (a).

(c) Let L/K be a finite Galois extension. If $\varepsilon > 0$, then $\psi_{L/K}(\varepsilon) > 0$, so $\mathrm{Gal}(L/K)^\varepsilon = \mathrm{Gal}(L/K)_{\lceil \psi_{L/K}(\varepsilon) \rceil} \subseteq \mathrm{Gal}(L/K)_1$. Since the image of P_K in $\mathrm{Gal}(L/K)$ is precisely $\mathrm{Gal}(L/K)_1$ it follows that $G_K^{0+} \subseteq P_K$. But we see more: there exists an $\varepsilon_0 > 0$, such that $\mathrm{Gal}(L/K)^{\varepsilon_0} = \mathrm{Gal}(L/K)_1$, so the closed subgroups $G_K^{0+} \subseteq P_K$ have the same images in every finite quotient, which shows that they are equal. (Recall that a closed subgroup H of a profinite group $\varprojlim_N G/N$ is also profinite and isomorphic to $\varprojlim_N H/(N\cap H))$

□

DEFINITION 4.77. For convenience, we say that a P_K*-module* is a $\mathbb{Z}[1/p]$-module M, together with a morphism $\rho : P_K \to \mathrm{Aut}_{\mathbb{Z}}(M)$, which factors through a finite discrete quotient. A morphism of P_K-modules is a morphism of $\mathbb{Z}[1/p]$-modules compatible with the action of P_K.

PROPOSITION 4.78 ([**27**, Ch. I, Prop. 1.1]). *For readability, write* $G := G_K$ *and* $P := P_K$. *Let* M *be a* P*-module.*

(a) *There exists a unique decomposition* $M = \bigoplus_{x\in\mathbb{R}_{\geq 0}} M(x)$ *of* P*-modules, such that*
   (i) $M(0) = M^P$
   (ii) $M(x)^{G^x} = 0$ *for* $x > 0$.
   (iii) $M(x)^{G^y} = M(x)$ *for* $y > x$.
(b) $M(x) = 0$ *for all but finitely many* $x \in \mathbb{R}_{\geq 0}$.
(c) *For every* $x \in \mathbb{R}_{\geq 0}$, *the assignment* $M \mapsto M(x)$ *is an exact endofunctor on the category of* P*-modules.*
(d) $\mathrm{Hom}_P(M(x), M(y)) = 0$ *unless* $x = y$.

PROOF. Let $\rho : P \to \mathrm{Aut}_{\mathbb{Z}}(M)$ be the representation implicit in saying that M is a P-module and let $H = \mathrm{im}(\rho)$. By assumption, H is a finite discrete group. For $x \in \mathbb{R}_{\geq 0}$, let $H(x) := \rho(G_K^x)$, and $H(x+) := \rho(G_K^{x+}) = \bigcup_{y>x} H(y)$. In particular, $H = H(0+)$. Note that for every $x \in \mathbb{R}_{\geq 0}$, $H(x)$ and $H(x+)$ are normal subgroups of H.

For every $x \in \mathbb{R}_{\geq 0}$, we define elements of $\mathbb{Z}[1/p][H]$

$$\pi(x) := \frac{1}{|H(x)|} \sum_{h\in H(x)} h \quad \text{and} \quad \pi(x+) := \frac{1}{|H(x+)|} \sum_{h\in H(x+)} h$$

We have seen such elements before: They define splittings of the projections $\mathbb{Z}[1/p][H(x)] \twoheadrightarrow \mathbb{Z}[1/p]$, $h \mapsto 1$, and similarly for $H(x+)$.

LEMMA 4.79. *Almost all the elements* $\pi(x+)(1-\pi(x))$ *are* 0, *and the nonzero elements of the set* $\{\pi(x+)(1-\pi(x)) | x > 0\} \cup \{\pi(0+)\}$ *form a set of orthogonal, central idempotents in* $\mathbb{Z}[1/p][H]$, *whose sum is* 1.

PROOF. An easy calculation shows that for all $x \geq 0$, $\pi(x)$ and $\pi(x+)$ are idempotents. They are central as the groups $H(x)$, $H(x+)$ are normal subgroups of H. It follows that the same is true for $\pi(x+)(1-\pi(x))$ and $\pi(0+)$. For the orthogonality, note that $\pi(x+)\pi(x) = \pi(x)$, so $\pi(x+)(1-\pi(x)) = \pi(x+) - \pi(x)$. For $y > x$ we see that $\pi(x)\pi(y) = \pi(x)$, and we compute

$$\begin{aligned}(\pi(x+) - \pi(x))(\pi(y+) - \pi(y)) &= \pi(x+)\pi(y+) - \pi(x+)\pi(y)\\ &= \pi(x+) - \pi(x+)\pi(y),\end{aligned}$$

which is 0 if $\pi(y) \neq \pi(x)$. If $\pi(x) = \pi(y)$, then $H(x) = H(y)$, so $H(x) = H(x+)$ and hence $\pi(x+) - \pi(x) = 0$.

Since H is a finite group, we have $H(x) = H(x+)$ for almost all $x \in \mathbb{R}_{\geq 0}$, thus only finitely many $\pi(x+)(1-\pi(x))$ are nonzero.

Finally, let us compute the sum of the idempotents:

$$\sum_{x>0} \pi(x+)(1-\pi(x)) + \pi(0+) = \sum_{n=1}^{r} (\pi(x_n+) - \pi(x_n)) + \pi(0+) \tag{4.5}$$

where $x_1 < x_2 < \ldots < x_r$ are the finitely many elements of $\mathbb{R}_{>0}$ where $H(x_n+) \neq H(x_n)$. Note that in this case

$$H(0+) = H(x_1), \quad H(x_{n-1}+) = H(x_n), n \in [2,r], \quad H(x_r+) = \{1\}.$$

It follows that the coefficient of $h \in H(x_n) \setminus H(x_n+)$ in (4.5) is

$$\underbrace{-\frac{1}{|H(x_n)|} + \frac{1}{|H(x_{n-1}+)|}}_{=0} \underbrace{-\frac{1}{|H(x_{n-1})|} + \frac{1}{|H(x_{n-2}+)|}}_{=0} + \ldots = 0$$

A similar computation shows that the coefficient of 1 in (4.5) is 1, so the proof is complete. □

From the lemma we obtain the decomposition of M that we were looking for: For $x > 0$, $M(x) := \{m \in M | \pi(x+)(1-\pi(x))m = m\}$, and $M(0) := \{m \in M | \pi(0+)m = m\}$. By construction, this is a decomposition into finitely many subrepresentations.

We verify that it has the desired properties. For property (i), we just have to observe that $H(0+) = \mathrm{im}(P)$, as $P = G_K^{0+}$. For (iii), let x_n be defined as in the previous paragraph, i.e. such that $H(x_n) \supsetneq H(x_n+)$. For $m \in M(x_n)$, we know by definition that $m = \pi(x_n+)m - \pi(x_n)m$. Thus, if $y > x_n$, then $H(y) \subseteq H(x_n+)$, and hence m is invariant under $H(y)$, as $\pi(y)\pi(x) = \pi(x)$ if $y \geq x$. But this also means that $m = \pi(x_n+)m - \pi(x_n)m = m - \pi(x_n)m$, so $\pi(x_n)m = 0$. This shows that $M(x_n)^{G^{x_n}} = 0$, so (ii) also is verified.

To prove that the decomposition of M is unique, let $M = \bigoplus_{x \geq 0} M'(x)$ be a second decomposition with the properties (i) - (iii). Clearly, $M'(x) \neq 0$ if and only if x is one of the $x_1, \ldots, x_r$ (or $x = 0$) from above, and the same is true for $M(x)$. We know that $M'(0) = M^P = M(0)$. For $i \in \{1, \ldots, r\}$, let $m \in M'(x_i)$. Then $\pi(x_i)m \in M'(x_i)^{G^{x_i}} = 0$ and $\pi(y)m = m$ for $y > x_i$. We can use the first decomposition to uniquely write $m = m_0 + m_1 + \ldots + m_r$ with $m_0 \in M^P$ and $m_j \in M(x_j)$. Note that $\pi(x_i)m_j = 0$ for $i \leq j$ and $\pi(x_i)m_j = m_j$ for $i > j$. Thus,

$$0 = \pi(x_i)m = m_0 + \ldots + m_{i-1}.$$

If $i = r$ is the final jump, we immediately see that $m = m_r$. If $i < r$, then

$$m = \pi(x_{i+1})m = m_0 + \ldots + m_i.$$

Again we see $m = m_i$. We conclude that $M'(x_i) \subseteq M(x_i)$. The converse inclusion follows by symmetry.

We already remarked that the decomposition of M only has finitely many nonzero components, so it remains to check (c) and (d). If M, N are two P-modules and $\varphi : M \to N$ an H-equivariant map, then φ restricts to $\varphi(x) : M(x) \to N(x)$. This is obvious for $x = 0$. If $x_i > 0$ is one of the finitely many elements with the property that $H(x_i) \supsetneq H(x_i+)$, then for $m \in M(x_i)$, we immediately see that $\pi(x_i)\varphi(m) = 0$, and $\pi(x_i+)\varphi(m) = \varphi(m)$. Hence $\varphi(m) \in N(x_i)$.

This also implies (d): if $y \neq x$, then in the decomposition $M(x) = \bigoplus_{y \in \mathbb{R}_{\geq 0}} (M(x))(y)$ of the P-module $M(x)$, we have $(M(x))(y) = 0$ if $y \neq x$. Thus there are no nonzero P-morphisms $M(y) \to M(x)$, unless $y = x$.

To see that the functor $M \mapsto M(x)$ is exact, let $0 \to M' \to M \to M'' \to 0$ be a short exact sequence of P-modules. Clearly $M'(x) \subseteq M(x)$, and $M(x) \twoheadrightarrow M''(x)$, as the map $M \to M''$ is surjective, and $M(y) \to M''(x)$ is the zero map if $y \neq x$. To see that the sequence $0 \to M'(x) \to M(x) \to M''(x) \to 0$ is exact in the middle, we can use the same argument: M' maps surjectively onto the kernel of $M \to M''$. □

DEFINITION 4.80. If M is a P_K-module, then the decomposition from Proposition 4.78 is called the *break decomposition* of M.

COROLLARY 4.81. *Let A be a $\mathbb{Z}[1/p]$-algebra, and M an A-module, on which $P = P_K$ acts A-linearly through a finite quotient, i.e. the representation $P \to \mathrm{Aut}_{\mathbb{Z}}(M)$ factors through a finite quotient and through $\mathrm{Aut}_A(M) \subseteq \mathrm{Aut}_{\mathbb{Z}}(M)$.*

(a) *In the break decomposition $M = \bigoplus_{x \in \mathbb{R}_{\geq 0}} M(x)$, every $M(x)$ is an A-submodule of M.*
(b) *If B is an A-algebra, then the break decomposition of $B \otimes_A M$ is $\bigoplus_{x \in \mathbb{R}_{\geq 0}} B \otimes_A M(x)$.*
(c) *If the $\mathbb{Z}[1/p]$-algebra A is local and noetherian, and M a free A-module of finite rank, then every $M(x)$ is free of finite rank.*

PROOF. (a) If $a \in A$, then multiplication by a induces a P-equivariant endomorphism of M, hence by Proposition 4.78, (c), multiplication by a maps $M(x)$ to $M(x)$ for every $x \in \mathbb{R}_{\geq 0}$.
(b) This is clear by construction: the idempotents $\pi(x), \pi(x+)$ used in the construction of the decomposition also exist in $A[P]$ and $B[P]$.
(c) If M is a free A-module of finite rank, then if A is noetherian, its A-submodule $M(x)$ is projective of finite rank. Hence if A is local, $M(x)$ is free of finite rank.

□

DEFINITION 4.82. If A is a local noetherian $\mathbb{Z}[1/p]$-algebra and M a free A-module of finite rank on which P_K acts A-linearly through a finite quotient, then the number $\mathrm{rank}_A M(x)$ is called *multiplicity of x*. If $\mathrm{rank}_A M(x) > 0$, then x is called *break of M*.

The real number

$$\mathrm{Swan}(M) := \sum_{x \geq 0} x \, \mathrm{rank}_A(M(x))$$

is called the *Swan conductor of M*.

REMARK 4.83. (a) With this definition, it is obvious that $\mathrm{Swan}(M) = 0$ if and only if the action of P_K on M is trivial, i.e. if and only if $M^{P_K} = M$.
(b) If B is an A-algebra, then $\mathrm{Swan}(M) = \mathrm{Swan}(M \otimes_A B)$ by Corollary 4.81.
(c) $\mathrm{Swan}(M)$ is additive in short exact sequences by Proposition 4.78, (c).

Now we return to representations of the larger groups G_K^0 or G_K.

LEMMA 4.84. *Let M be a $\mathbb{Z}[1/p]$-module on which G_K^0 (resp. G_K) acts such that the restricted action of P_K on M factors through a finite quotient. Then the break decomposition $M = \bigoplus_{x \geq 0} M(x)$ is a decomposition of G_K^0-modules (resp. G_K-modules).*

PROOF. We only do the proof for G_K; the argument for G_K^0 is identical. Let $H \subseteq \mathrm{Aut}_{\mathbb{Z}}(M)$ denote the finite image of P_K, and as in the proof of Proposition 4.78, for $x > 0$ let $H(x)$ be the image of G_K^x, and for $x \geq 0$ let $H(x+)$ be the image of G_K^{x+}. We again get idempotents $\pi(x)$ and $\pi(x+)$, and $\pi(0+)$. As P_K is a normal subgroup of G_K, the elements in the image of G_K in $\mathrm{Aut}_{\mathbb{Z}}(M)$ commute with the $\pi(x)$, $\pi(x+)$:

$$g\pi(x+) = \left(\frac{1}{|H(x+)|}\sum_{h\in H(x+)} ghg^{-1}\right)g = \pi(x+)g,$$

and analogously for $\pi(x)$. As $M(x) = \{m \in M|\pi(x+)(1-\pi(x))m = m\}$, the claim follows. □

4.4.2. *Integrality.* Now let us return to ℓ-adic representations. We denote by K a complete discretely valued field with perfect residue field k of characteristic $p > 0$. We write $G_K = \mathrm{Gal}(K^{\mathrm{sep}}/K)$ with respect to a fixed separable closure K^{sep} of K, and $P_K = G_K^{0+}$ for its wild inertia group, see the discussion preceding Definition 4.68 and Lemma 4.76.

Let E be a finite extension of $\mathbb{Q}_\ell$ with ℓ a prime different from p.

DEFINITION 4.85. If V is a finite dimensional E-vector space and $\rho : G_K \to \mathrm{GL}(V)$ a continuous representation, then we know by Lemma 4.70 that the restriction of ρ to P_K factors through a finite group. By Lemma 4.84 and Corollary 4.81 the break decomposition

$$V = \bigoplus_{x\in\mathbb{R}_{\geq 0}} V(x)$$

is a decomposition of continuous E-representations of G_K. (For the continuity observe that ρ factors through $\bigoplus_{x\geq 0}\mathrm{GL}(V(x))$, which carries the subspace topology in $\mathrm{GL}(V)$).

The real number $\mathrm{Swan}(V)$ is called the *Swan conductor of V*. By definition, $\mathrm{Swan}(V) = 0$ if and only if V is tame.

If V is a finite dimensional E-vector space and $\rho : G_K \to \mathrm{GL}(V)$ a continuous representation, then according to Lemma 4.67 there exists a free finite rank $\mathcal{O}_E$-submodule $\mathcal{V} \subseteq V$ with $\mathrm{rank}_{\mathcal{O}_E}\mathcal{V} = \dim_E V$, such that ρ factors through $\mathrm{GL}(\mathcal{V}) \subseteq \mathrm{GL}(V)$. We then get an $\mathcal{O}_E$-linear, G_K-stable break decomposition

$$\mathcal{V} = \bigoplus_{x\geq 0}\mathcal{V}(x)$$

and Corollary 4.81 shows that $V(x) = \mathcal{V}(x)\otimes_{\mathcal{O}_E} E$. It follows that $\mathrm{Swan}(V) = \mathrm{Swan}(\mathcal{V})$.

Similarly, if $\mathbb{F}_\lambda$ is the (finite) residue field of E and $\overline{\mathcal{V}} := \mathcal{V}\otimes_{\mathcal{O}_E}\mathbb{F}_\lambda$ denotes the reduction of $\mathcal{V}$, then we obtain an $\mathbb{F}_\lambda$-linear, G_K-stable break decomposition

$$\overline{\mathcal{V}} = \bigoplus_{x\geq 0}\overline{\mathcal{V}}(x).$$

Again it follows from Corollary 4.81 that $\overline{\mathcal{V}}(x) = \mathcal{V}(x)\otimes_{\mathcal{O}_E}\mathbb{F}_\lambda$, and hence that $\mathrm{Swan}(V) = \mathrm{Swan}(\mathcal{V}) = \mathrm{Swan}(\overline{\mathcal{V}})$.

The final goal of this section is to prove that $\mathrm{Swan}(V)$ is an integer.

THEOREM 4.86. *If V is a finite dimensional E-vector space, and $\rho : G_K \to \mathrm{GL}(V)$ a continuous representation, then* $\mathrm{Swan}(V) = b(V)$ *(Definition* 4.72*).*

REMARK 4.87. For $\mathrm{Swan}(V)$ to be an integer, it is necessary that V really is a representation not just of P_K but (at least) of G_K^0. For representations of P_K it can still be shown that their Swan conductors are rational numbers, see [**25**, p. 214].

PROOF OF THEOREM 4.86. We compute the number $\operatorname{Swan}(\overline{\mathcal{V}})$, where $\overline{\mathcal{V}} = \mathcal{V} \otimes_{\mathcal{O}_E} \mathbb{F}_\lambda$ with $\mathcal{V}$ a free $\mathcal{O}_E$-module with continuous action $\rho_{\mathcal{O}_E} : G_K \to \mathrm{GL}(\mathcal{V})$ such that ρ factors through $\rho_{\mathcal{O}_E}$. As the residue field $\mathbb{F}_\lambda$ of E is a finite field, $\mathrm{GL}(\overline{\mathcal{V}})$ is a finite group, and consequently G_K acts on $\overline{\mathcal{V}}$ through a finite quotient G, which corresponds to a finite Galois extension L/K. If $\overline{\mathcal{V}} = 0$ then there is nothing to do, so assume $\overline{\mathcal{V}} \neq 0$.

Let $x \in \mathbb{R}_{\geq 0}$ be an element such that $\overline{\mathcal{V}}(x) \neq 0$. Then $G^x \neq G^{x+\varepsilon}$ for all $\varepsilon > 0$ and hence $G_{\lceil \psi_{L/K}(x) \rceil} \neq G_{\lceil \psi_{L/K}(x)+\varepsilon \rceil}$ for all $\varepsilon > 0$, so $\psi_{L/K}(x) \in \mathbb{Z}_{\geq 0}$. By the computation following Definition 3.44, this means that

$$x = \varphi_{L/K}(\psi_{L/K}(x)) = \sum_{i=1}^{\psi_{L/K}(x)} \frac{|G_i|}{|G_0|}.$$

On the other hand, for an integer $i \in [1, \psi_{L/K}(x)]$ we have $G^x = G_{\psi_{L/K}(x)} \subseteq G_i$, so

$$\overline{\mathcal{V}}(x)^{G_i} \subseteq \overline{\mathcal{V}}(x)^{G^x} = 0.$$

If $i > \psi_{L/K}(x)$, then $\varphi_{L/K}(i) > x$, and hence $\overline{\mathcal{V}}(x)^{G_i} = \overline{\mathcal{V}}(x)$. Consequently we compute

$$\dim_{\mathbb{F}_\lambda}(\overline{\mathcal{V}}(x)/\overline{\mathcal{V}}(x)^{G_i}) = \begin{cases} \dim_{\mathbb{F}_\lambda} \overline{\mathcal{V}}(x) & i \leq \psi_{L/K}(x) \\ 0 & i > \psi_{L/K}(x), \end{cases}$$

and thus

$$\begin{aligned} \operatorname{Swan}(\overline{\mathcal{V}}(x)) &= x \dim_{\mathbb{F}_\lambda}(\overline{\mathcal{V}}(x)) \\ &= \sum_{i=1}^{\infty} \frac{|G_i|}{|G_0|} \dim_{\mathbb{F}_\lambda}(\overline{\mathcal{V}}(x)/\overline{\mathcal{V}}(x)^{G_i}) \end{aligned}$$

Both sides of this equation are additive with respect to direct sums, so we finally get

$$\operatorname{Swan}(\overline{\mathcal{V}}) = \sum_{i=1}^{\infty} \frac{|G_i|}{|G_0|} \dim_{\mathbb{F}_\lambda}(\overline{\mathcal{V}}/\overline{\mathcal{V}}^{G_i}).$$

This shows that

$$\operatorname{Swan}(\overline{\mathcal{V}}) = b(V),$$

see Proposition 4.74. In particular, note that the left-hand side and hence the right-hand side is independent of the choice of G. □

EXAMPLE 4.88. We continue Example 3.33. Consider the field $K := k((x))$, where k is an algebraically closed field of characteristic $p > 0$. Let $L := K[t]/(t^p - t - x^{-m})$. The Galois group of this extension is $G = \mathbb{F}_p$.

Assume that $(m, p) = 1$, pick a prime $\ell \neq p$, and an arbitrary nontrivial 1-dimensional representation $\chi : \mathbb{F}_p \to \overline{\mathbb{Q}}_\ell^\times$. The reduction $\overline{\chi} : \mathbb{F}_p \to \overline{\mathbb{F}}_\ell^\times$ is nontrivial, and we compute $\operatorname{Swan}(\chi)$: In Example 3.33 we saw that $G_0 = G_1 = \ldots = G_m \supsetneq G_{m+1} = 0$. It follows that

$$\operatorname{Swan}(\chi) = \operatorname{Swan}(\overline{\chi}) = \sum_{i=1}^{\infty} \frac{|G_i|}{|G_0|} \dim_{\mathbb{F}_\ell}(\overline{\chi}/\overline{\chi}^{G_i}) = \sum_{i=1}^{m} 1 = m$$

Part 2. Generalities about étale topology and ℓ-adic sheaves

5. Recollection on étale morphisms

We recall the definition of an étale morphism and state some of its properties without proof. The general references for this section are [**19**] and [**22**], see also [**39**, I, §3].

DEFINITION 5.1. Let A be a ring (commutative with 1 as always). We say that an A-algebra B is *étale* (or that B is étale over A) if B is finitely presented as an A-algebra and one of the following equivalent conditions is satisfied:

(a) B is formally étale over A, i.e for all A-algebras C and all nilpotent ideals $I \subseteq C$ the map
$$\mathrm{Hom}_{A\text{-alg}}(B, C) \to \mathrm{Hom}_{A\text{-alg}}(B, C/I), \quad \varphi \mapsto \varphi \bmod I$$
is bijective.

(b) For all A-algebras C and all ideals $I \subseteq C$ with $I^2 = 0$ the map
$$\mathrm{Hom}_{A\text{-alg}}(B, C) \to \mathrm{Hom}_{A\text{-alg}}(B, C/I), \quad \varphi \mapsto \varphi \bmod I$$
is bijective.

(c) B is flat as an A-module and is unramified, i.e. $\Omega^1_{B/A} = 0$.

(d) B is flat as an A-module and for each prime ideal $\mathfrak{q} \subseteq B$ the natural map $k(\mathfrak{p}) = A_{\mathfrak{p}}/\mathfrak{p}A_{\mathfrak{p}} \to B_{\mathfrak{q}}/\mathfrak{p}B_{\mathfrak{q}}$ is a separable extension of fields, where $\mathfrak{p} := A \cap \mathfrak{q}$.

(e) If $B = A[x_1, \dots, x_n]/I$ is a presentation of B, then for all prime ideals $\mathfrak{p} \subseteq A[x_1, \dots, x_n]$ with $\mathfrak{p} \supseteq I$, there exist polynomials $f_1, \dots, f_n \in I$ such that
$$I_{\mathfrak{p}} = (f_1, \dots, f_n) \subseteq A[x_1, \dots, x_n]_{\mathfrak{p}} \quad \text{and} \quad \det\left(\frac{\partial f_i}{\partial x_j}\right)_{i,j} \notin \mathfrak{p}.$$

DEFINITION 5.2. A morphism of schemes $f : X \to Y$ is *étale* if for any point $x \in X$ with image $y = f(x) \in Y$ there exist open neighborhoods $V = \operatorname{Spec} B$ of x and $U = \operatorname{Spec} A$ of y such that f restricts to a morphism $f_{|V} : V \to U$ and the induced map $A \to B$ makes B into an étale A-algebra.

EXAMPLE 5.3.

(a) Let k be a field and A a k-algebra. Then A is étale over k iff $A \cong \prod_{i \in I} L_i$, where I is some finite index set and L_i/k is a finite separable extension for every $i \in I$.

(b) Open immersions are étale.

(c) In the situation of Definition 3.11 the A-algebra B is étale iff L/K is unramified.

PROPOSITION 5.4 ([**19**, (17.3.3), (17.3.4), (17.7.3)]).

(a) *The composition of étale morphisms is étale.*

(b) *Let $Y' \to Y$ be a morphism of schemes. If $f : X \to Y$ is étale, then the base change map $X \times_Y Y' \to Y'$ is étale. The converse is true if $Y' \to Y$ is faithfully flat and quasi-compact (the latter condition is satisfied if e.g. Y' is noetherian.)*

(c) *If $f : X \to S$ and $g : Y \to S$ are étale, so is any S-morphism $X \to Y$.*

PROPOSITION 5.5 ([**22**, Exp. I, Cor 5.4]). *Consider a diagram*

$$\begin{array}{ccc} X & \xrightarrow{\quad f,g \quad} & Y \\ & {\scriptstyle a}\searrow \quad \swarrow{\scriptstyle b} & \\ & S, & \end{array}$$

where X, Y, S are locally noetherian and connected, b is separated and étale and $a = b \circ f = b \circ g$. Assume there exists a point $x \in X$ with $f(x) = g(x) =: y$ and that f and g induce the same morphism $k(y) \to k(x)$ on the residue fields. Then $f = g$.

6. The étale fundamental group

In this section let X be a connected noetherian scheme. We partly follow [**22**] and [**52**].

6.1. Definition and basic properties.

DEFINITION 6.1. (a) An *étale covering* of X is a surjective finite étale morphism $f : Y \to X$. The category of such finite étale morphisms is denoted $\mathbf{F\acute{E}t}_X$.

(b) A *geometric point of* X is a morphism $\bar{x} : \operatorname{Spec}\Omega \to X$, with Ω an algebraically closed field. Note that giving $\bar{x}$ is equivalent to giving a point x of X and an embedding of the residue field $k(x)$ into Ω.

(c) If $f : Y \to X$ is a finite étale cover and $\bar{x} : \operatorname{Spec}\Omega_1 \to X$, $\bar{y} : \operatorname{Spec}\Omega_2 \to Y$ geometric points, we say that $\bar{y}$ *lies over* $\bar{x}$ if there exists an algebraically closed field Ω containing both Ω_1 and Ω_2, such that the diagram

$$\begin{array}{ccccc} \operatorname{Spec}\Omega & \longrightarrow & \operatorname{Spec}\Omega_2 & \xrightarrow{\bar{y}} & Y \\ & \searrow & & & \downarrow{\scriptstyle f} \\ & & \operatorname{Spec}\Omega_1 & \xrightarrow{\bar{x}} & X \end{array}$$

commutes. For the geometric point $\operatorname{Spec}\Omega_2 \xrightarrow{\bar{y}} Y \xrightarrow{f} X$ we also write $f(\bar{y})$.

(d) If $\bar{x} : \operatorname{Spec}\Omega \to X$ is a geometric point of X, we denote by $\operatorname{Fib}_{\bar{x}}$ the functor

$$\operatorname{Fib}_{\bar{x}} : \mathbf{F\acute{E}t}_X \to \mathbf{FiniteSet},\ (Y \to X) \mapsto \operatorname{Hom}_X(\operatorname{Spec}\Omega, Y).$$

Note that $\operatorname{Fib}_{\bar{x}}(Y \to X)$ can be naturally identified with the finite set underlying the geometric fiber $Y_{\bar{x}} := Y \times_X \operatorname{Spec}\Omega$.

(e) If $\bar{x}$ is a geometric point of X, then the group $\operatorname{Aut}(\operatorname{Fib}_{\bar{x}})$ is called the *étale fundamental group of X with respect to the base point $\bar{x}$*, and it is denoted $\pi_1^{\text{ét}}(X, \bar{x})$. When confusion is unlikely, we will also write $\pi_1(X, \bar{x})$ for the étale fundamental group.

Grothendieck defined the étale fundamental group in [**22**], and we will summarize its main properties in this section.

THEOREM 6.2 (Grothendieck ([**22**, V.7])). *Let X be a connected noetherian scheme and $\bar{x}$ a geometric point. The étale fundamental group $\pi_1^{\acute{e}t}(X, x)$ is a profinite group and the functor $\operatorname{Fib}_{\bar{x}}$ induces an equivalence*

$$\mathbf{F\acute{E}t}_X \xrightarrow{\cong} \left\{\textit{finite sets with continuous } \pi_1^{\acute{e}t}(X, \bar{x})\textit{-action}\right\}.$$

In particular, open subgroups of $\pi_1^{\acute{e}t}(X, \bar{x})$ correspond to (pointed) finite étale coverings, and open normal subgroups U correspond to (pointed) Galois coverings with Galois group $\pi_1^{\acute{e}t}(X, \bar{x})/U$ (see below).

This theorem has been extended to the non-noetherian case in [**52**].

DEFINITION 6.3. A finite étale covering $f : Y \to X$ is called *Galois covering* if Y is connected and if $\mathrm{Aut}_X(Y)$ acts transitively on $\mathrm{Fib}_{\bar{x}}(Y)$. If f is Galois, then $\mathrm{Aut}_X(Y)$ is called the *Galois group of f*.

The property of being Galois is independent of the choice of the base point by Proposition 5.5. By the same proposition $\mathrm{Aut}_X(Y)$ is a finite group, and if Y is connected then $\mathrm{Aut}_X(Y)$ acts transitively on $\mathrm{Fib}_{\bar{x}}(Y)$ if and only if $|\mathrm{Aut}_X(Y)| = \deg(Y/X)$.

Alternatively, let G be a finite group, and consider the scheme $G_X := \coprod_{g\in G} X_{(g)}$, where $X_{(g)} = X$. This is the constant X-group scheme associated with G. If $f : Y \to X$ is a finite étale covering, then a homomorphism $G \to \mathrm{Aut}_X(Y)$ gives rise to an action $\varphi : G_X \times_X Y \to Y$. If Y is connected, then f is a Galois covering with group G if and only if the morphism

$$(\mathrm{id}, \varphi) : G_X \times_X Y \to Y \times_X Y$$

is an isomorphism, i.e. if X is a *G-torsor*.

PROPOSITION 6.4 ([**46**, §1.5], see also [**52**, Prop. 5.3.9]). *Let X be a connected noetherian scheme and $f : Y \to X$ a finite étale covering. There exists a Galois covering $g : P \to X$, factoring through f.*

If Y is connected then one finds g, such that any other Galois covering $g' : P' \to X$ which factors through f also factors through g. More succinctly, we have the following diagram:

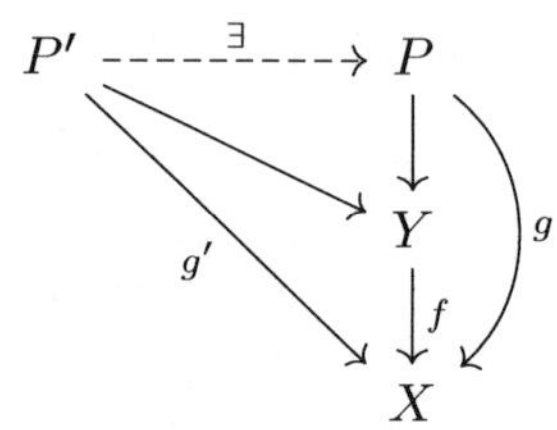

EXAMPLE 6.5. Let K be a field and $X = \mathrm{Spec}\, K$. Giving a finite étale covering $f : Y \to X$ is equivalent to giving a finite étale K-algebra A, and setting $Y = \mathrm{Spec}\, A$. The K-algebra A is a finite product of separable extensions of K, see Example 5.3. Hence Y is connected if and only if A itself is a separable field extension of K, and $f : Y \to X$ is a Galois covering if and only if A/K is a Galois extension.

To see that $\pi_1^{\acute{e}t}(X, \bar{x})$ is a profinite group in a natural way, Grothendieck proceeds as follows. Let Gal_X be the set of Galois coverings of X. Let I be an index set, so that we can write $\mathrm{Gal}_X = \{f_\alpha : P_\alpha \to X | \alpha \in I\}$. We define an ordering on I by writing $\alpha \le \beta$ if and only if there exists an X-morphism $P_\beta \to P_\alpha$. By Proposition 6.4 this system is directed: For $\alpha, \beta \in I$ every connected component of the fiber product $P_\alpha \times_X P_\beta$ is an étale covering of X. By Proposition 6.4 we find $f_\gamma : P_\gamma \to X$ factoring through $P_\alpha \times_X P_\beta$, so $\alpha, \beta \le \gamma$. Now we fix arbitrary geometric points $p_\alpha \in \mathrm{Fib}_{\bar{x}}(P_\alpha)$. By Proposition 5.5, for $\alpha \le \beta \in I$, there exists a *unique* map $\varphi_{\beta,\alpha} : P_\beta \to P_\alpha$, mapping p_β to p_α. The system $(P_\alpha, \varphi_{\beta,\alpha})$ is a directed projective system. Note that its inverse limit exists in the category of schemes because the maps $\varphi_{\beta,\alpha}$ are finite, and hence affine ([**19**, Prop. 8.2.3]), but in general it is not étale over X, as it is usually not locally of finite type. Now let $f : Y \to X$ be an arbitrary finite étale covering. For every $\alpha \in I$ we obtain a map $\mathrm{Hom}_X(P_\alpha, Y) \to \mathrm{Fib}_{\bar{x}}(Y)$, $g \mapsto g(p_\alpha)$. These maps give rise to a map

$$\varinjlim_I \mathrm{Hom}(P_\alpha, Y) \to \mathrm{Fib}_{\bar{x}}(Y),$$

which is easily seen to be a bijection. In Grothendieck's words, $\mathrm{Fib}_{\bar{x}}$ is *strictly pro-representable.*

Moreover, it is not difficult to see that every automorphism of $\mathrm{Fib}_{\bar{x}}$ comes from a unique automorphism of the projective system $(P_\alpha, \varphi_{\alpha,\beta})$: If $\psi \in \mathrm{Aut}(\mathrm{Fib}_{\bar{x}})$ then for every $\alpha \in I$, there exists a unique automorphism of P_α mapping p_α to $\psi(p_\alpha)$, as the P_α are Galois. It follows that

$$\pi_1^{\text{ét}}(X,\bar{x}) = \mathrm{Aut}(\mathrm{Fib}_{\bar{x}}) = \mathrm{Aut}\left(\varinjlim_I \mathrm{Hom}_X(P_\alpha, -)\right) = \varprojlim \mathrm{Aut}_X(P_\alpha)^{\mathrm{opp}}, \tag{6.1}$$

where $(-)^{\mathrm{opp}}$ denotes the opposite group.

EXAMPLE 6.6. (a) Let us bring the above construction down to earth: Let K be a field and write $X = \operatorname{Spec} K$. Choosing a geometric point $\bar{x}$ of $\operatorname{Spec} K$ amounts to choosing an embedding of K into an algebraically closed field Ω. Without loss of generality we take $\Omega = K^{\mathrm{alg}}$, an algebraic closure of K. The Galois coverings P_α of X in the above construction correspond to finite Galois extensions L_α. Fixing a geometric point p_α lying over $\bar{x}$ is equivalent to fixing a K-embedding of L_α into K^{alg}. The transition maps $\varphi_{\alpha,\beta}$ then correspond to embeddings $L_\alpha \subseteq L_\beta \subseteq K^{\mathrm{alg}}$, so the projective system $(P_\alpha, \varphi_{\alpha,\beta})$ corresponds to the inductive system of finite Galois extension of K *contained in* K^{alg}.

Finally, (6.1) shows that

$$\pi_1^{\text{ét}}(X,\bar{x}) = \varprojlim \mathrm{Aut}_X(\operatorname{Spec} L_\alpha)^{\mathrm{opp}} = \varprojlim_{\substack{L \subseteq K^{\mathrm{alg}} \\ L/K \text{ finite Galois}}} \mathrm{Aut}_K(L) = \mathrm{Gal}(K^{\mathrm{alg}}/K).$$

(b) If X is a connected normal noetherian scheme, write $K(X)$ for its function field, i.e. for the stalk at its generic point. Fix an algebraic closure $K(X)^{\mathrm{alg}}$, and let $\bar{\eta} : \operatorname{Spec} K(X)^{\mathrm{alg}} \to X$ denote the associated geometric point of X. If $K(X)^{\mathrm{unr}}$ is the composite of all finite Galois extensions $L/K(X)$, such that the normalization of X in L is étale over X, then $K(X)^{\mathrm{unr}}$ is an (usually infinite) Galois extension of $K(X)$ and there is a natural isomorphism

$$\mathrm{Gal}(K(X)^{\mathrm{unr}}/K(X)) \cong \pi_1^{\text{ét}}(X,\bar{\eta}).$$

PROPOSITION 6.7 (Dependence on base point). *Again X is noetherian and connected. If $\bar{x}$ and $\bar{x}'$ are two geometric points of X, then there is a continuous isomorphism*

$$\pi_1^{\text{ét}}(X,\bar{x}) \cong \pi_1^{\text{ét}}(X,\bar{x}'),$$

which is canonical up to inner automorphism.

Proving the proposition amounts to constructing an isomorphism of the inverse system $(P_\alpha, \varphi_{\alpha,\beta})$ constructed using $\bar{x}$ and the analogous inverse system constructed using $\bar{x}'$, which is not too difficult.

From the Definition of the fundamental group we easily see that it is functorial for schemes equipped with a geometric point.

PROPOSITION 6.8 (Functoriality). *If $f : X' \to X$ is a morphism of connected noetherian schemes, and $\bar{x}' : \operatorname{Spec}\Omega \to X'$ a geometric point, then f induces a continuous homomorphism of groups*

$$\pi^{\acute{e}t}(X', \bar{x}') \to \pi_1^{\acute{e}t}(X, f(\bar{x}')).$$

PROOF. Let $Z \to X$ be a finite étale covering. Then $Z' = Z \times_X X' \to X'$ is an étale covering, and $\operatorname{Fib}_{\bar{x}'}(Z') = \operatorname{Fib}_{f(\bar{x}')}(Z)$. It follows that any automorphism of $\operatorname{Fib}_{\bar{x}'}$ induces an automorphism of $\operatorname{Fib}_{f(\bar{x}')}$. □

EXAMPLE 6.9. Let X be connected and normal with function field $K(X)$. Fix an algebraic closure $K(X)^{\mathrm{alg}}$ of $K(X)$, and write $\bar{\eta}$ for the associated geometric point. With this setup Example 6.6 shows that

$$\operatorname{Gal}(K(X)^{\mathrm{alg}}/K(X)) = \pi_1^{\text{ét}}(\operatorname{Spec}(K(X)), \bar{\eta}) \to \pi_1^{\text{ét}}(X, \bar{\eta})$$

is surjective.

6.2. Complex varieties. If X is a scheme of finite type over the field of complex numbers $\mathbb{C}$, then to X one associates a complex analytic space X^{an}. Its underlying set is $X(\mathbb{C})$. If X is a smooth complex variety, then X^{an} is a complex manifold. In this section we summarize how the étale fundamental group of X and the fundamental group of the topological space X^{an} are related. For details and proofs we refer the reader to [**22**, Exp. XII].

If $f : Y \to X$ is a finite étale covering, then $f^{\mathrm{an}} : Y^{\mathrm{an}} \to X^{\mathrm{an}}$ is a covering space of X^{an} with finite fibers. This defines a functor from the category $\mathbf{F\acute{E}t}_X$ to the category of covering spaces of X^{an} with finite fibers. The Riemann Existence Theorem ([**22**, Exp. XII, Thm. 5.1]) states that this functor is in fact an equivalence. We will use the following reformulation in terms of fundamental groups.

THEOREM 6.10. *Let X be a connected finite type $\mathbb{C}$-scheme and $x \in X(\mathbb{C})$. Then there is a natural continuous isomorphism*

$$\pi_1(X^{an}, x)^\wedge \xrightarrow{\cong} \pi_1^{\acute{e}t}(X, x),$$

where the left-hand side is the profinite completion *of the fundamental group of the topological space X^{an}, based at x.*

One uses this theorem to prove:

COROLLARY 6.11. *Let X be a smooth complex curve of genus g. Define $U := X \setminus \{x_1, \dots x_n\}$ as complement of finitely many closed points. If $\bar{u} \in U$ is a geometric point, then the étale fundamental group $\pi_1^{\acute{e}t}(U, \bar{u})$ is isomorphic to the profinite completion of the group*

$$\langle a_1, \dots, a_g, b_1, \dots, b_g, c_1, \dots, c_n \,|\, [a_1, b_1] \cdot \ldots \cdot [a_g, b_g] c_1 \cdot \ldots \cdot c_n = 1 \rangle .$$

More generally the corollary is true over any algebraically closed field of characteristic 0, [**22**, Exp. XIII, Cor. 2.12].

Similarly, since the fundamental group of a smooth quasi-projective complex variety is a finitely generated group (this can be deduced from the curve case via Lefschetz type theorems), we obtain:

COROLLARY 6.12. *If X is a smooth complex quasi-projective variety, and $x \in X(\mathbb{C})$, then $\pi_1^{\acute{e}t}(X,x)$ is topologically finitely generated.*

With considerable effort, the theory of specialization maps for fundamental groups allows to generalize the corollary:

THEOREM 6.13. *Let k be an algebraically closed field and X a smooth quasi-projective k-variety with geometric point $\bar{x}$. Then $\pi_1^{\acute{e}t}(X,\bar{x})$ is topologically finitely generated in each of the following two cases:*

- $\mathrm{char}(k) = 0$ *(implicit in* [**22**, p.291], *see also* [**30**, Thm. 1]).
- $\mathrm{char}(k) > 0$ *and X is proper* ([**22**, Exp. X, Thm. 2.9]).

The picture is different for non-proper varieties in positive characteristic:

EXAMPLE 6.14 ([**18**]). If k is an algebraically closed field of characteristic $p > 0$, U a smooth affine curve over k and $\bar{u}$ a geometric point of U, then the maximal pro-p-quotient of $\pi_1^{\text{ét}}(U,\bar{u})$ is a free pro-p-group on $\#k$ generators. This can be seen by cohomological methods: One checks that $H^i_{\text{ét}}(U,\mathbb{F}_p) = H^i(\pi_1^{\text{ét}}(U,\bar{u}),\mathbb{F}_p)$, for all $i \geq 0$, where the right-hand side is continuous group cohomology. Since U is affine, we have $H^2_{\text{ét}}(U,\mathbb{F}_p) = 0$. The numbers of generators and relations of the maximal pro-p-quotient of $\pi_1^{\text{ét}}(U,\bar{u})$ can be described in terms of the $\mathbb{F}_p$-dimension of the groups H^1 and H^2 above ([**51**, §4.]).

6.3. Tame coverings. The pathologies that appear in characteristic $p > 0$, exemplified by Example 6.14, can partly be remedied by restricting the attention to *tame* coverings.

DEFINITION 6.15.
(a) ([**21**, Def. 2.2.2]) Let X be a normal connected scheme and $U \subseteq X$ an open subset. If $f : V \to U$ is a finite étale covering, then f is said to be *tamely ramified along* $X \smallsetminus U$, if and only if for every codimension 1 point $\eta \in X \smallsetminus U$, the extension $K(X) \subseteq K(V)$ is tamely ramified with respect to the discrete valuation defined by η.
(b) ([**29**, Sec. 4]) If U is a normal connected separated scheme of finite type over a field, then a finite étale covering $f : V \to U$ is said to be *tame* if for all open immersions $U \hookrightarrow X$ with X a proper connected normal k-scheme, f is tamely ramified with respect to $X \smallsetminus U$.
If U is not connected, then $f : V \to U$ is tame if the induced coverings of the components of U are tame.
(c) The category of tame coverings of U is denoted $\mathbf{F\acute{E}t}_U^{\text{tame}}$. It is a full subcategory of $\mathbf{F\acute{E}t}_U$.

REMARK 6.16.
(a) The normality assumption guarantees that the local ring $\mathcal{O}_{X,\eta}$ at a codimension 1 point is a discrete valuation ring.
(b) If U is a scheme over a field of characteristic 0, then every finite étale covering is tame.
(c) In Definition 6.15, (b), note that there always exists a normal connected *proper* k-scheme containing U: By Nagata's compactification theorem ([**36**], [**11**]) there exists a proper X containing U, which we can then normalize.

(d) In Definition 6.15, (b), if $\dim U = 1$, then there exists a *unique* normal proper curve X containing U. So in this case tameness of coverings of U only has to be checked on this single compactification.

In [**22**, Exp. XIII, §.2] Grothendieck defined the notion of a tame covering, but only with respect to a normal crossings divisor (see Definition 6.17 and Theorem 6.18, (c) below). In characteristic 0, the famous theorem of Hironaka on resolution of singularities states that if U is smooth then one can always find a *smooth* proper variety X containing U, such that $X \setminus U$ is a strict normal crossings divisor. Proving the analogous statement in positive characteristic is an open problem. Theorem 6.18 below shows that Definition 6.15, (b), agrees with Grothendieck's notion, whenever a comparison is possible.

DEFINITION 6.17. Let X be a locally noetherian scheme and D an effective Cartier divisor on X.

(a) D is said to have *strict normal crossings* if for every $x \in \operatorname{supp} D$ the following hold:
- $\mathcal{O}_{X,x}$ is a regular local ring.
- There exists a regular sequence $t_1, \dots, t_n$ and $m \in \{1, \dots, n\}$, such that the ideal $\mathcal{O}_X(-D)_x \subseteq \mathcal{O}_{X,x}$ is generated by $t_1 \cdot \ldots \cdot t_m$. In other words: locally, D looks like the intersection of coordinate hyperplanes.

(b) D is said to have *normal crossings*, if for every $x \in \operatorname{supp} D$ there exists an étale morphism $f : V \to X$, such that $x \in \operatorname{im}(f)$ and such that f^*D has strict normal crossings. In other words: étale locally, D looks like the intersection of coordinate hyperplanes.

THEOREM 6.18 (Kerz-Schmidt, [**29**]). *Let $f : V \to U$ be a finite étale covering of regular separated schemes over a field k. The following statements are equivalent:*

(a) *f is tame.*

(b) *For any regular k-curve C and any morphism $\varphi : C \to U$, the induced covering $V \times_U C \to C$ is tamely ramified along $\overline{C} \setminus C$, where $\overline{C}$ is the unique normal proper k-curve containing C as an open subscheme.*

If there exists an open immersion $U \hookrightarrow X$ with X a smooth, proper, separated, finite type k-scheme, such that $X \setminus U$ is a normal crossings divisor, then (a) *and* (b) *are equivalent to*

(c) *f is tamely ramified along $X \setminus U$.*

6.4. The tame fundamental group. With essentially the same method as used in Section 6.1 one proves:

THEOREM 6.19 ([**29**]). *Let U be a finite type, connected, separated, normal k-scheme, and $\bar{u}$ a geometric point of U. Denote by $\mathrm{Fib}_{\bar{u}}^{\mathrm{tame}}$ the restriction of the functor $\mathrm{Fib}_{\bar{u}}$ to $\mathbf{F\acute{E}t}_U^{\mathrm{tame}}$. Then the group $\pi_1^{\mathrm{tame}}(U, \bar{u}) := \mathrm{Aut}(\mathrm{Fib}_{\bar{u}}^{\mathrm{tame}})$ is profinite, and $\mathrm{Fib}_{\bar{u}}^{\mathrm{tame}}$ induces an equivalence*

$$\mathbf{F\acute{E}t}_U^{\mathrm{tame}} \xrightarrow{\cong} \left\{\textit{finite sets with continuous } \pi_1^{\mathrm{tame}}(U, \bar{u})\textit{-action}\right\}.$$

DEFINITION 6.20. The profinite group $\pi_1^{\mathrm{tame}}(U, \bar{u})$ from the theorem is called the *tame fundamental group of U* with respect to the base point $\bar{u}$.

It is not difficult to check the following basic properties from the definitions.

PROPOSITION 6.21. *Let k be a field and U, $\bar{u}$ as in Theorem 6.19. Then*

(a) *There is a canonical continuous surjective map*

$$\pi_1^{ét}(U,\bar{u}) \twoheadrightarrow \pi_1^{\mathrm{tame}}(U,\bar{u}). \tag{6.2}$$

corresponding to the inclusion $\mathbf{FÉt}_U^{\mathrm{tame}} \subseteq \mathbf{FÉt}_U$.

(b) *The kernel of (6.2) can be described as follows: If X is a normal compactification of U and $\eta \in X \smallsetminus U$ a codimension 1 point, denote by K_η^{sh} the fraction field of the strict henselisation of the discrete valuation ring $\mathcal{O}_{X,\eta}$. Fix an algebraic closure $\overline{K_\eta^{\mathrm{sh}}}$. Then there is a morphism $G := \mathrm{Gal}(\overline{K_\eta^{\mathrm{sh}}}/K_\eta^{\mathrm{sh}}) \to \pi_1^{ét}(U,\bar{u})$, canonical up to inner automorphism of $\pi_1^{ét}(U,\bar{u})$ by Proposition 6.7 and Proposition 6.8. Denote by P_η the image of the wild ramification group of G. The kernel of (6.2) is the smallest closed normal subgroup $\mathbf{P}$ of $\pi_1^{ét}(U,\bar{u})$ containing P_η, where η runs through all codimension 1 points on all normal compactifications X of U.*

(c) *If $\operatorname{char} k = p \geq 0$, then the map (6.2) induces an isomorphism*

$$\pi_1^{ét}(U,\bar{u})^{(p')} \xrightarrow{\cong} \pi_1^{\mathrm{tame}}(U,\bar{u})^{(p')}$$

where $(-)^{(p')}$ denotes the maximal pro-prime-to-p-quotient.

PROOF. We sketch the argument.

(a) As before let $\{P_\alpha | \alpha \in I\}$ be the set of Galois coverings of U. We saw above Example 6.5 that fixing points $p_\alpha \in \mathrm{Fib}_{\bar{u}}(P_\alpha)$ defines unique morphisms $\varphi_{\beta,\alpha} : P_\beta \to P_\alpha$ for $\beta \geq \alpha$, and that the projective system $(P_\alpha, \varphi_{\beta,\alpha})$ induces an isomorphism $\pi_1^{ét}(U,\bar{u}) \cong \varprojlim_{\alpha\in I} \mathrm{Aut}_U(P_\alpha)^{\mathrm{opp}}$. If I^{tame} denotes the subset of I such that $\alpha \in I^{\mathrm{tame}}$ if and only if $P_\alpha \to U$ is tame, then I^{tame} is also a directed set (exercise!), and $\pi_1^{\mathrm{tame}}(U,\bar{u}) = \varprojlim_{\alpha\in I^{\mathrm{tame}}} \mathrm{Aut}_U(P_\alpha)^{\mathrm{opp}}$. The claim follows.

(b) Just for this proof, if G is a profinite group, we write $\mathbf{Set}(G)$ for the category of finite sets with continuous G-action. The categories $\mathbf{Set}(\pi_1^{ét}(U,\bar{u})/\mathbf{P})$ and $\mathbf{Set}(\pi_1^{\mathrm{tame}}(U,\bar{u}))$ are strictly full subcategories of $\mathbf{Set}(\pi_1^{ét}(U,\bar{u}))$, and our goal is to show that they are they are identical.

In other words, we need to show that if $f : V \to U$ is a finite connected étale covering, then f is tame if and only if $\mathbf{P}$ acts trivially on $\mathrm{Fib}_{\bar{u}}(V)$. In fact we may assume f to be Galois étale.

Let $K(U)$, $K(V)$ denote the function fields of U and V. Note that $\mathrm{Gal}(K(V)/K(U)) = \mathrm{Aut}_U(V)^{\mathrm{opp}}$ is (noncanonically) isomorphic to $\mathrm{Fib}_{\bar{u}}(V)$ in $\mathbf{Set}(\pi_1^{ét}(U,\bar{u}))$. If $\mathbf{P}$ acts trivially on $\mathrm{Fib}_{\bar{u}}(V)$, then the image of the group P_η in $\mathrm{Gal}(K(V)/K(U))$ is trivial for all codimension 1 points $\eta \in X \smallsetminus U$ in all normal compactifications X of U. But this means that $K(V)/K(U)$ is tamely ramified at η.

Conversely, if f is tame, then $K(V)/K(U)$ is tame with respect to all codimension 1 points η in $X \smallsetminus U$ in some normal compactification X of U. This means that the image of P_η is trivial in $\mathrm{Gal}(K(V)/K(U))$. But $\mathrm{Gal}(K(V)/K(U)) = \mathrm{Aut}_U(V)^{\mathrm{opp}}$ is a quotient of $\pi_1^{ét}(U,\bar{u})$, so the image of $\mathbf{P}$ in $\mathrm{Gal}(K(V)/K(U))$ is trivial, and thus $\mathbf{P}$ acts trivially on $\mathrm{Fib}_{\bar{u}}(V)$.

(c) This follows from the previous part, as the groups P_η are pro-p-groups.

□

Using the fact that every curve can be lifted to characteristic 0 together with Corollary 6.11, Grothendieck proves the following structure theorem for curves.

THEOREM 6.22 ([**22**, Exp. XIII, Cor. 2.12]). *If k is an algebraically closed field of characteristic $p \geq 0$, X a smooth projective curve of genus g over k and $U = X \smallsetminus \{x_1, \dots, x_n\}$, with $x_1, \dots, x_n$ closed points, then for any geometric point $\bar{u}$ of U, the tame fundamental group $\pi_1^{\mathrm{tame}}(U, \bar{u})$ is a quotient of the profinite completion of the group*

$$\langle a_1, \dots, a_g, b_1, \dots, b_g, c_1, \dots, c_n \,\|\, [a_1, b_1] \cdot \ldots \cdot [a_g, b_g] c_1 \cdot \ldots \cdot c_n = 1 \rangle. \tag{6.3}$$

Moreover, the maximal prime-to-p-quotient of $\pi_1^{\mathrm{tame}}(U, \bar{u})$ is isomorphic to the pro-prime-to-p-completion of (6.3).

7. Étale sheaves

In this section we recall the definition of an étale sheaf on a scheme and give some basic constructions.

CONVENTION 7.1. Throughout this section schemes are assumed to be *separated* and *noetherian.*

7.1.1. *Étale neighborhoods.* Let X be a scheme. Recall that a *geometric point* of X is a morphism $\bar{x} = \operatorname{Spec} k \to X$, where k is an algebraically closed field. If x is the image of $\bar{x} \to X$, we say that $\bar{x}$ is a geometric point over x or that x is the center of $\bar{x} \to X$. By abuse of notation we also denote by $\bar{x}$ the morphism $\bar{x} \to X$. An *étale neighborhood* of $\bar{x}$ is a diagram

$$\begin{array}{ccc} & & U \\ & \nearrow & \downarrow{\scriptstyle u} \\ \bar{x} & \longrightarrow & X, \end{array}$$

where u is étale. A morphism between two étale neighborhoods of $\bar{x}$ is given by the obvious commutative diagram. Note that the opposite category of étale neighborhoods of $\bar{x}$ is filtered. (Indeed: If $U_* = (\bar{x} \to U \to X)$ and $V_* = (\bar{x} \to V \to X)$ are two étale neighborhoods, their cartesian product $\bar{x} \to U \times_X V \to X$ is another étale neighborhood mapping to U_* and V_*. If $f, g : U_* \to V_*$ are two morphisms of étale neighborhoods, then the connected component U_0 of U containing the center of $\bar{x}$ naturally defines an étale neighborhood U_{0*} of $\bar{x}$ with a morphism $j : U_{0,*} \to U_*$, and by Proposition 5.5 we have $f \circ j = g \circ j$.)

7.1.2. *Étale sheaves.* Let X be a scheme. We denote by $(\text{ét}/X)$ the category with objects the étale X-schemes and morphisms the X-morphisms (these are automatically étale by Proposition 5.4). Let A be a ring. Then an *étale presheaf* of A-modules on X is a functor

$$\mathcal{F} : (\text{ét}/X)^{\mathrm{op}} \to (A\text{-mod}).$$

In case $A = \mathbb{Z}$, we will simply say that $\mathcal{F}$ is an étale presheaf on X or is a presheaf on $X_{\text{ét}}$. Let $\bar{x} \to X$ be a geometric point. The stalk of a presheaf of A-modules $\mathcal{F}$ on $X_{\text{ét}}$ at $\bar{x}$ is given by

$$\mathcal{F}_{\bar{x}} := \varinjlim_{\bar{x} \to U} \mathcal{F}(U),$$

where the limit is over the étale neighborhoods of $\bar{x}$. (One can take just one neighborhood for each isomorphism class to avoid set theoretical problems.)

We say $\mathcal{F}$ is an *étale sheaf* of A-modules on X or a *sheaf of A-modules* on $X_{\text{ét}}$, if it is a presheaf of A-modules on $X_{\text{ét}}$ and for any étale map $U \to X$ and all families $\{u_i : U_i \to U \mid i \in I\}$ of étale maps with $\bigcup_i u_i(U_i) = U$ the sequence

$$0 \to \mathcal{F}(U) \to \prod_i \mathcal{F}(U_i) \rightrightarrows \prod_{i,j} \mathcal{F}(U_i \times_U U_j) \tag{7.1}$$

is exact. Recall that the exactness of this sequence means that (1) an element $s \in \mathcal{F}(U)$ is zero iff $s_{|U_i} = 0$ for all i and (2) for a collection of elements $s_i \in \mathcal{F}(U_i)$, $i \in I$, there exists an element $s \in \mathcal{F}(U)$ with $s_{|U_i} = s_i$ iff $s_{i|U_i \times_U U_j} = s_{j|U_i \times_U U_j}$ for all $i, j \in I$. (In particular, if $\{U_i \to U\}$ is a Zariski open cover this is just the usual sheaf condition.) A morphism between (pre-)sheaves of A-modules on $X_{\text{ét}}$ is a natural transformation of functors.

If P is a presheaf of A-modules on $X_{\text{ét}}$ then there exits a sheaf of A-modules aP on $X_{\text{ét}}$ and a morphism $P \to aP$ such that any morphism $P \to \mathcal{F}$ with $\mathcal{F}$ a sheaf on X_{et} factors uniquely as $P \to aP \to \mathcal{F}$. The sheaf aP is called the *sheaf associated to* P or *the sheafification of* P and it satisfies $(aP)_{\bar{x}} = P_{\bar{x}}$ for any geometric point $\bar{x} \to X$. The category of presheaves of A-modules on $X_{\text{ét}}$ is abelian. Hence the category of sheaves of A-modules on $X_{\text{ét}}$ is also abelian. (The cokernel of a morphism of sheaves is given by the sheafification of the cokernel in the bigger category of presheaves.) A sequence of sheaves of A-modules $0 \to \mathcal{F}' \to \mathcal{F} \to \mathcal{F}'' \to 0$ is exact if and only if the sequence of A-modules $0 \to \mathcal{F}'_{\bar{x}} \to \mathcal{F}_{\bar{x}} \to \mathcal{F}''_{\bar{x}} \to 0$ is exact for all geometric points $\bar{x} \to X$.

REMARK 7.2. Let $\mathcal{F}$ be a presheaf on $X_{\text{ét}}$. Then $\mathcal{F}$ is a sheaf iff the sequence (7.1) is exact in the following two situations: (1) $U \to X$ is an étale map and $U = \bigcup_i U_i$ is a Zariski open cover (i.e. the family $\{u_i\}$ is given by the open immersions $\{U_i \hookrightarrow U\}$) and (2) $U \to X$ is an étale map with U affine and the family $\{u_i\}$ consists just of one affine, surjective and étale map $U' \to U$, see e.g. [**39**, II, Prop. 1.5.].

EXAMPLE 7.3. Let X be a scheme.

(a) Let M be an A-module. For $U \to X$ an étale map, denote by $\pi_0(U)$ the set of connected components of U (it is finite by Convention 7.1). Then $U \mapsto M^{\pi_0(U)}$ defines a sheaf of A-modules on $X_{\text{ét}}$. It is called the *constant étale sheaf associated to* M and denoted by M_X or simply by M again.

(b) The assignment $U \mapsto \Gamma(U, \mathcal{O}_U)$ defines a sheaf of $\Gamma(X, \mathcal{O}_X)$-modules on $X_{\text{ét}}$. (Indeed, by Remark 7.2 it suffices to show that for a faithfully flat morphism of rings $\varphi : A \to B$, the sequence $0 \to A \xrightarrow{\varphi} B \xrightarrow{d} B \otimes_A B$ is exact, where $d(b) = 1 \otimes b - b \otimes 1$. Since φ is faithfully flat this is equivalent to proving that the sequence is exact when tensored with B. Thus we have to show that $0 \to B \xrightarrow{d_0} B^{\otimes_A 2} \xrightarrow{d_1} B^{\otimes_A 3}$ is exact, where $d_0(b) = 1 \otimes b$ and $d_1(b_1 \otimes b_2) = b_1 \otimes 1 \otimes b_2 - 1 \otimes b_1 \otimes b_2$. Define $s_0 : B^{\otimes 2} \to B$ by $s_0(b_1 \otimes b_2) = b_1 b_2$ and $s_1 : B^{\otimes 3} \to B^{\otimes 2}$, $s_1(b_1 \otimes b_2 \otimes b_3) = b_1 \otimes b_2 b_3$. Then $\mathrm{id}_B = s_0 \circ d_0$ and $\mathrm{id}_{B^{\otimes 2}} = d_0 \circ s_0 + s_1 \circ d_1$, hence the exactness.) The underlying sheaf of abelian groups on $X_{\text{ét}}$ is denoted by $\mathbb{G}_{a,X}$ or simply $\mathbb{G}_a$ if it is clear that we view it as a sheaf on $X_{\text{ét}}$.

(c) The assignment $U \mapsto \Gamma(U, \mathcal{O}_U)^\times$ defines a sheaf on $X_{\text{ét}}$. This follows directly from b) above. We denote this sheaf by $\mathbb{G}_{m,X}$ or simply $\mathbb{G}_m$.

(d) Let $n \geq 1$ be a natural number. We define the sheaf μ_n on $X_{\text{ét}}$ as the kernel of multiplication by n on $\mathbb{G}_m$, i.e. we have an exact sequence

$$0 \to \mu_n \to \mathbb{G}_m \xrightarrow{\cdot n} \mathbb{G}_m.$$

Explicitly $\mu_n(U) = \{a \in \Gamma(U, \mathcal{O}_U) \,|\, a^n = 1\}$. If R is a ring we also write by abuse of notation $\mu_n(R) := \{a \in R \,|\, a^n = 1\}$. We elaborate a bit.

(i) Assume n is invertible in $\mathcal{O}_X$ then

$$0 \to \mu_n \to \mathbb{G}_m \xrightarrow{\cdot n} \mathbb{G}_m \to 0$$

is an exact sequence of sheaves on $X_{\text{ét}}$. Only the surjectivity has to be proved and this follows from the fact that for all U and all $a \in \mathbb{G}_m(U)$ the map $\operatorname{Spec} \mathcal{O}_U[t]/(t^n - a) \to U$ is étale.

(ii) Let $R \neq 0$ be a ring in which n is invertible and in which the polynomial $t^n - 1$ decomposes, i.e.

$$t^n - 1 = (t - \zeta_1)\cdots(t - \zeta_n) \quad \text{in } R[t]. \tag{7.2}$$

Then $\{\zeta_1, \ldots, \zeta_n\}$ is a cyclic subgroup of order n of $R^\times$. Indeed, let K be a field and $\varphi : R \to K$ a morphism. Then φ induces a morphism $\{\zeta_1, \ldots, \zeta_n\} \to \mu_n(K)$. This morphism is injective. (Else differentiating (7.2) would give $n\varphi(\zeta_i)^{n-1} = 0$ for some i, which is absurd since $n, \varphi(\zeta_i) \in K^\times$.) Thus this morphism is bijective and we know that $\mu_n(K)$ is cyclic of order n.

Furthermore, any morphism $R \to A$ into a *local* ring A induces a bijection $\{\zeta_1, \ldots, \zeta_n\} = \mu_n(A)$. Indeed, if A is a local R-algebra any $a \in A$ with $a^n = 1$ is of the form $a = \zeta_i + x$ for some i and $x \in \mathfrak{m}_A$ the maximal ideal of A. We obtain $1 = (\zeta_i + x)^n = 1 + xu$ with $u \in A^\times$, i.e. $x = 0$. This proves the bijection.

(iii) Assume R is as in (ii) above and X is an R-scheme. Then the choice of a cyclic generator $\zeta \in R$ of $\{\zeta_1, \ldots, \zeta_n\}$ defines an isomorphism

$$\mu_n \cong (\mathbb{Z}/n\mathbb{Z})_X \quad \text{on } X_{\text{ét}}.$$

Explicitly for all étale maps $U \to X$ we have an isomorphism

$$(\mathbb{Z}/n\mathbb{Z})^{\pi_0(U)} \xrightarrow{\simeq} \mu_n(U), \quad (\bar{m}_i)_{i \in \pi_0(U)} \mapsto (\zeta^{m_i})_{i \in \pi_0(U)}.$$

It suffices to prove this Zariski (or even étale) locally and hence it follows from (ii) above.

Notice that the statements (i) - (iii) are wrong if n is not invertible on X.

(e) Let G be a commutative group scheme, i.e. a commutative group object in the category of schemes. Then the assignment $U \mapsto G(U) := \operatorname{Hom}(U, G)$ defines a sheaf of abelian groups on $X_{\text{ét}}$, see e.g. [**22**, VIII, Thm. 5.2]. In fact, it is not hard to see that the Examples (b) - (d) are of this form.

7.1.3. *Direct and inverse image.* Let $\pi : X \to Y$ be a morphism of schemes and A a ring. If $\mathcal{F}$ is a sheaf of A-modules on $X_{\text{ét}}$, then we obtain a sheaf of A-modules $\pi_*\mathcal{F}$ on $Y_{\text{ét}}$ via

$$\pi_*\mathcal{F}(V) = \mathcal{F}(V \times_Y X), \quad V \to Y \text{ étale}.$$

We call this sheaf the *direct image of $\mathcal{F}$ under π*. We get a left exact functor from the category of A-modules on $X_{\text{ét}}$ to the category of A-modules on $Y_{\text{ét}}$, $\mathcal{F} \mapsto \pi_*(\mathcal{F})$.

This functor has a left adjoint denoted by π^*, i.e. for each $\mathcal{F}$ on $X_{\text{ét}}$ and $\mathcal{G}$ on $Y_{\text{ét}}$ there is an isomorphism of abelian groups

$$\operatorname{Hom}(\mathcal{G}, \pi_*\mathcal{F}) \cong \operatorname{Hom}(\pi^*\mathcal{G}, \mathcal{F}),$$

which is functorial in $\mathcal{F}$ and $\mathcal{G}$. We call $\pi^*\mathcal{G}$ the *inverse image of $\mathcal{G}$ under π*. If $\bar{x}$ is a geometric point on X and we denote $\pi(\bar{x})$ the geometric point of Y given by the composition $\bar{x} \to X \to Y$, then we have

$$(\pi^*\mathcal{G})_{\bar{x}} = \mathcal{G}_{\pi(\bar{x})}. \tag{7.3}$$

It follows that π^* is an exact functor. By abuse of notation we also write $\mathcal{G}_{|X} := \pi^*\mathcal{G}$. In case $u : U \to Y$ is étale $u^*\mathcal{G}$ coincides with the restriction of the functor $\mathcal{G}$ to the category $(\text{ét}/U)$. If $\pi : X \to Y$ is any morphism and $u : U \to Y$ is étale and we denote by $\pi_1 : X \times_Y U \to U$ and $u_1 : X \times_Y U \to X$ the projection maps then we clearly have

$$u^*\pi_*\mathcal{F} = \pi_{1*}u_1^*\mathcal{F}. \tag{7.4}$$

EXAMPLE 7.4. (a) Let $\pi : X \to Y$ be a morphism of schemes and M an A-module. Then (with the notation from Example 7.3, (a))

$$\pi^*(M_Y) = M_X.$$

Indeed, for an étale map $V \to Y$ there is a natural morphism

$$M_Y(V) = M^{\pi_0(V)} \to M^{\pi_0(V \times_Y X)} = \pi_*(M_X)(V).$$

(If V is connected it is the diagonal.) This induces a morphism of sheaves $M_Y \to \pi_*(M_X)$ on $Y_{\text{ét}}$, by adjunction also a morphism $\pi^*(M_Y) \to M_X$. It suffices to check that this latter map is an isomorphism on the stalks at the geometric points of X. This follows directly from (7.3).

(b) In general $\pi_*(M_X)$ is not isomorphic to M_Y. For example, if $i : X \hookrightarrow Y$ is a closed immersion then $i_*(M_X)$ is zero in the stalks at geometric points whose center lies in $Y \setminus X$.

7.1.4. *Action of a finite group.* Let G be a finite group acting on a scheme X, i.e. we are given a group homomorphism $G \to \operatorname{Aut}(X)$. Assume G acts admissibly on X, i.e. X is a union of open affines $U = \operatorname{Spec} A$ such that the action of G restricts to an action on U. (This is e.g. the case if X is quasi-projective over an affine scheme). Then we can form the quotient $\pi : X \to X/G$, where X/G is defined by glueing the schemes $U/G := \operatorname{Spec} A^G$, for U as above. We have $\operatorname{Hom}(X, Y)^G = \operatorname{Hom}(X/G, Y)$ for all schemes Y, where G acts on $\operatorname{Hom}(X, Y)$ via precomposition. In general, X/G might not be noetherian, which violates our Convention 7.1. Therefore, by convention we will only consider finite group actions on X for which the quotient X/G is actually noetherian. This is e.g. the case if X is of finite type over some (noetherian) base S and G acts via S-automorphisms. See [**22**, Exp.V, 1].

An étale sheaf of A-modules on (X, G) is an étale sheaf of A-modules $\mathcal{F}$ on X together with morphisms

$$\mathcal{F}(\sigma) : \mathcal{F} \to \sigma^*\mathcal{F}, \quad \sigma \in G,$$

such that $\mathcal{F}(1_G) = \mathrm{id}_{\mathcal{F}}$ and $\mathcal{F}(\tau\sigma) = \tau^*(\mathcal{F}(\sigma)) \circ \mathcal{F}(\tau)$. In particular, the $\mathcal{F}(\sigma)$'s are isomorphisms. Sometimes we just say $\mathcal{F}$ is a sheaf with G-action on X and it is understood that this action is compatible with the given action on X. For any geometric point $\bar{x} \to X$ we obtain isomorphisms $\mathcal{F}_{\bar{x}} \xrightarrow{\cong} \mathcal{F}_{\sigma(\bar{x})}$, where $\sigma(\bar{x})$ denotes the composition $\bar{x} \to X \xrightarrow{\sigma} X$. Also notice that we have natural isomorphisms $\sigma_*\sigma^*\mathcal{F} \cong \mathcal{F} \cong \sigma^*\sigma_*\mathcal{F}$ induced by adjunction; hence to give a map $\mathcal{F}(\sigma)$ as above is equivalent to give a map $\sigma_*\mathcal{F} \to \mathcal{F}$.

Let $\pi : X \to X/G$ be as above. The action of G on $\mathcal{F}$ induces maps $\sigma_*\mathcal{F} \to \mathcal{F}$ and applying π_* we obtain a map $\pi_*\mathcal{F} = \pi_*\sigma_*\mathcal{F} \to \pi_*\mathcal{F}$. Thus G acts on $\pi_*\mathcal{F}$ (where we equip X/G with the trivial G-action). We can therefore form the sheaf of G-invariant elements $(\pi_*\mathcal{F})^G$ whose section on $U \to X/G$ (étale map) are given by

$$(\pi_*\mathcal{F})^G(U) = \{a \in \mathcal{F}(U \times_{X/G} X) \,|\, \sigma \cdot a = a \text{ for all } \sigma \in G\},$$

where we abbreviate the notation $\sigma \cdot a := \mathcal{F}(\sigma)(a)$.

7.1.5. *Extension by zero.* Let $j : U \hookrightarrow X$ be an open immersion and denote by $i : Z \hookrightarrow X$ the closed immersion of the complement of U (equipped with some scheme structure). Let $\mathcal{F}$ be a sheaf of A-modules on U. Then we define the *extension by zero* of $\mathcal{F}$ as

$$j_!\mathcal{F} := \operatorname{Ker}(j_*\mathcal{F} \to i_*i^*j_*\mathcal{F}),$$

where $j_*\mathcal{F} \to i_*i^*j_*\mathcal{F}$ is the adjunction map for (i_*, i^*). For a point $x \in X$ and $\bar{x}$ a geometric point over x we have

$$(j_!\mathcal{F})_{\bar{x}} = \begin{cases} \mathcal{F}_{\bar{x}}, & \text{if } x \in U, \\ 0, & \text{if } x \in Z. \end{cases}$$

It follows that $j_!$ is an exact functor from the category A-modules on $U_{\text{ét}}$ to the category of A-modules on $X_{\text{ét}}$. It is left adjoint to j^* i.e. there is a functorial isomorphism

$$\operatorname{Hom}(j_!\mathcal{G}, \mathcal{F}) \cong \operatorname{Hom}(\mathcal{G}, j^*\mathcal{F}),$$

where $\mathcal{G}$ is a sheaf on $U_{\text{ét}}$ and $\mathcal{F}$ a sheaf on $X_{\text{ét}}$.

DEFINITION 7.5. Let X be a scheme, A a noetherian ring and M an A-module. Then we call an étale sheaf $\mathcal{F}$ of A-modules on X *locally constant with stalk M* if there exists a family $\{u_i : U_i \to X\}$ of étale maps with $\bigcup_i u_i(U_i) = X$ such that $\mathcal{F}_{|U_i} \cong M_{U_i}$ for all i.

PROPOSITION 7.6. *In the situation of Definition 7.5 assume M is finite (as a set). Then $\mathcal{F}$ is represented by a finite étale group scheme $X_{\mathcal{F}}$ over X, i.e. there is a finite étale morphism $X_{\mathcal{F}} \to X$ such that*

$$\mathcal{F} \cong \operatorname{Hom}_X(-, X_{\mathcal{F}}) \quad \textit{on } (\textit{ét}/X).$$

Assume X is connected, then it follows that there exists a connected finite étale Galois cover $P \to X$ which trivializes $\mathcal{F}$, i.e. $\mathcal{F}_{|P} \cong M_P$.

Before we prove the proposition we recall the following statement from descent theory:

THEOREM 7.7 ([**22**, VIII, Thm. 2.1, Cor. 5.7] and [**19**, Cor. (17.7.3)]). *Let $f : U \to X$ be faithfully flat and quasi-compact. Let $V \to U$ be an affine morphism such that there exists an isomorphism $\gamma : p_1^*V \xrightarrow{\cong} p_2^*V$ satisfying*

$$p_{13}^*(\gamma) \circ p_{12}^*(\gamma) = p_{23}^*(\gamma),$$

where $p_i : U \times_X U \to U$ and $p_{ij} : U \times_X U \times_X U \to U \times_X U$ denote the projection on the respective factors, $p_i^* V$ denotes the pullback of V along p_i and $p_{ij}^*(\gamma)$ denotes the pullback of γ along p_{ij}.

Then there exists an affine morphism $Y \to X$ such that $V = Y \times_X U$. (One says V/U descends *to Y/X.) Furthermore if $V \to U$ is finite or étale, then so is $Y \to X$.*

PROOF OF PROPOSITION 7.6. We find a finite family $\{u_i : U_i \to X\}$ of étale maps such that $u_i^* \mathcal{F} \cong M_{U_i}$. The constant sheaf M_{U_i} is represented by the U_i-group scheme $U_i \times M$ (= disjoint union over $|M|$ copies of U_i with group action induced by the one on M). Then $U_i \times M \to U_i$ is a finite étale U_i-group scheme representing $u_i^* \mathcal{F}$. Set $U := \bigsqcup_i U_i$. Then we have an étale and surjective (in particular quasi-compact by Convention 7.1) morphism $U \to X$ and a finite étale U-group scheme $U \times M$ representing $\mathcal{F}_{|U}$. Denote by $p_1, p_2 : U \times_X U \to U$ the two projections. The natural isomorphism $p_1^*(\mathcal{F}_{|U}) \cong p_2^*(\mathcal{F}_{|U})$ induces a gluing data $p_1^*(U \times M) \cong p_2^*(U \times M)$ and hence the finite étale U-group scheme $U \times M$ descends to a finite étale map $X_{\mathcal{F}} \to X$, see Theorem 7.7. By construction and the sheaf property $X_{\mathcal{F}}$ represents $\mathcal{F}$ over X.

If X is connected, then any connected component of $X_{\mathcal{F}}$ is still a finite étale covering of X. We take a finite étale Galois covering $P \to X$ which factors over all the connected components of $X_{\mathcal{F}}$; this can be achieved as in the discussion below Example 6.5. Then $X_{\mathcal{F}} \times_X P \cong P \times M$ represents $\mathcal{F}_{|P}$, i.e. $\mathcal{F}_{|P} \cong M_P$. □

7.1.6. *Constructible sheaves.* Let X be a scheme. A subset $Z \subseteq X$ is called *locally closed* if it is the intersection of an open and a closed subset in X. We equip such a Z with the reduced scheme structure and obtain an immersion $i : Z \hookrightarrow X$.

Let A be a noetherian ring which is torsion (i.e. $mA = 0$ for some natural number m) and $\mathcal{F}$ a sheaf of A-modules on $X_{\text{ét}}$. Then we say that $\mathcal{F}$ is *constructible* if there exist finite type A-modules $M_1, \ldots, M_n$ and locally closed subsets $X_1, \ldots, X_n \subseteq X$ such that

$$X = \bigsqcup_i X_i \quad \text{and} \quad \mathcal{F}_{|X_i} \text{ is locally constant with stalks } M_i.$$

We have (see [**5**, IX, Prop. 2.6, Prop. 2.9]):

(a) The category of constructible sheaves of A-modules is abelian.
(b) If $0 \to \mathcal{F}' \to \mathcal{F} \to \mathcal{F}'' \to 0$ is a short exact sequence of étale sheaves of A-modules, and $\mathcal{F}', \mathcal{F}''$ are constructible, then so is $\mathcal{F}$.
(c) An étale sheaf of A-modules is constructible iff it is a noetherian object in the category of étale sheaves of A-module, i.e., any ascending sequence of submodules becomes stationary.

REMARK 7.8. Let $\pi : X \to Y$ be a morphism and $\mathcal{G}$ a constructible sheaf of A-modules on $Y_{\text{ét}}$, then $\pi^* \mathcal{G}$ is a constructible A-module on $X_{\text{ét}}$. (This follows directly from Example 7.4.) If π is proper and $\mathcal{F}$ is a constructible A-module on $X_{\text{ét}}$, then $\pi_* \mathcal{F}$ is a constructible A-module on $Y_{\text{ét}}$, see [**5**, Exp. XIV, Thm 1.1].

8. ℓ-adic sheaves

In this section we recall the notions of constructible and lisse sheaves with ℓ-adic coefficients. We show that lisse sheaves correspond to continuous representations of the fundamental group. The references for this section are [**12**, Arcata, II and Rapport, 1.2], [**13**, (1.1)] and [**20**, Exp. V, VI]. *We keep Convention* 7.1.

DEFINITION 8.1. Let R be a complete local DVR with maximal ideal $\mathfrak{m}$. Assume $R/\mathfrak{m}$ has characteristic $\ell > 0$. Let X be a scheme.

(a) A *constructible R-sheaf* on X is a projective system of R-modules $\mathcal{F} = (\mathcal{F}_n)_{n\geq 1}$ on $X_{\text{ét}}$ satisfying the following two properties:
   (i) $\mathfrak{m}^n \cdot \mathcal{F}_n = 0$ and $\mathcal{F}_n$ is a constructible $R/\mathfrak{m}^n$-module on $X_{\text{ét}}$, see 7.1.6.
   (ii) $\mathcal{F}_n = \mathcal{F}_{n+1} \otimes_{R/\mathfrak{m}^{n+1}} R/\mathfrak{m}^n$, for all $n \geq 1$.

(b) A *lisse* R-sheaf on X is a constructible R-sheaf $\mathcal{F} = (\mathcal{F}_n)$ such that each $\mathcal{F}_n$ is a locally constant sheaf of $R/\mathfrak{m}^n$-modules.

EXAMPLE 8.2. (a) Let R be as above and $\mathcal{F}$ a locally constant sheaf of finitely generated $R/\mathfrak{m}^{n_0}$-modules on $X_{\text{ét}}$. Then we can view $\mathcal{F}$ also as a lisse R-sheaf $(\mathcal{F}_n)$, via $\mathcal{F}_n := \mathcal{F}$ for $n > n_0$ and $\mathcal{F}_n := \mathcal{F} \otimes_{R/\mathfrak{m}^{n_0}} R/\mathfrak{m}^n R$, for $n \leq n_0$.

(b) Let $\pi : X \to Y$ be a proper morphism and $\mathcal{F} = (\mathcal{F}_n)$ a constructible R-sheaf on X. We want to define the constructible sheaf $\pi_*\mathcal{F}$. The first try is to consider the system $(\pi_*(\mathcal{F}_n))$. But this might not satisfy (ii). To repair this we have to use the following two facts (see [**20**, Exp. VI, Lem 2.2.2] and [**20**, Exp. V, Prop. 3.2.3]):
   (i) The system $(\pi_*(\mathcal{F}_n))$ satisfies the Mittag-Leffler-Artin-Rees condition, i.e., there exists a natural number r_0 such that for all n and all $r \geq r_0$ we have
   $$\text{image}(\pi_*\mathcal{F}_{n+r} \to \pi_*\mathcal{F}_n) = \text{image}(\pi_*\mathcal{F}_{n+r_0} \to \pi_*\mathcal{F}_n) =: I_n.$$
   (ii) There exists a natural number s_0 such that for all n and all $s \geq s_0$ we have the natural isomorphism
   $$I_{n+s} \otimes_R R/\mathfrak{m}^n \xrightarrow{\simeq} I_{n+s_0} \otimes_R R/\mathfrak{m}^n.$$

   Then we define $\pi_*\mathcal{F} = (I_{n+s_0} \otimes_R R/\mathfrak{m}^n)_n$. There is a natural morphism of projective systems $\pi_*\mathcal{F} \to (\pi_*(\mathcal{F}_n))$ which is an *AR-isomorphism*, i.e., if we denote by (K_n) and (C_n) the kernel and cokernel, respectively, then there exists a natural number N such that transition maps $K_{n+N} \to K_n$ and $C_{n+N} \to C_n$ are zero, for all $n \geq 1$. (Here we can take $N = r_0 + s_0$.)

(c) Let ℓ be an invertible prime on X. We have surjections of $\mathbb{Z}_\ell$-sheaves on $X_{\text{ét}}$, $\mu_{\ell^{n+1}} \to \mu_{\ell^n}$, $a \mapsto a^\ell$. The corresponding projective system of $\mathbb{Z}_\ell$-modules on $X_{\text{ét}}$ is denoted by $\mathbb{Z}_\ell(1)$,
   $$\mathbb{Z}_\ell(1) := (\mu_{\ell^n}).$$
   It follows from Example 7.3, (d), (iii), that $\mathbb{Z}_\ell(1)$ is a lisse $\mathbb{Z}_\ell$-sheaf on X. But notice that in general there is no étale covering $\{U_i \to X\}$ such that $\mu_{\ell^n|U_i} = (\mathbb{Z}/\ell^n)_{U_i}$ for all n (e.g. $X = \operatorname{Spec}\mathbb{Q}$). This shows that lisse sheaves are in general not locally constant projective systems of sheaves.

8.0.1. *The category of lisse R-sheaves.* Let $R, \mathfrak{m}, \ell$ be as in Definition 8.1 and X a scheme. Let $\mathcal{F} = (\mathcal{F}_n), \mathcal{G} = (\mathcal{G}_n)$ be two constructible R-sheaves on X. Then we have morphisms of R-modules
$$\text{Hom}(\mathcal{F}_{n+1}, \mathcal{G}_{n+1}) \to \text{Hom}(\mathcal{F}_n, \mathcal{G}_n), \quad \varphi \mapsto \varphi \otimes R/\mathfrak{m}^n.$$
The category of constructible (resp. lisse) R-sheaves is the category with morphisms given by
$$\text{Hom}(\mathcal{F}, \mathcal{G}) = \varprojlim_n \text{Hom}(\mathcal{F}_n, \mathcal{G}_n).$$

It is an abelian category, see [**20**, V, Thm 5.2.3].

8.0.2. *The category of* $\bar{\mathbb{Q}}_\ell$*-sheaves.* Let X be a scheme and ℓ an invertible prime on X. Fix an algebraic closure $\bar{\mathbb{Q}}_\ell$ of $\mathbb{Q}_\ell$.

(a) Let E be a *finite* field extension of $\mathbb{Q}_\ell$ inside $\bar{\mathbb{Q}}_\ell$ and R the integral closure of $\mathbb{Z}_\ell$ in E. The category of constructible E-sheaves on X is by definition the localization of the category of constructible R-sheaves localized with respect to the full-subcategory of torsion sheaves. This means the following:

Objects: There is an essentially surjective functor

$$(\text{constr. } R\text{-sheaves}) \to (\text{constr. } E\text{-sheaves}), \quad \mathcal{F} \mapsto \mathcal{F} \otimes_R E$$

Morphisms: $\mathrm{Hom}(\mathcal{F} \otimes_R E, \mathcal{G} \otimes_R E) = \mathrm{Hom}(\mathcal{F}, \mathcal{G}) \otimes_R E$.

We say that a constructible E-sheaf $\mathcal{F} \otimes_R E$ is *lisse* if there exists an étale covering $\{U_i \to X\}$ and lisse R-sheaves $\mathcal{F}_i$ on U_i such that $\mathcal{F}_{|U_i} \otimes_R E \cong \mathcal{F}_i \otimes_R E$.

(b) If $E'/E/\mathbb{Q}_\ell$ are finite extensions inside $\bar{\mathbb{Q}}_\ell$, then there is a natural functor (constr. E-sheaves) $\to$ (constr. E'-sheaves) on objects given by

$$\mathcal{F} \otimes_R E \mapsto (\mathcal{F} \otimes_R E) \otimes_E E' = \mathcal{F} \otimes_R E'.$$

The category of constructible $\bar{\mathbb{Q}}_\ell$-sheaves is the inductive 2-limit over the categories of constructible E-sheaves, $E \subseteq \bar{\mathbb{Q}}_\ell$. This means the following:

Objects: For all finite extensions $E/\mathbb{Q}_\ell$ inside $\bar{\mathbb{Q}}_\ell$ there are functors

$$(\text{constr. } E\text{-sheaves}) \to (\text{constr. } \bar{\mathbb{Q}}_\ell\text{-sheaves}), \quad \mathcal{F} \mapsto \mathcal{F} \otimes_E \bar{\mathbb{Q}}_\ell.$$

These satisfy:

(a) Each object in (constr. $\bar{\mathbb{Q}}_\ell$-sheaves) is isomorphic to an object of the form $\mathcal{F} \otimes_E \bar{\mathbb{Q}}_\ell$, for some constructible E-sheaf $\mathcal{F}$.

(b) If E'/E is a finite extension and $\mathcal{F} \in$ (constr. E-sheaves), then there is a canonical isomorphism

$$\mathcal{F} \otimes_E \bar{\mathbb{Q}}_\ell \cong (\mathcal{F} \otimes_E E') \otimes_{E'} \bar{\mathbb{Q}}_\ell.$$

Morphisms: For $\mathcal{F}, \mathcal{G} \in$ (constr. E-sheaves) we have

$$\mathrm{Hom}(\mathcal{F} \otimes_E \bar{\mathbb{Q}}_\ell, \mathcal{G} \otimes_E \bar{\mathbb{Q}}_\ell) = \mathrm{Hom}(\mathcal{F}, \mathcal{G}) \otimes_E \bar{\mathbb{Q}}_\ell.$$

A *lisse* $\bar{\mathbb{Q}}_\ell$*-sheaf* on X is a constructible $\bar{\mathbb{Q}}_\ell$-sheaf which is locally of the form $\mathcal{F} \otimes_E \bar{\mathbb{Q}}_\ell$, with $\mathcal{F}$ a lisse E-sheaf.

DEFINITION 8.3. Let X be a scheme and ℓ an invertible prime on X.

(a) Let R be a complete local DVR with maximal ideal $\mathfrak{m}$ and residue characteristic $\ell > 0$. We say that a lisse R-sheaf $\mathcal{F} = (\mathcal{F}_n)$ is *constant* or *trivial* if there exists a finitely generated R-module M such that we have an isomorphism of projective systems $(\mathcal{F}_n) \cong ((M \otimes_R R/\mathfrak{m}^n)_X)$, where the index X denotes the constant sheaf associated to the module, see Example 7.3, (a). In this case we also write $\mathcal{F} = M_X$ or just $\mathcal{F} = M$.

(b) Let E be a finite extension of $\mathbb{Q}_\ell$ with ring of integers R. We say that a lisse E-sheaf $\mathcal{F}$ is *constant* or *trivial* if there exists a finite dimensional E-vector space V and an R-lattice $M \subseteq V$ such that $\mathcal{F} = M_X \otimes_R E$ in the above notations. As lisse E-sheaf $M_X \otimes_R E$ depends (up to isomorphism) only on V. We write $\mathcal{F} = V_X$ or just $\mathcal{F} = V$.

(c) We say that a lisse $\bar{\mathbb{Q}}_\ell$-sheaf is *constant* or *trivial* if there exists a finite dimensional $\bar{\mathbb{Q}}_\ell$-vector space V, a finite extension $E/\mathbb{Q}_\ell$ and a finite dimensional E-vector space V_E such that $V = V_E \otimes_E \bar{\mathbb{Q}}_\ell$ and $\mathcal{F} = V_{E,X} \otimes_E \bar{\mathbb{Q}}_\ell$. This is independent of the choice of E and V_E. We write $\mathcal{F} = V_X$ or just $\mathcal{F} = V$.

CONVENTION 8.4. Let X be a scheme and assume ℓ is a prime number which is invertible on X. We say that a ring A is an *ℓ-adic coefficient ring* if it is either $\bar{\mathbb{Q}}_\ell$, a finite extension of $\mathbb{Q}_\ell$ or equal to R or $R/\mathfrak{m}^n$, $n \geq 1$, where R is a complete local DVR finite over $\mathbb{Z}_\ell$ with maximal ideal $\mathfrak{m}$. We can therefore speak about constructible (resp. lisse, resp. constant) A-sheaves on X. (In case $A = R/\mathfrak{m}^n$ a lisse A-sheaf $\mathcal{F}$ is just a locally constant and constructible sheaf of A-modules on $X_{\text{ét}}$; by Example 8.2, (a), we can view $\mathcal{F}$ also as a lisse R-sheaf.) We can write any constructible A-sheaf in the form $(\mathcal{F}_n) \otimes_R A$ with R as above and $(\mathcal{F}_n)$ a constructible R-sheaf.

8.0.3. *Tate twist.* Let X be a scheme. For $n \geq 1$ and $i \geq 0$ denote by

$$\mathbb{Z}/\ell^n(i)$$

the sheaf on $X_{\text{ét}}$ associated to

$$U \mapsto \underbrace{\mu_{\ell^n}(U) \otimes_{\mathbb{Z}/\ell^n} \ldots \otimes_{\mathbb{Z}/\ell^n} \mu_{\ell^n}(U)}_{i\text{-times}}$$

and by

$$\mathbb{Z}/\ell^n(-i)$$

the sheaf on $X_{\text{ét}}$ associated to

$$U \mapsto \mathrm{Hom}_U(\mathbb{Z}/\ell^n(i)_{|U}, \mathbb{Z}/\ell^n_{|U}).$$

Let A be an ℓ-adic coefficient ring and $\mathcal{F}$ a constructible A-sheaf. Then we define the i-th Tate twist of $\mathcal{F}$

$$\mathcal{F}(i), \quad i \in \mathbb{Z},$$

as follows: Take a DVR R finite over $\mathbb{Z}_\ell$ and a constructible R-sheaf $(\mathcal{F}_n)_n$ such that $\mathcal{F} = (\mathcal{F}_n) \otimes_R A$. Notice that the $\mathcal{F}_n$ are in particular sheaves of $\mathbb{Z}/\ell^n$-modules on $X_{\text{ét}}$. We define $\mathcal{F}_n(i)$ as the sheaf on $X_{\text{ét}}$ associated to $U \mapsto \mathcal{F}_n(U) \otimes_{\mathbb{Z}/\ell^n} \mathbb{Z}/\ell^n(i)(U)$ and define

$$\mathcal{F}(i) := (\mathcal{F}_n(i)) \otimes_R A.$$

This does not depend on the choice of R and $(\mathcal{F}_n)$. It is a constructible A-sheaf and it is lisse if $\mathcal{F}$ is. We have $\mathcal{F}(i)(j) = \mathcal{F}(i+j)$, for $i, j \in \mathbb{Z}$.

8.0.4. *Stalk of a lisse A-sheaf.* Let $\bar{x} \to X$ be a geometric point of X and ℓ a prime number which is invertible on X.

(a) Let R be a complete local DVR with finite residue field of characteristic ℓ and $\mathcal{F} = (\mathcal{F}_n)$ a lisse R-sheaf on X. We define the stalk of $\mathcal{F}$ at $\bar{x}$ to be

$$\mathcal{F}_{\bar{x}} := \varprojlim_n \mathcal{F}_{n,\bar{x}}.$$

(b) Let E be a finite extension of $\mathbb{Q}_\ell$ and $\mathcal{F}$ a lisse E-sheaf. We define the stalk of $\mathcal{F}$ at $\bar{x}$ to be

$$\mathcal{F}_{\bar{x}} := (\varprojlim_n \mathcal{F}'_{n,\bar{x}}) \otimes_R E,$$

where R is the ring of integers of E over $\mathbb{Q}_\ell$ and $\mathcal{F}' = (\mathcal{F}'_n)$ is a lisse R-sheaf defined on some étale neighborhood of $\bar{x}$ such that locally around $\bar{x}$ we have $\mathcal{F} = \mathcal{F}' \otimes_R E$. This definition is independent of the choice of $\mathcal{F}'$ (up to isomorphism).

(c) Let $\mathcal{F}$ be a lisse $\bar{\mathbb{Q}}_\ell$-sheaf. We define the stalk of $\mathcal{F}$ at $\bar{x}$ to be

$$\mathcal{F}_{\bar{x}} := (\varprojlim_n \mathcal{F}'_{n,\bar{x}}) \otimes_R \bar{\mathbb{Q}}_\ell,$$

where R is the ring of integers of a finite extension E over $\mathbb{Q}_\ell$ and $\mathcal{F}' = (\mathcal{F}'_n)$ is a lisse R-sheaf defined on some étale neighborhood of $\bar{x}$ such that locally around $\bar{x}$ we have $\mathcal{F} = (\mathcal{F}' \otimes_R E) \otimes_E \bar{\mathbb{Q}}_\ell$. This definition is independent of the choice of E and $\mathcal{F}'$ (up to isomorphism).

Thus, all together, if A is an ℓ-adic coefficient ring the stalk $\mathcal{F}_{\bar{x}}$ of a lisse A-sheaf $\mathcal{F}$ at a geometric point $\bar{x}$ is defined. Furthermore (with the above notation) $\varprojlim_n \mathcal{F}'_{n,\bar{x}}$ is a finite type R-module. (In fact it is separated for the $\mathfrak{m}$-adic topology and R is an $\mathfrak{m}$-adically complete ring, thus it is generated by any lift of a system of generators of $(\varprojlim_n \mathcal{F}'_{n,\bar{x}})/\mathfrak{m} = \mathcal{F}'_{1,\bar{x}}$, see e.g. [**38**, Thm 8.4].) Hence $\mathcal{F}_{\bar{x}}$ is a finite type A-module for general A.

We say that a lisse A-sheaf $\mathcal{F}$ is *free* if its stalks are free A-modules. In this case and if X is connected the rank of $\mathcal{F}$ – denoted by $\mathrm{rk}(\mathcal{F})$ – is by definition the rank of $\mathcal{F}_{\bar{x}}$ at some (and hence any) geometric point of X.

8.0.5. *ℓ-adic representations.* Let A be an ℓ-adic coefficient ring (Convention 8.4), which is *not* $\bar{\mathbb{Q}}_\ell$. Let M be a finitely generated A-module, then we can equip it with the ℓ-adic topology as follows: There is a DVR R, finite over $\mathbb{Z}_\ell$, such that A is either the fraction field or a quotient of R. We can find a finitely generated R-module M' such that $M = M' \otimes_R A$ and we define the topology on M in such a way that the image of $\{m + \ell^n \cdot M'\}_{n\geq 0}$ in M is a system of open neighborhoods of $m \in M$. This topology is independent of the choice of M'. (Indeed, if A is a finite ring this is just the discrete topology and otherwise M' is an R-lattice in M and for any other R-lattice M'' we find integers $a, b \geq 1$ such that $\ell^a \cdot M' \subseteq M''$ and $\ell^b \cdot M'' \subseteq M'$; hence M' and M'' define the same topology.) In particular the A-module of A-linear endomorphisms $\mathrm{End}_A(M)$ also carries the ℓ-adic topology and the composition of endomorphisms is a continuous operation. Now let $\mathrm{Aut}_A(M)$ be the group of A-linear automorphisms $\mathrm{Aut}_A(M)$. We have an inclusion $\mathrm{Aut}_A(M) \to \mathrm{End}_A(M) \times \mathrm{End}_A(M)$, $\sigma \mapsto (\sigma, \sigma^{-1})$. We put the product topology on $\mathrm{End}_A(M) \times \mathrm{End}_A(M)$ and the subspace topology on $\mathrm{Aut}_A(M)$. In this way $\mathrm{Aut}_A(M)$ becomes a topological group (compare to Definition 4.66).

Let X be a connected scheme and $\bar{x} \to X$ a geometric point. We define an *A-representation of* $\pi_1(X, \bar{x})$ to be a continuous group homomorphism

$$\pi_1(X, \bar{x}) \to \mathrm{Aut}_A(M),$$

where M is a finitely generated A-module.

By definition, a *$\bar{\mathbb{Q}}_\ell$-representation of* $\pi_1(X, \bar{x})$ is a group homomorphism $\rho_V : \pi_1(X, \bar{x}) \to \mathrm{Aut}_{\bar{\mathbb{Q}}_\ell}(V)$, where V is a finite dimensional $\bar{\mathbb{Q}}_\ell$-vector space, such that there exists a finite field extension $E/\mathbb{Q}_\ell$, a finite dimensional E-vector space V_0, an isomorphism $V \cong V_0 \otimes_E \bar{\mathbb{Q}}_\ell$ and an E-representation $\rho_{V_0} : \pi_1(X, \bar{x}) \to \mathrm{Aut}_E(V_0)$ such that ρ_V is equal to the composition

$$\rho_V : \pi_1(X, \bar{x}) \xrightarrow{\rho_{V_0}} \mathrm{Aut}_E(V_0) \xrightarrow{\otimes_E \bar{\mathbb{Q}}_\ell} \mathrm{Aut}_{\bar{\mathbb{Q}}_\ell}(V).$$

THEOREM 8.5. *Let X be a connected scheme and $\bar{x}$ a geometric point. Let A be an ℓ-adic coefficient ring. Then there is a natural equivalence of categories*

$$(\textit{lisse } A\textit{-sheaves on } X) \xrightarrow{\simeq} (A\textit{-representations of } \pi_1(X,\bar{x})), \quad \mathcal{F} \mapsto \mathcal{F}_{\bar{x}}.$$

Furthermore, let $\pi : X' \to X$ be a finite étale Galois cover with Galois group G. Then the above equivalence induces an equivalence between the following subcategories

$$(\textit{lisse } A\textit{-sheaves on } X,\ \textit{constant on } X') \xrightarrow{\simeq} (\textit{finitely gen. } A[G]\textit{-modules}).$$

If A is finite then under this equivalence a finitely generated $A[G]$-module M corresponds to the lisse A-sheaf given by $(\pi_* M_{X'})^G$ (see 7.1.4 and (8.1) below for how we view $M_{X'}$ as a sheaf on (X',G)).

If A is infinite and M is a finitely generated $A[G]$-module, then there exists a complete local DVR R, finite over $\mathbb{Z}_\ell$ and contained in A, and a finitely generated $R[G]$-module N such that $M = N \otimes_R A$. In this case M corresponds to the lisse A-sheaf $((\pi_* N_{X'})^G \otimes_R R/\mathfrak{m}^n)_n \otimes_R A$.

PROOF. Let (P_α) be an inverse system of finite étale Galois coverings of X with maps $\bar{x} \to P_\alpha$ making the diagram

$$\begin{array}{ccc} & & P_\alpha \\ & \nearrow & \downarrow \\ \bar{x} & \longrightarrow & X, \end{array}$$

commutative and such that an element of $\pi_1(X,\bar{x})$ is the same as a compatible system of X-automorphisms of (P_α) fixing $\bar{x}$, i.e. such that $\pi_1^{\text{ét}}(X,\bar{x}) = \varprojlim \operatorname{Aut}_X(P_\alpha)^{\text{opp}}$ (see (6.1)).

First case: A is a finite ring. Let $\mathcal{F}$ be a lisse A-sheaf. Given an étale neighborhood $U \to X$ of $\bar{x}$ we set $U_\alpha := U \times_X P_\alpha$ and obtain an inverse system (U_α) of étale neighborhoods of $\bar{x}$. An element $\sigma \in \pi_1(X,\bar{x})$ induces a compatible system of X-automorphisms of (U_α). Therefore such an element induces a map

$$\mathcal{F}(U) \to \mathcal{F}(U_\alpha) \xrightarrow{\sigma^*} \mathcal{F}(U_\alpha) \to \mathcal{F}_{\bar{x}}.$$

Taking the limit over all étale neighborhoods U of $\bar{x}$ we get a morphism $\sigma^* : \mathcal{F}_{\bar{x}} \to \mathcal{F}_{\bar{x}}$. We obtain in this way the structure of a $\pi_1(X,\bar{x})$-representation on $\mathcal{F}_{\bar{x}}$. Clearly this construction is functorial.

For the other direction let M be an A-representation of $\pi_1(X,\bar{x})$. The group $\operatorname{Aut}_A(M)$ is finite hence there is a finite étale Galois cover $\pi : P = P_\alpha \to X$ with Galois group G such that $\pi_1(X,\bar{x}) \to \operatorname{Aut}_A(M)$ factors over G. We consider the constant sheaf M_P as a sheaf on (P,G) (see 7.1.4) by defining the map $M_P(\sigma) : \sigma_* M_P \to M_P$ ($\sigma \in G$) via

$$M_P(V \times_{P,\sigma} P) = M \xrightarrow{\sigma^{-1}\cdot} M = M_P(V) \tag{8.1}$$

where $V \to P$ is étale with V connected. Then set

$$\mathcal{F}_M := \pi_*(M_P)^G.$$

We claim that $\mathcal{F}_M$ is étale locally isomorphic to the constant sheaf M and that this construction is independent of the choice of $\pi : P \to X$. Then $\mathcal{F}_M$ is a lisse A-sheaf and the assignment $M \mapsto \mathcal{F}_M$ is functorial. To prove the claim let $V \to P$ be an étale map and assume V is connected. We have $V \times_X P \cong V \times G$. For $\sigma \in G$, the automorphism $\mathrm{id}_V \times \sigma$ ($\sigma \in G$) on $V \times_X P$ translates via this isomorphism into the automorphism $\mathrm{id} \times \sigma \cdot$ of $V \times G$. Hence

$$\begin{aligned}\mathcal{F}_M(V) &= \{a \in \pi_*(M_P)(V) \mid M_P(\sigma)(a) = a \text{ for all } \sigma \in G\} \\ &= \{(a_\tau) \in M \times G \mid \sigma^{-1} \cdot a_{\sigma \cdot \tau} = a_\tau \text{ for all } \sigma, \tau \in G\} \cong M,\end{aligned}$$

where the last isomorphism is given by $M \ni m \mapsto (\sigma \cdot m)_\sigma \in M \times G$. Thus $\mathcal{F}_{M|P} = M_P$. Moreover if we choose another $P' \to X$ we obtain a sheaf $\mathcal{F}'_M$ in the same way. Then we find $P'' \to X$ dominating P' and P and we construct $\mathcal{F}''_M$. There is a natural map $\mathcal{F}_M \to \mathcal{F}''_M$ induced by $\pi_* M_P \to \pi''_*(M_{P|P''}) = \pi''_* M_{P''}$ and by the above we obtain

$$\mathcal{F}_{M|P''} = M_{P''} = \mathcal{F}''_{M|P''}.$$

Thus the natural map $\mathcal{F}_M \to \mathcal{F}''_M$ is an isomorphism étale locally, hence is an isomorphism globally. Same with $\mathcal{F}'_M$. This shows that $\mathcal{F}_M$ is independent of the choice of $P \to X$.

We defined two functors $\mathcal{F} \mapsto \mathcal{F}_{\bar{x}}$ and $M \mapsto \mathcal{F}_M$ and we have to show that they are inverse to each other. It is straightforward to check $\mathcal{F}_{M,\bar{x}} = M$ (as $\pi_1(X,\bar{x})$-representations). Now let $\mathcal{F}$ be a lisse A-sheaf and $\mathcal{F}_{\bar{x}} = M$ its associated representation. By Proposition 7.6 there exists a finite étale Galois cover $\pi : P \to X$ with Galois group G such that $\pi^*\mathcal{F} \cong M_P$. (In particular the representation M of $\pi_1(X,\bar{x})$ factors over G.) Then

$$\mathcal{F}_M = \pi_*(\pi^*\mathcal{F})^G = \mathcal{F}.$$

Here the first equality holds by definition; for the second, first observe that there is a canonical map $\mathcal{F} \to \pi_*(\pi^*\mathcal{F})^G$ (induced by adjunction); it suffices to check that it is an isomorphism when restricted to P which follows from $\mathcal{F}_{M|P} \cong M_P$. This finishes the proof in the case A is finite.

Second case: $A = R$ is a DVR finite over $\mathbb{Z}_\ell$. Let $\mathfrak{m}$ be the maximal ideal of R. A lisse R-sheaf is a projective system $\mathcal{F} = (\mathcal{F}_n)$ as in Definition 8.1. By the first case each $\mathcal{F}_n$ gives rise to a $\pi_1(X,\bar{x})$-representation $\mathcal{F}_{n,\bar{x}}$ which factors over some finite quotient. Clearly they fit together to give a continuous $\pi_1(X,\bar{x})$-representation on $\mathcal{F}_{\bar{x}} = \varprojlim_n \mathcal{F}_{n,\bar{x}}$. We already remarked in 8.0.4 that $\mathcal{F}_{\bar{x}}$ is actually a finitely generated R-module. This gives the functor $\mathcal{F} \mapsto \mathcal{F}_{\bar{x}}$ from the statement. To construct a functor in the other direction, let M be a R-representation of $\pi_1(X,\bar{x})$. We obtain induced representations on $M_n := M \otimes_R R/\mathfrak{m}^n$ which by the first case correspond to lisse $R/\mathfrak{m}^n$-modules $\mathcal{F}_{M_n}$. It is straightforward to check that we obtain in this way a projective system $\mathcal{F}_M = (\mathcal{F}_{M_n})$ which defines a lisse R-sheaf and that the functor $M \mapsto \mathcal{F}_M$ thus defined is inverse to $\mathcal{F} \mapsto \mathcal{F}_{\bar{x}}$.

Third case: A is a finite field extension of $\mathbb{Q}_\ell$ or equal to $\bar{\mathbb{Q}}_\ell$. We get the functor $\mathcal{F} \mapsto \mathcal{F}_{\bar{x}}$ in an analogous way as above. For the functor in the other direction let V be an A-representation of $\pi_1(X,\bar{x})$. Then by Lemma 4.67 we find a DVR R finite over $\mathbb{Z}_\ell$ and contained in A and a finitely generated R-submodule M of V which has a continuous $\pi_1(X,\bar{x})$-action and satisfies $M \otimes_R A \cong V$ as $A[\pi_1(X,\bar{x})]$-modules. By the second case above M gives rise to a lisse R-sheaf $\mathcal{F}_M$ and we define $\mathcal{F}_V := \mathcal{F}_M \otimes_R A$. One checks that this construction is independent of the choice of M and defines a functor $M \mapsto \mathcal{F}_M$ which is inverse to the functor from the statement of the theorem. This finishes the proof. □

EXAMPLE 8.6. Let k be a finite field with $q = p^n$ elements and ℓ a prime different from p. Then the lisse $\bar{\mathbb{Q}}_\ell$-sheaves of rank 1 on $\operatorname{Spec} k$ correspond to homomorphisms $\operatorname{Gal}(\bar{k}/k) \to \bar{\mathbb{Q}}_\ell^\times$ which factor over a continuous homomorphism $\chi : \operatorname{Gal}(\bar{k}/k) \to E^\times$, with E a finite extension of $\mathbb{Q}_\ell$. Let us see what this is. We have an isomorphism of topological groups $\widehat{\mathbb{Z}} \xrightarrow{\cong} \operatorname{Gal}(\bar{k}/k)$ under which 1 is mapped to the q-power Frobenius $x \mapsto x^q$. Let R be the ring of integers of E with maximal ideal $\mathfrak{m} = (\pi)$. Then a homomorphism χ as above factors over a lattice $\frac{1}{\pi^n}R$, i.e. is induced by a continuous group homomorphism $\chi_R : \widehat{\mathbb{Z}} \to \operatorname{Aut}_R(\frac{1}{\pi^n}R) = R^\times$, see Lemma 4.67. The topological group $\widehat{\mathbb{Z}}$ is the free profinite group on 1 generator, so giving a continuous map from $\widehat{\mathbb{Z}}$ to a profinite group G is the same as giving an element of G. Here $R^\times$ is profinite and hence a homomorphism χ_R as above corresponds to the choice of an element of $R^\times$.

All together we see that (up to isomorphism) a lisse $\bar{\mathbb{Q}}_\ell$-sheaf on $\operatorname{Spec} k$ of rank 1 corresponds uniquely to an element of $\bar{\mathbb{Z}}_\ell^\times$, the integral closure of $\mathbb{Z}_\ell$ in $\bar{\mathbb{Q}}_\ell$.

EXAMPLE 8.7. Let X be a connected scheme, $\bar{x} \to X$ a geometric point and ℓ a prime number invertible on X. We have the lisse sheaf $\mathbb{Z}_\ell(1)$ on X (see 8.0.3) at our disposal. The corresponding representation of $\pi_1(X, \bar{x})$ is the following: In the algebraically closed field $k(\bar{x})$ take a family of elements $\zeta_n \in k(\bar{x})$, $n \geq 1$, such that ζ_n is an ℓ^n-th primitive root of unity and $\zeta_{n+1}^\ell = \zeta_n$ for all n. Let $P \to X$ be a finite étale Galois cover with Galois group G_P and a map $\bar{x} \to P$ inducing $\bar{x} \to X$. Denote by x_P the image of $\bar{x} \to P$. If $H^0(P, \mathcal{O}_P)$ contains an ℓ^n-th primitive root of unity, then it also contains a unique ℓ^n-th primitive root of unity which maps to ζ_n under the composition $H^0(P, \mathcal{O}_P) \to \mathcal{O}_{P,x_P} \to k(x_P) \hookrightarrow k(\bar{x})$ (see Example 7.3, (d)); we denote it again by $\zeta_n \in H^0(P, \mathcal{O}_P)$. If $\zeta_n \in H^0(P, \mathcal{O}_P)$, then for $\sigma \in G_P$ the element $\sigma(\zeta_n)$ is again an ℓ^n-th primitive root of unity. Hence

$$\sigma(\zeta_n) = \zeta_n^{r_{n,\sigma}}, \quad \text{for some } r_{n,\sigma} \in (\mathbb{Z}/\ell^n\mathbb{Z})^\times.$$

Let n_P be the maximal n such that $\zeta_n \in H^0(P, \mathcal{O}_P)$. We obtain a group homomorphism

$$G_P \to (\mathbb{Z}/\ell^{n_P}\mathbb{Z})^\times, \quad \sigma \mapsto r_{n_P,\sigma}.$$

This map is independent of the choice of the ζ_n. If $P' \to X$ is another such Galois cover which factors via $P' \to P$ then we obtain a commutative diagram

$$\begin{array}{ccc} G_{P'} & \longrightarrow & (\mathbb{Z}/\ell^{n_{P'}})^\times \\ \downarrow & & \downarrow \\ G_P & \longrightarrow & (\mathbb{Z}/\ell^{n_P})^\times, \end{array}$$

with vertical maps the natural surjections. Therefore we can take the inverse limit over an inverse system of finite étale Galois covers (P_α) pro-representing the fiber functor $F_{\bar{x}}$ and obtain a $\mathbb{Z}_\ell$-representation

$$\pi_1(X, \bar{x}) \to \mathbb{Z}_\ell^\times.$$

It is the representation corresponding to the lisse sheaf $\mathbb{Z}_\ell(1)$.

EXAMPLE 8.8. We can use Theorem 8.5 to construct new lisse sheaves which are related to Example 3.33. For example let k be a perfect field of characteristic $p > 1$ containing the field with $q = p^n$ elements $\mathbb{F}_q$ and $m \geq 1$ a natural number. Then the natural inclusion $k[x] \subseteq k[x][t]/(t^q - t - x^m) := B$ defines an étale Galois covering $\pi : X := \operatorname{Spec} B \to \mathbb{A}^1_k = \operatorname{Spec} k[x]$. (This map is étale by Definition 5.1, (e), since $\partial(t^q - t - x^m)/\partial t = -1$.) The Galois group is isomorphic to $\mathbb{F}_q$, where $a \in \mathbb{F}_q$ acts via $\bar{t} \mapsto \bar{t} + a$. Let $\bar{\eta} \to X$ be a geometric point over the generic point of X, we obtain

$$\mathbb{F}_q = \operatorname{coker}(\pi_1(X, \bar{\eta}) \hookrightarrow \pi_1(\mathbb{A}^1_k, \bar{\eta})).$$

Now let ℓ be a prime different from p and A an ℓ-adic coefficient ring and take a group homomorphism

$$\psi : (\mathbb{F}_q, +) \to A^\times.$$

Then the composition

$$\pi_1(\mathbb{A}^1_k, \bar{\eta}) \twoheadrightarrow \mathbb{F}_q \xrightarrow{\psi} A^\times = \operatorname{Aut}_A(A)$$

is an A-representation of $\pi_1(\mathbb{A}^1_k, \bar{\eta})$ of rank 1. Hence it defines a free lisse A-sheaf $\mathcal{L}_{m,\psi}$ of rank 1 on $\mathbb{A}^1_k$. Its sheaf theoretic description is as follows: If A is finite, then denote by L_ψ the sheaf on $(X, \mathbb{F}_q)$ (see 7.1.4) whose underlying sheaf on X is the constant sheaf A and where the action of $a \in \mathbb{F}_q$ is given by multiplication with $\psi(a)$. Concretely, if $\bar{x} \to X$ is a geometric point we get an isomorphism

$$L_{\psi,\bar{x}} = A \to A = L_{\psi,\bar{x}+a}, \quad b \to \psi(a)b,$$

where $\bar{x} + a$ is the composition $\bar{x} \to X \xrightarrow{+a} X$. Then $\mathcal{L}_\psi = \pi_*(L_\psi)^{\mathbb{F}_q}$. One obtains the description for general (infinite) A by the usual limit procedure.

EXAMPLE 8.9. One can generalize the above example for the case $m = 1$ as follows (see e.g. [**12**, Sommes trig. 1.]): Let $\mathbb{F}_q$ be the finite field with $q = p^n$ elements and G a commutative algebraic group scheme over $\mathbb{F}_q$. Let $F : G \to G$ be the Frobenius morphism (it is given by $\mathcal{O}_G \to \mathcal{O}_G$, $x \mapsto x^p$). Then $F^n = (F \circ \ldots \circ F)$ (n-times) is an $\mathbb{F}_q$-endomorphism of G. We obtain a finite étale Galois covering (the *Lang isogeny*)

$$F - \operatorname{id}_G : G \to G$$

whose kernel and Galois group is the finite group $G(\mathbb{F}_q)$. Let A be an ℓ-adic coefficient ring and $\chi : G(\mathbb{F}_q) \to A^\times$ a homomorphism. Then the composition

$$\pi_1(G, \bar{\eta}) \twoheadrightarrow \operatorname{coker}(\pi_1(G, \bar{\eta}) \xrightarrow{F-\mathrm{id}} \pi_1(G, \bar{\eta})) = G(\mathbb{F}_q) \xrightarrow{\chi} A^\times$$

is an A-representation of $\pi_1(G, \bar{\eta})$. Hence we obtain a lisse rank 1 sheaf $\mathcal{L}_\chi$ on G. In case $G = \mathbb{A}^1_{\mathbb{F}_q}$ we obtain the example above for $m = 1$.

9. ℓ-adic cohomology

In this section we recall without proofs the main properties of étale cohomology on a scheme over an algebraically closed field with coefficients in a lisse ℓ-adic sheaf. The references for this section are [**5**], [**12**] and [**20**]. See also [**17**].

CONVENTION 9.1. Let k be a perfect field of characteristic $p \geq 0$ and ℓ a prime number which is invertible in k. Throughout this section a k-scheme is a scheme which is *separated* and of finite type over k.

9.1.1. *Étale cohomology.* Let $f : X \to Y$ be a morphism of k-schemes.

(a) Let A be a torsion ring. The category of A-modules on $X_{\text{ét}}$ has enough injectives. Therefore we can define the i-th higher direct image of an A-module $\mathcal{F}$ on $X_{\text{ét}}$ under f as the i-th right derived functor of f_*, i.e.

$$R^i f_* \mathcal{F} := H^i(f_* \mathcal{I}^\bullet),$$

where $\mathcal{F} \to \mathcal{I}^\bullet$ is an injective resolution of A-modules on $X_{\text{ét}}$.

Choose a compactification of f (using Nagata's theorem [**11**], [**36**]), i.e. a proper morphism $\bar{f} : \bar{X} \to Y$ together with a dominant open immersion $j : X \hookrightarrow \bar{X}$ such that $f = \bar{f} \circ j$. Then we define

$$R^i f_! \mathcal{F} := R^i \bar{f}_* j_! \mathcal{F}.$$

This definition is independent of the choice of the compactification, see [**12**, Arcata, IV, (5.3)].

In case k is *algebraically closed* and $\pi : X \to \operatorname{Spec} k$ is the structure map we also write

$$H^i(X, \mathcal{F}) := \Gamma(\operatorname{Spec} k, R^i \pi_* \mathcal{F}) \quad \text{and} \quad H^i_c(X, \mathcal{F}) := H^i(\bar{X}, j_! \mathcal{F})$$

and call it the i-th cohomology and the i-th *cohomology with compact support* of $\mathcal{F}$, respectively. (The A-module $H^i(X, \mathcal{F})$ is in this case also equal to the i-th right derived functor of the global section functor $\Gamma(X, -)$.)

(b) Assume k is algebraically closed. Let A be an ℓ-adic coefficient ring (in the sense of Convention 8.4) and let $\mathcal{F}$ be a constructible A-sheaf on $X_{\text{ét}}$. Then we find a DVR R which is finite over $\mathbb{Z}_\ell$ such that A is an R-algebra, and a constructible R-sheaf $(\mathcal{F}_n)_n$ such that $\mathcal{F} = (\mathcal{F}_n) \otimes_R A$. We define the i-th étale cohomology group of $\mathcal{F}$, respectively the i-th étale cohomology group with compact support of $\mathcal{F}$ as

$$H^i(X, \mathcal{F}) := (\varprojlim_n H^i(X, \mathcal{F}_n)) \otimes_R A, \quad H^i_c(X, \mathcal{F}) = (\varprojlim_n H^i_c(X, \mathcal{F}_n)) \otimes_R A.$$

This definition is independent of the choice of R and the constructible R-sheaf $(\mathcal{F}_n)$, see [**20**, VI, 2.2].

REMARK 9.2. Let $\bar{k}$ be an algebraic closure of k, A a finite ℓ-adic coefficient ring, $\pi : X \to \operatorname{Spec} k$ a k-scheme and $\mathcal{F}$ a constructible A-sheaf on X. Then $R^i \pi_* \mathcal{F}$ and $R^i \pi_! \mathcal{F}$ are lisse A-sheaves on $\operatorname{Spec} k$ and the corresponding $\operatorname{Gal}(\bar{k}/k)$-representations are equal to $H^i(X \otimes_k \bar{k}, \mathcal{F})$ and $H^i_c(X \otimes_k \bar{k}, \mathcal{F})$, respectively. (This follows from the definition of the stalk at $\operatorname{Spec} \bar{k} \to \operatorname{Spec} k$ and étale base change, cf. (7.4).)

9.1.2. In the following we assume that k is *algebraically closed*. Let $f : X \to Y$ be a morphism of k-schemes, A an ℓ-adic coefficient ring and $\mathcal{F}$ a constructible A-sheaf on $X_{\text{ét}}$. We list some properties:

(a) If A is finite (i.e. $A = R/\mathfrak{m}^n$ for some DVR R finite over $\mathbb{Z}_\ell$), then $R^i f_! \mathcal{F}$ and $R^i f_* \mathcal{F}$ are constructible on Y, see [**12**, Arcata, IV, Thm (6.2)] and [**12**, Th. finitude, Thm 1.1]. If A is general then $H^i(X, \mathcal{F})$ and $H^i_c(X, \mathcal{F})$ are finitely generated A-modules, cf. [**20**, VI, 2.2].

(b) Let $0 \to \mathcal{F}' \to \mathcal{F} \to \mathcal{F}'' \to 0$ be a short exact sequence of constructible A-sheaves. Then there is a long exact sequence

$$\cdots \to H^i(X, \mathcal{F}) \to H^i(X, \mathcal{F}'') \to H^{i+1}(X, \mathcal{F}') \to \cdots,$$

and similarly with H^i replaced by H^i_c. (If A is finite this follows immediately from the general theory of right derived functors and the fact that $j_!$ is exact. For general A notice that the groups $H^i(X,\mathcal{F}_n)$ are finite by (a); hence any projective system of subquotients of $(H^i(X,\mathcal{F}_n))$ satisfies the Mittag-Leffler condition; hence the long exact sequence of pro-groups gives a long exact sequence in the limit; further if A is not finite, then it is flat over R and hence tensoring with A over R is exact.)

(c) Assume $f : X \to Y$ is finite. If A is finite then $R^i f_*\mathcal{F} = 0$ for all $i \geq 1$, see [**12**, Arcata, II, Prop. (3.6)]. This implies (via a Leray spectral sequence argument) that for general A we have $H^i(X,\mathcal{F}) = H^i(Y, f_*\mathcal{F})$.

(d) Assume X is a proper k-scheme and $j : U \hookrightarrow X$ is an open subscheme with complement $i : Z \hookrightarrow X$. Then there is a long exact sequence

$$\cdots \to H^i_c(U, j^*\mathcal{F}) \to H^i(X,\mathcal{F}) \to H^i(Z, i^*\mathcal{F}) \to H^{i+1}_c(U, j^*\mathcal{F}) \to \cdots.$$

If we write $\mathcal{F} = (\mathcal{F}_n) \otimes_R A$ then this sequence is induced as above by the exact sequences $0 \to j_! j^*\mathcal{F}_n \to \mathcal{F}_n \to i_* i^*\mathcal{F}_n \to 0$, for $n \geq 1$.

(e) Assume X is affine. Then $H^i(X,\mathcal{F}) = 0$ for all $i > \dim X$, see [**5**, XIV, Cor. 3.2].

(f) We have $H^i(X,\mathcal{F}) = 0$ for all $i > 2\dim X$, see [**5**, X, Cor. 4.3].

(g) Assume $k = \mathbb{C}$ and X is smooth over $\mathbb{C}$. Then

$$H^i(X,\mathbb{Z}/n\mathbb{Z}) \cong H^i(X(\mathbb{C}),\mathbb{Z}/n\mathbb{Z}) \quad \text{and} \quad H^i_c(X,\mathbb{Z}/n\mathbb{Z}) \cong H^i_c(X(\mathbb{C}),\mathbb{Z}/n\mathbb{Z}),$$

where $n \geq 1$ and the right-hand side of the two isomorphisms is the singular cohomology (resp. with compact supports) of the complex manifold $X(\mathbb{C})$. See [**5**, XI, Thm. 4.4] and [**12**, Arcata, IV, Thm (6.3)]

(h) *Cohomology with support.* Let $j : U \hookrightarrow X$ be an open subset of X with complement $i : Z \hookrightarrow X$. If A is a finite ℓ-adic coefficient ring and $\mathcal{F}$ a constructible A-sheaf on X we define

$$i^!\mathcal{F} := \operatorname{Ker}(i^*\mathcal{F} \to i^* j_* j^*\mathcal{F}).$$

We obtain a left exact functor $\mathcal{F} \mapsto \Gamma_Z(X,\mathcal{F}) := \Gamma(Z, i^!\mathcal{F})$. Its right derived functor is denoted by $H^i_Z(X,\mathcal{F})$.

If A is a general ℓ-adic coefficient ring, we write $\mathcal{F} = (\mathcal{F}_n) \otimes_R A$ and set

$$H^i_Z(X,\mathcal{F}) := (\varprojlim_n H^i_Z(X,\mathcal{F}_n)) \otimes_R A.$$

We have an exact sequence

$$\ldots \to H^i(X,\mathcal{F}) \to H^i(U,\mathcal{F}_{|U}) \to H^{i+1}_Z(X,\mathcal{F}) \to \cdots.$$

(In case A is finite this is the usual localization sequence. For general A it follows from this and (a) that $H^i_Z(X,\mathcal{F}_n)$ is finite and hence as above taking the inverse limit over the sequences for $\mathcal{F}_n$ and tensoring with A gives the exact sequence in general.)

REMARK 9.3. Let X be a smooth $\mathbb{C}$-scheme. Notice that 9.1.2, (g) also implies

$$H^i(X,\mathbb{Z}_\ell) \cong H^i(X(\mathbb{C}),\mathbb{Z}) \otimes_{\mathbb{Z}} \mathbb{Z}_\ell, \quad \text{for all primes } \ell,$$

where $\mathbb{Z}_\ell$ on the left-hand side denotes the constant lisse $\mathbb{Z}_\ell$-sheaf $(\mathbb{Z}/\ell^n)_n$. Indeed, $H^i(X(\mathbb{C}),\mathbb{Z})$ is a finitely generated $\mathbb{Z}$-module, hence

$$H^i(X(\mathbb{C}),\mathbb{Z}) \otimes_{\mathbb{Z}} \mathbb{Z}_\ell = \varprojlim_n (H^i(X(\mathbb{C}),\mathbb{Z}) \otimes_{\mathbb{Z}} \mathbb{Z}/\ell^n).$$

On the other hand the exact sequence $0 \to \mathbb{Z} \xrightarrow{\cdot \ell^n} \mathbb{Z} \to \mathbb{Z}/\ell^n \to 0$ yields an exact sequence

$$0 \to H^i(X(\mathbb{C}),\mathbb{Z}) \otimes_{\mathbb{Z}} \mathbb{Z}/\ell^n \to H^i(X(\mathbb{C}),\mathbb{Z}/\ell^n) \to {}_{\ell^n}H^{i+1}(X(\mathbb{C}),\mathbb{Z}) \to 0,$$

where we denote by ${}_{\ell^n}(-)$ the kernel of the multiplication by ℓ^n. Using (g) and taking $\varprojlim_n$ we arrive at a short exact sequence

$$0 \to H^i(X(\mathbb{C}),\mathbb{Z}) \otimes_{\mathbb{Z}} \mathbb{Z}_\ell \to H^i(X,\mathbb{Z}_\ell) \to N \to 0,$$

where N consists of the sequences (a_n) with $a_n \in H^{i+1}(X(\mathbb{C}),\mathbb{Z})$, $\ell^n \cdot a_n = 0$, and such that $a_{n-1} = \ell \cdot a_n$. In particular all the a_n are ℓ-power-torsion. Thus there exists an $n_0 \geq 1$ such that $\ell^{n_0} \cdot a_n = 0$ for all n (since $H^{i+1}(X(\mathbb{C}),\mathbb{Z})$ is finitely generated over $\mathbb{Z}$). Hence $a_n = \ell^{n_0} \cdot a_{n+n_0} = 0$, i.e. $N = 0$.

THEOREM 9.4 (Poincaré duality). *Let X be a smooth k-scheme of pure dimension d and $\bar{k}$ an algebraic closure of k. Let A be an ℓ-adic coefficient ring and $\mathcal{F}$ a free lisse A-sheaf. Then for all $i \in \mathbb{Z}$ there is a natural (i.e. functorial in $\mathcal{F}$) and* $\mathrm{Gal}(\bar{k}/k)$*-equivariant isomorphism*

$$H^{2d-i}(X \otimes_k \bar{k}, \mathcal{F}^\vee(d)) \xrightarrow{\simeq} H^i_c(X \otimes_k \bar{k}, \mathcal{F})^\vee.$$

Here (d) is the Tate twist 8.0.3, $H^i_c(X \otimes_k \bar{k}, \mathcal{F})^\vee := \mathrm{Hom}_A(H^i_c(X \otimes_k \bar{k}, \mathcal{F}), A)$ *and $\mathcal{F}^\vee$ is defined as follows: write $\mathcal{F} = (\mathcal{F}_n) \otimes_R A$ with a free lisse R-sheaf $(\mathcal{F}_n)$ (this is possible by Theorem* 8.5 *and Lemma* 4.67 *), then $\mathcal{F}^\vee := (\mathcal{H}om(\mathcal{F}_n, R/\mathfrak{m}^n)) \otimes_R A$ is a free lisse A-sheaf.*

PROOF. If A is finite this is [**5**, XVIII, (3.2.6.2)]. (In *loc. cit.* it is done for $A = \mathbb{Z}/\ell^n$, but since for any DVR R with maximal ideal $\mathfrak{m}$, the ring $R/\mathfrak{m}^n$ is self injective, the same argument works.) In general we can assume that A is flat over R and write $\mathcal{F} = (\mathcal{F}_n) \otimes_R A$ with $(\mathcal{F}_n)$ a free lisse R-sheaf. Since $\mathcal{F}_n = \mathcal{F}_{n+1} \otimes R/\mathfrak{m}^n$ we have $\varinjlim_m \mathcal{H}om(\mathcal{F}_m, R/\mathfrak{m}^n) = \mathcal{H}om(\mathcal{F}_n, R/\mathfrak{m}^n)$. Therefore the Poincaré isomorphism for the $\mathcal{F}_n$ yields a $\mathrm{Gal}(\bar{k}/k)$-equivariant isomorphism

$$H^{2d-i}(X \otimes_k \bar{k}, \mathcal{F}^\vee(d)) \xrightarrow{\simeq} (\varprojlim_n \varinjlim_m \mathrm{Hom}_R(H^i_c(X \otimes_k \bar{k}, \mathcal{F}_m), R/\mathfrak{m}^n)) \otimes_R A,$$

where we use that taking cohomology commutes with direct limits. Set $H_m := H^i_c(X \otimes_k \bar{k}, \mathcal{F}_m)$ and $H := \varprojlim_m H^i_m$. By [**20**, VI, Lem. 2.2.2] and [**20**, V, Prop. 3.2.3] there exists an $r_0 \geq 1$ such that for $r \geq r_0$ and $m \geq 1$ we have

$$\mathrm{im}(H_{m+r} \to H_m) = \mathrm{im}(H_{m+r_0} \to H_m) =: H'_m$$

and there exists an $s_0 \geq 1$ such that for $s \geq s_0$ and $n \geq 1$ we have an isomorphism $H \otimes_R R/\mathfrak{m}^n \xrightarrow{\simeq} H'_{n+s} \otimes_R R/\mathfrak{m}^n$. Hence

$$\begin{aligned}\varinjlim_m \mathrm{Hom}_R(H_m, R/\mathfrak{m}^n) &\cong \varinjlim_m \mathrm{Hom}_R(H'_m, R/\mathfrak{m}^n)\\ &\cong \varinjlim_m \mathrm{Hom}_{R/\mathfrak{m}^n}(H'_m \otimes_R R/\mathfrak{m}^n, R/\mathfrak{m}^n)\\ &\cong \mathrm{Hom}_{R/\mathfrak{m}^n}(H \otimes_R R/\mathfrak{m}^n, R/\mathfrak{m}^n)\\ &\cong \mathrm{Hom}_R(H, R/\mathfrak{m}^n).\end{aligned}$$

We obtain

$$\varprojlim_n \varinjlim_m \mathrm{Hom}_R(H_m, R/\mathfrak{m}^n) = \varprojlim_n \mathrm{Hom}_R(H, R/\mathfrak{m}^n) = \mathrm{Hom}_R(H, R).$$

Since R is noetherian, H is actually a finitely presented R-module and since A is flat over R we obtain $\mathrm{Hom}_R(H,R)\otimes_R A \cong \mathrm{Hom}_A(H\otimes_R A, A)$. Putting all the isomorphisms together we obtain the statement. □

COROLLARY 9.5. *Let X be a smooth affine k-scheme of pure dimension d, A an ℓ-adic coefficient ring and $\mathcal{F}$ a free lisse A-sheaf on X. Then*

$$H^i_c(X\otimes_k \bar{k}, \mathcal{F}) = 0, \quad \textit{for all } i < d.$$

PROOF. This follows from Poincaré duality and 9.1.2, (e). □

THEOREM 9.6 (Purity). *Let $i: Y \hookrightarrow X$ be a closed immersion of pure codimension c between smooth k-schemes, A an ℓ-adic coefficient ring and $\mathcal{F}$ a lisse A-sheaf on X. Then there is a natural $\mathrm{Gal}(\bar{k}/k)$-equivariant isomorphism*

$$H^j(Y\otimes_k \bar{k}, i^*\mathcal{F}) \xrightarrow{\simeq} H^{j+2c}_{Y\otimes_k\bar{k}}(X\otimes_k\bar{k}, \mathcal{F}(c)), \quad \textit{for all } j,$$

where (c) on the right-hand side denotes the Tate twist. In particular $H^j_{Y\otimes_k\bar{k}}(X\otimes_k \bar{k},\mathcal{F}) = 0$ for all $j < 2c$.

PROOF. If A is finite we have $R^j i^!\mathcal{F} = 0$, $j \neq 2c$, and $R^{2c}i^!\mathcal{F} = i^*\mathcal{F}(-c)$, by [**5**, XVI, Cor 3.8, Rem. 3.10, a)]. Therefore the statement follows in this case from the local-global spectral sequence

$$E_2^{a,b} = H^a(X\otimes_k\bar{k}, i_*R^b i^!\mathcal{F}(c)) \Rightarrow H^*_{Y\otimes_k\bar{k}}(X\otimes_k\bar{k},\mathcal{F}(c)).$$

For general A applying $\varprojlim_n$ and $\otimes_R A$ gives the assertion. □

THEOREM 9.7 (Lefschetz trace formula, [**12**, Cycle, Cor 3.7]). *Let X be a smooth proper scheme over an algebraically closed field k of characteristic $p > 0$ and $\ell \neq p$ a prime number. Let $f: X \to X$ be a k-morphism and assume that the graph Γ_f of f and the diagonal Δ_X intersect properly in $X\times X$ (i.e. the intersection scheme $\Gamma_f\cap\Delta_X$ is either empty or zero dimensional). Then*

$$(\Gamma_f\cdot\Delta_X) = \sum_i (-1)^i\,\mathrm{Tr}(f^*|H^i(X,\mathbb{Q}_\ell)).$$

Here the left-hand side is the degree of the intersection product $\Gamma_f\cdot\Delta_X$, concretely

$$(\Gamma_f\cdot\Delta_X) := \sum_{x\in\Gamma_f\cap\Delta_X} \mathrm{length}(\mathcal{O}_{X,x}/I_f), \tag{9.1}$$

where $I_f\subseteq\mathcal{O}_{X,x}$ is the ideal generated by the elements $a - f^*(a)$, $a\in\mathcal{O}_{X,x}$.

LEMMA 9.8. *Let k be an algebraically closed field, X an integral k-scheme with generic point η, E a finite extension of $\mathbb{Q}_\ell$ with ring of integers R and $\mathcal{F}$ a constructible E-sheaf on X. Suppose there is a finite group G acting on $\mathcal{F}$, i.e. there is a constructible R-sheaf $\mathcal{F}' = (\mathcal{F}_n)$ such that the $\mathcal{F}_n$ form a projective system of $R[G]$-modules and $\mathcal{F} = \mathcal{F}'\otimes_R E$. Then the G-invariant sections $\mathcal{F}_n^G$ form a projective system which induces a constructible R-sheaf denoted by $(\mathcal{F}')^G$. We define the constructible E-sheaf*

$$\mathcal{F}^G := (\mathcal{F}')^G\otimes_R E.$$

Then

$$H^i(X,\mathcal{F}^G) = H^i(X,\mathcal{F})^G.$$

PROOF. Set $N := |G|$. For all $n_1 \geq n$ we have $\mathcal{F}_{n_1} \otimes_R R/\mathfrak{m}^n = \mathcal{F}_n$ by definition. We obtain natural maps

$$(\mathcal{F}_{n_1})^G \otimes_R R/\mathfrak{m}^n \to \mathcal{F}_n^G, \tag{9.2}$$

the cokernel of which is killed by N. (Indeed, it suffices to check this on the stalks. If we take a local section $s_n \in \mathcal{F}_n^G$ we can lift it to a local section $s \in \mathcal{F}_{n_1}$. Then $s' := \sum_{\sigma \in G} \sigma(s)$ is a G-invariant section of $\mathcal{F}_{n_1}^G$ which modulo $\mathfrak{m}^n$ equals $N \cdot s_n$.) Now the stalks of $\mathcal{F}_n^G$ are finite (as sets). It follows that the images of the maps (9.2) become stationary for $n_1 \to \infty$. We define $(\mathcal{F}')_n^G \subseteq \mathcal{F}_n^G$ to be the intersection over the images of all these maps. Then $(\mathcal{F}')_n^G$ is a constructible $R/\mathfrak{m}^n$-subsheaf of $\mathcal{F}_n^G$ and by construction $(\mathcal{F}')_{n+1}^G \otimes_R R/\mathfrak{m}^n = (\mathcal{F}')_n^G$. We obtain the constructible R-sheaf $(\mathcal{F}')^G := ((\mathcal{F}')_n^G)$ from the first part of the Lemma. Since the cokernel of the inclusion $(\mathcal{F}')_n^G \hookrightarrow \mathcal{F}_n^G$ is killed by N we obtain:

$$H^i(X, \mathcal{F}^G) \overset{\text{defn}}{=} (\varprojlim_n H^i(X, (\mathcal{F}')_n^G)) \otimes_R E = (\varprojlim_n H^i(X, \mathcal{F}_n^G)) \otimes_R E.$$

Let $\mathcal{F}_n \to \mathcal{I}^\bullet$ be a resolution of $\mathcal{F}_n$ by injective $R/\mathfrak{m}^n[G]$-modules on $X_{\text{ét}}$. The sheaf of G-invariants $(\mathcal{I}^j)^G$ is an injective sheaf of $R/\mathfrak{m}^n$-modules for all j. (Since the functor of taking G-invariants from the category of $R/\mathfrak{m}^n[G]$-modules on $X_{\text{ét}}$ to the category of $R/\mathfrak{m}^n$-modules on $X_{\text{ét}}$ has an exact left adjoint, namely the functor which to an $R/\mathfrak{m}^n$-module M associates the $R/\mathfrak{m}^n[G]$-module given by the same M with trivial G-action.) Further $H^0(\mathcal{I}^{\bullet G}) = \mathcal{F}_n^G$ and $H^j(\mathcal{I}^{\bullet G})$ is N-torsion for all $j \neq 0$, by a similar argument as above. It follows that kernel and cokernel of $H^i(X, \mathcal{F}_n^G) \to H^i(\Gamma(X, \mathcal{I}^{\bullet G}))$ are killed by N. On the other hand $H^i(\Gamma(X, \mathcal{I}^{\bullet G})) = H^i(\Gamma(X, \mathcal{I}^\bullet)^G)$ and kernel and cokernel of the natural map $H^i(\Gamma(X, \mathcal{I}^\bullet)^G) \to H^i(\Gamma(X, \mathcal{I}^\bullet))^G = H^i(X, \mathcal{F}_n)^G$ are killed by N. All together we obtain a natural map

$$H^i(X, \mathcal{F}_n^G) \to H^i(X, \mathcal{F}_n)^G$$

whose kernel and cokernel are killed by $2N$. Taking the limit and tensoring with E we arrive at the statement. □

Part 3. Ramification theory of ℓ-adic sheaves

10. Grothendieck-Ogg-Shafarevich

In this section we explain Grothendieck's proof of the Grothendieck-Ogg-Shafarevich formula, partly following Katz. References for this section are [**20**, Exp. X], [**27**, Ch. 2], see also [**43**].

10.0.1. Throughout this section we fix the following notation:

- k is a perfect field of characteristic $p > 0$ and $\bar{k}$ an algebraic closure.
- ℓ is a prime number different from p.
- C is a smooth proper and geometrically connected curve over k.
- $U \subseteq C$ is a *strict* open subset (in particular it is affine).
- $K = k(C)$ is the function field of C, $\bar{K}$ an algebraic closure and $K^{\text{sep}} \subseteq \bar{K}$ a separable closure. We denote by $\eta : \operatorname{Spec} K \to C$ the generic point of C and by $\bar{\eta} : \operatorname{Spec} \bar{K} \to C$ the induced geometric point over η.
- For a closed point $x \in C$ we denote by K_x the completion of K with respect to the valuation $v_x : K^\times \to \mathbb{Z}$ corresponding to x and by K_x^{sep} a separable closure. For all x we choose an embedding $\iota_x : K^{\text{sep}} \hookrightarrow K_x^{\text{sep}}$ over K.

- $G = \mathrm{Gal}(K^{\mathrm{sep}}/K)$ is the absolute Galois group of K and for a closed point $x \in C$ we denote by $D_x := D_x^{\iota_x}$ the image of the inclusion $\mathrm{Gal}(K_x^{\mathrm{sep}}/K_x) \hookrightarrow G$ induced by ι_x, i.e. D_x is the decomposition subgroup of G with respect to ι_x. We denote by $P_x \subseteq I_x \subseteq D_x$ the wild inertia and inertia subgroups, respectively. (Notice that if we take a different K-embedding $\iota_{x,1} : K^{\mathrm{sep}} \hookrightarrow K_x^{\mathrm{sep}}$, then there exists a $\sigma \in G$ such that $\iota_{x,1} = \iota_x \circ \sigma$; thus the resulting decomposition group is conjugate to D_x, i.e. $D_x^{\iota_{x,1}} = \sigma^{-1} D_x \sigma$. Same with I_x, P_x. Hence D_x, P_x, I_x are only well-defined up to conjugation by elements in G.)

10.1.1. Let A be an ℓ-adic coefficient ring (Convention 8.4) and $\mathcal{F}$ a free lisse A-module on U. By Theorem 8.5, we can identify $\mathcal{F}$ with a representation of $\pi_1(U, \bar{\eta})$ on $\mathcal{F}_{\bar{\eta}}$. The natural surjection $G \to \pi_1(U, \bar{\eta})$ hence induces a G-action on $\mathcal{F}_{\bar{\eta}}$, which we can further restrict to a P_x-action, $x \in C$. The action of P_x on $\mathcal{F}_{\bar{\eta}}$ factors over a finite quotient of P_x, see Lemma 4.70. Thus the Swan conductor of the P_x-representation on $\mathcal{F}_{\bar{\eta}}$ is defined (Definition 4.82) and denoted by

$$\mathrm{Swan}_x(\mathcal{F}).$$

We observe that $\mathrm{Swan}_x(\mathcal{F})$ is independent of the chosen embedding $\iota_x : K^{\mathrm{sep}} \hookrightarrow K_x^{\mathrm{sep}}$. Indeed, if we choose a different embedding there is a $\sigma \in G$ such that we have to consider the restriction of the G-action on $\mathcal{F}_{\bar{\eta}}$ to the group $\sigma^{-1} P_x \sigma$. But then σ maps the break decomposition of $\mathcal{F}_{\bar{\eta}}$ (Definition 4.80) with respect to $\sigma^{-1} P_x \sigma$ isomorphically to the break decomposition of $\mathcal{F}_{\bar{\eta}}$ with respect to P_x. Hence we obtain the same number $\mathrm{Swan}_x(\mathcal{F})$, when we compute it with respect to the $\sigma^{-1} P_x \sigma$-action.

We recall the following facts (see Remark 4.83, Theorem 4.86):

(a) Let A' be an ℓ-adic coefficient ring and $A \to A'$ a ring homomorphism. Then for all $x \in C$

$$\mathrm{Swan}_x(\mathcal{F} \otimes_A A') = \mathrm{Swan}_x(\mathcal{F}).$$

(b) $\mathrm{Swan}_x(\mathcal{F}) \in \mathbb{Z}_{\geq 0}$.

(c) $\mathrm{Swan}_x(\mathcal{F}) = 0 \iff \mathcal{F}$ is tame at x, i.e. $(\mathcal{F}_{\bar{\eta}})^{P_x} = \mathcal{F}_{\bar{\eta}}$. In particular, $\mathrm{Swan}_x(\mathcal{F}) = 0$ for all $x \in U$.

(d) Let $0 \to \mathcal{F}' \to \mathcal{F} \to \mathcal{F}'' \to 0$ be an exact sequence of free lisse A-sheaves on U. Then

$$\mathrm{Swan}_x(\mathcal{F}) = \mathrm{Swan}_x(\mathcal{F}') + \mathrm{Swan}_x(\mathcal{F}'').$$

10.1.2. Recall that the compactly supported Euler characteristic of a lisse $\bar{\mathbb{Q}}_\ell$-sheaf $\mathcal{F}$ on U is defined to be

$$\begin{aligned}\chi_c(\bar{U}, \mathcal{F}) &= \sum_i (-1)^i \dim_{\bar{\mathbb{Q}}_\ell} H_c^i(\bar{U}, \mathcal{F}) \\ &= -\dim_{\bar{\mathbb{Q}}_\ell} H_c^1(\bar{U}, \mathcal{F}) + \dim_{\bar{\mathbb{Q}}_\ell} H_c^2(\bar{U}, \mathcal{F}),\end{aligned}$$

where $\bar{U} = U \otimes_k \bar{k}$ and we abuse notation denoting the pullback of $\mathcal{F}$ to $\bar{U}$ still by $\mathcal{F}$.

The *Grothendieck-Ogg-Shafarevich* formula is the following theorem.

THEOREM 10.1. *Let $\mathcal{F}$ be a lisse $\bar{\mathbb{Q}}_\ell$-sheaf on U. Then*

$$\chi_c(\bar{U}, \mathcal{F}) = \mathrm{rk}(\mathcal{F}) \cdot \chi_c(\bar{U}, \bar{\mathbb{Q}}_\ell) - \sum_{x \in C \setminus U} [k(x) : k] \cdot \mathrm{Swan}_x(\mathcal{F}).$$

Outline of the proof: First we give a more general definition of the compactly supported Euler characteristic which works for any free lisse A-sheaf $\mathcal{F}$, where A is any ℓ-adic coefficient ring, and does not change when we replace $\mathcal{F}$ by $\mathcal{F} \otimes_A A'$, for any $A \to A'$. Using this and the corresponding fact for the Swan conductor we can reduce the theorem to a statement about lisse $\mathbb{F}_\lambda$-sheaves, where $\mathbb{F}_\lambda$ is a finite extension of $\mathbb{F}_\ell$. The corresponding representation has finite monodromy group G'. Using facts from the representation theory in mixed characteristic explained in Section 4.3.2 we find a lisse E-sheaf $\mathcal{F}'$, where E is a finite extension of $\mathbb{Q}_\ell$, which has the same finite monodromy group G' and its compactly supported Euler characteristic and its Swan conductors are equal to those of $\mathcal{F}$. Thus we are reduced to the case where $\mathcal{F}$ is a lisse $\bar{\mathbb{Q}}_\ell$-sheaf on U which is trivialized by a finite étale cover $U' \to U$. In this case we can prove the theorem via a direct computation relying on the Lefschetz trace formula, the Hurwitz genus formula and the results from Section 4.

After developing some preliminaries, we will give a detailed proof at the end of this subsection.

LEMMA 10.2. *Let $\bar{x} \in C \otimes_k \bar{k}$ be a closed point mapping to $x \in C$. Let A be an ℓ-adic coefficient ring and $\mathcal{F}$ a lisse A-sheaf on U. Then*

$$\mathrm{Swan}_x(\mathcal{F}) = \mathrm{Swan}_{\bar{x}}(\mathcal{F}),$$

where here as everywhere we abuse notation and write $\mathcal{F}$ on the right-hand side instead of $\mathcal{F}_{|U \otimes_k \bar{k}}$.

PROOF. The point $\bar{x}$ corresponds to a prime ideal in $k(x) \otimes_k \bar{k}$ which by Hensel's Lemma corresponds uniquely to a prime ideal in $K_x \otimes_k \bar{k}$. Denote by L the completion of the function field of $U \otimes_k \bar{k}$ with respect to $\bar{x}$; it is the localization of $K_x \otimes_k \bar{k}$ with respect to the prime ideal corresponding to $\bar{x}$. Then L is a complete discrete valuation field with residue field $\bar{k}$. Since k is perfect, the canonical inclusion $K_x \hookrightarrow L$ is unramified. It follows that L is isomorphic to the maximal unramified extension of K_x inside K_x^{sep}, hence its absolute Galois group is isomorphic to the inertia group I_x. Hence the restriction to I_x of the representation corresponding to $\mathcal{F}$ is isomorphic to the representation corresponding to $\mathcal{F}_{|U \otimes_k \bar{k}}$. The claim follows. □

LEMMA 10.3. *Assume $k = \bar{k}$. Let A be an ℓ-adic coefficient ring and $\mathcal{F}$ a lisse A-sheaf on U. Then there exists a two term complex of A-modules*

$$C(\mathcal{F}) : C^1(\mathcal{F}) \to C^2(\mathcal{F})$$

with the following properties:

(a) *$H^i(C(\mathcal{F})) = H^i_c(U, \mathcal{F})$, all i.*
(b) *If $\mathcal{F}$ is free then $C^i(\mathcal{F})$ is a free A-module of finite rank.*
(c) *The functor $\mathcal{F} \mapsto C(\mathcal{F})$ is exact.*
(d) *If A' is an ℓ-adic coefficient ring and $A \to A'$ a ring homomorphism and $\mathcal{F}$ is free, we have a canonical isomorphism $C(\mathcal{F}) \otimes_A A' \cong C(\mathcal{F} \otimes_A A')$.*

PROOF. We are following the proof given in [**27**, Lem. 2.2.7]. Denote by $j : U \hookrightarrow C$ the inclusion of U into C. Pick once and for all a closed point $P \in U$. It is a closed subscheme of U of codimension 1 which is smooth over k. Thus the long exact localization sequence 9.1.2, (h) together with purity (Theorem 9.6) gives a long exact sequence

$$\dots \to H^i(C, j_!\mathcal{F}) \to H^i(C \smallsetminus P, (j_!\mathcal{F})_{|C\smallsetminus P}) \to H^{i-1}(P, (j_!\mathcal{F})(-1)_{|P}) \to \dots.$$

By definition $H^i(C, j_!\mathcal{F}) = H^i_c(U, \mathcal{F})$, which vanishes for $i \neq 1, 2$ by Corollary 9.5. By 9.1.2, (e), we have

$$H^i(P, (j_!\mathcal{F})(-1)_{|P}) = 0, \quad \text{for } i \neq 0. \tag{10.1}$$

From this and 9.1.2, (e) we get

$$H^i(C \smallsetminus P, (j_!\mathcal{F})_{|C\smallsetminus P}) = 0, \quad \text{for } i \neq 1. \tag{10.2}$$

Set

$$C^1(\mathcal{F}) := H^1(C \smallsetminus P, (j_!\mathcal{F})_{|C\smallsetminus P}), \quad C^2(\mathcal{F}) := H^0(P, (j_!\mathcal{F})(-1)_{|P}).$$

By (10.1) and (10.2) the functors $\mathcal{F} \mapsto C^i(\mathcal{F})$ are exact and by the long exact sequence above we get a complex $C^1(\mathcal{F}) \to C^2(\mathcal{F})$ whose cohomology groups are equal to $H^i_c(U, \mathcal{F})$. It remains to prove (b) and (d). Notice that both statements are obvious if A is a field extension of $\mathbb{Q}_\ell$. Thus we may assume that there exists a DVR R finite over $\mathbb{Z}_\ell$ that surjects onto A. In this case we can represent $\mathcal{F}$ as a projective system $(\mathcal{F}_n)$ and by definition $C^i(\mathcal{F}) = \varprojlim_n C^i(\mathcal{F}_n)$. Since $C^i(\mathcal{F})$ is a finitely generated A-module it suffices to prove (b) for A finite, i.e. $A = R/\mathfrak{m}^n$. Furthermore let $\pi \in \mathfrak{m}$ be a local parameter of R. Then we have an exact sequence $0 \to \mathcal{F} \xrightarrow{\cdot\pi^n} \mathcal{F} \to \mathcal{F}_n \to 0$, which yields $C^i(\mathcal{F})/\pi^n \cong C^i(\mathcal{F}_n)$. If $R \to R'$ is a finite extension we get

$$C^i(\mathcal{F}) \otimes_R R' = \varprojlim_n (C^i(\mathcal{F}) \otimes_{R/\mathfrak{m}^n} R'/\mathfrak{m}'^n) = \varprojlim_n (C^i(\mathcal{F}_n) \otimes_{R/\mathfrak{m}^n} R'/\mathfrak{m}'^n).$$

Thus in (d) it suffices to consider A (and hence also A') finite.

So let A be a finite local ring with residue characteristic prime to p and $\mathcal{F}$ a lisse A-sheaf on U. If M is a finitely generated A-module we find a resolution $P_\bullet \to M$ by free finitely generated A-modules. We can view M and the P_j as constant sheaves on U and form the complex of lisse A-sheaves $\mathcal{F} \otimes_A P_\bullet$, which is augmented towards $\mathcal{F} \otimes_A M$. The j-th homology of this complex is by definition $\mathcal{T}or^A_j(\mathcal{F}, M)$, which is a lisse A-sheaf. Since $P_\bullet$ is a complex of free A-modules, we obtain $C^i(\mathcal{F} \otimes_A P_\bullet) = C^i(\mathcal{F}) \otimes_A P_\bullet$. All together the exactness of the C^i gives

$$C^i(\mathcal{T}or^A_j(\mathcal{F}, M)) = H_j(C^i(\mathcal{F} \otimes_A P_\bullet)) = H_j(C^i(\mathcal{F}) \otimes_A P_\bullet) = \mathcal{T}or^A_j(C^i(\mathcal{F}), M).$$

In case $\mathcal{F}$ is free, the left-hand side vanishes for $j \geq 1$ thus $C^i(\mathcal{F})$ is a finitely generated flat A-module, hence is free. Furthermore, the case $j = 0$ gives the base change statement. This finishes the proof. □

As an immediate consequence we get:

COROLLARY 10.4. *Assume $k = \bar{k}$. Let $\mathcal{F}$ be a free lisse A-sheaf on U and $C(\mathcal{F})$ the complex from Lemma 10.3. Then we define*

$$\chi_c(U, \mathcal{F}) := -\operatorname{rk}_A(C^1(\mathcal{F})) + \operatorname{rk}_A(C^2(\mathcal{F})).$$

We have:

(a) *If $H^i_c(U, \mathcal{F})$ is a free A-module, for $i = 1, 2$, then*

$$\chi_c(U, \mathcal{F}) = \sum_i (-1)^i \operatorname{rk}_A H^i_c(U, \mathcal{F}).$$

(b) *If A' is an ℓ-adic coefficient ring and $A \to A'$ a ring homomorphism, then*

$$\chi_c(U, \mathcal{F}) = \chi_c(U, \mathcal{F} \otimes_A A').$$

(c) *If $0 \to \mathcal{F}' \to \mathcal{F} \to \mathcal{F}'' \to 0$ is an exact sequence of free lisse A-sheaves on U, then*

$$\chi_c(U, \mathcal{F}) = \chi_c(U, \mathcal{F}') + \chi_c(U, \mathcal{F}'').$$

10.1.3. Assume $k = \bar{k}$. Let $U' \to U$ be a connected finite étale Galois covering with Galois group $G_{U'} = \mathrm{Gal}(K'/K)$, where $K' = k(U')$. We can extend it uniquely to a finite morphism $C' \to C$ between smooth proper and connected curves over k. Let $x' \in C' \setminus U'$ be a closed point lying over $x \in C \setminus U$ and $G_{U',x'} = \{\sigma \in G_{U'} \,|\, \sigma(x') = x'\} = \mathrm{Gal}(K'_{x'}/K_x)$ the decomposition group of $G_{U'}$ at x'. By Theorem 4.53 there exists a finitely generated and projective $\mathbb{Z}_\ell[G_{U',x'}]$-module

$$\mathrm{Sw}_{G_{U',x'}}$$

underlying the Swan representation of $G_{U',x'}$. We define

$$\mathrm{Sw}_{G_{U'},x} := \mathrm{Sw}_{G_{U',x'}} \otimes_{\mathbb{Z}_\ell[G_{U',x'}]} \mathbb{Z}_\ell[G_{U'}].$$

and we denote its character by

$$\mathrm{sw}_{G_{U'},x} : G_{U'} \to \mathbb{Z}_\ell. \tag{10.3}$$

PROPOSITION 10.5. *In the situation above we have the following properties:*

(a) *The definition of $\mathrm{Sw}_{G_{U'},x}$ is independent of the choice of x'/x.*

(b)

$$\mathrm{sw}_{G_{U'},x}(\sigma) = \begin{cases} \sum\limits_{\substack{y/x \\ \sigma(y)=y}} (1 - i_{G_{U',y}}(\sigma)), & \text{if } \sigma \neq 1, \\ \left(\sum\limits_{y/x} (1 + v_y(\mathfrak{D}_{C'/C}))\right) - |G_{U'}|, & \text{if } \sigma = 1, \end{cases}$$

where $y \in C'$ maps to x, $i_{G_{U',y}} : G_{U',y} \to \mathbb{Z}$ is the function defined in (3.5) *and $v_y(\mathfrak{D}_{C'/C})$ is the multiplicity at y of the different $\mathfrak{D}_{B/\mathcal{O}_{C,x}}$, where B is the integral closure of $\mathcal{O}_{C,x}$ in K', see Definition* 3.18.

(c) *Let R be a complete DVR which is finite over $\mathbb{Z}_\ell$ with maximal ideal $\mathfrak{m}$ and residue field $\mathbb{F}_\lambda$. Let $\mathcal{F}$ be a free lisse R-sheaf and assume that $(\mathcal{F} \otimes_R \mathbb{F}_\lambda)_{|U'}$ is trivial, i.e. we can view $\mathcal{F}_{\bar{\eta}} \otimes_R \mathbb{F}_\lambda$ as a $\mathbb{F}_\lambda[G_{U'}]$-module. Then*

$$\mathrm{Swan}_x(\mathcal{F}) = \mathrm{Swan}_x(\mathcal{F} \otimes_R \mathbb{F}_\lambda) = \dim_{\mathbb{F}_\lambda} \mathrm{Hom}_{\mathbb{F}_\lambda[G_{U'}]}(\mathrm{Sw}_{G_{U'},x} \otimes_{\mathbb{Z}_\ell} \mathbb{F}_\lambda, \mathcal{F}_{\bar{\eta}} \otimes_R \mathbb{F}_\lambda).$$

(d) *Let $\mathcal{F}$ be a lisse $\bar{\mathbb{Q}}_\ell$-sheaf on U such that $\mathcal{F}_{|U'}$ is trivial, i.e. $\mathcal{F}_{\bar{\eta}}$ is a $\bar{\mathbb{Q}}_\ell[G_{U'}]$-module. Then for all closed points $x \in C$*

$$\mathrm{Swan}_x(\mathcal{F}) = \frac{1}{|G_{U'}|} \sum_{\sigma \in G_{U'}} \mathrm{sw}_{G_{U'},x}(\sigma) \cdot \mathrm{Tr}(\sigma | \mathcal{F}_{\bar{\eta}}).$$

PROOF. Denote by $\iota_{x'} : G_{U',x'} \hookrightarrow G_{U'}$ the inclusion. By Theorem 4.64, (b), to prove (a) it suffices to check that the character of

$$\mathrm{Ind}_{\iota_{x'}}(\mathrm{Sw}_{G_{U',x'}}) = \mathrm{Sw}_{G_{U',x'}} \otimes_{\mathbb{Z}_\ell[G_{U',x'}]} \mathbb{Z}_\ell[G_{U'}]$$

is independent of the choice of x'. Hence it suffices to prove (b). To this end, notice that by Lemma 4.38 the character of $\mathrm{Ind}_{\iota_{x'}}(\mathrm{Sw}_{G_{U',x'}})$ is given by

$$G_{U'} \ni \sigma \mapsto \chi(\sigma) := \frac{1}{|G_{U',x'}|} \sum_{\substack{\tau \in G_{U'} \\ \tau\sigma\tau^{-1} \in G_{U',x'}}} \mathrm{sw}_{G_{U',x'}}(\tau\sigma\tau^{-1}),$$

where $\mathrm{sw}_{G_{U',x'}}$ is the Swan character of $G_{U',x'}$, see Definition 4.45.

First case: $\sigma \neq 1$. We have

$$\tau\sigma\tau^{-1} \in G_{U',x'} \Longleftrightarrow \sigma \in G_{U',\tau^{-1}(x')} = \tau^{-1} \circ G_{U',x'} \circ \tau$$

and

$$\mathrm{sw}_{G_{U',x'}}(\tau\sigma\tau^{-1}) = \mathrm{sw}_{G_{U',\tau^{-1}(x')}}(\sigma).$$

Since $G_{U'}$ acts transitively on the points over x, we have a bijection of sets $G_{U'}/G_{U',x'} \xrightarrow{1:1} \{y/x\} \subseteq C'$. Hence

$$\chi(\sigma) = \sum_{\substack{y/x \\ \sigma(y)=y}} \mathrm{sw}_{G_{U',y}}(\sigma).$$

The formula in (b) for $\sigma \neq 1$ thus follows from the definition of $\mathrm{sw}_{G_{U',y}}$, see Definition 4.45. (Notice that since $k = \bar{k}$ we have $G_{U',x'} = G^0_{U',x'}$.)

Second case: $\sigma = 1$. In this case

$$\chi(1) = \frac{|G_{U'}|}{|G_{U',x'}|} \cdot \mathrm{sw}_{G_{U',x'}}(1) = \sum_{y/x} \mathrm{sw}_{G_{U',y}}(1)$$
$$= \sum_{y/x} \Big(\big(\sum_{\sigma \neq 1} i_{G_{U',y}}(\sigma)\big) - (|G_{U',y}| - 1) \Big).$$

Since $\sum_{\sigma\neq 1} i_{G_{U',y}}(\sigma) = v_y(\mathfrak{D}_{C'/C})$ (see Proposition 3.25) and $\sum_{y/x} |G_{U',y}| = |G_{U'}|$ we obtain the formula for (b).

The formula in (c) follows directly from $\mathrm{Swan}_x(\mathcal{F}) = b(\mathrm{Res}_{\iota_{x'}}(\mathcal{F}_{\bar{\eta}}))$ (see Theorem 4.86), the definition of $b(-)$ in Definition 4.72 and that $\mathrm{Ind}_{\iota_{x'}}$ is left adjoint to $\mathrm{Res}_{\iota_{x'}}$, see (4.1).

Finally, let us prove (d). First notice that if P and V are two finite dimensional $\bar{\mathbb{Q}}_\ell$-representations of $G_{U'}$, then

$$\mathrm{Hom}_{\bar{\mathbb{Q}}_\ell[G_{U'}]}(P, V) = (P^\vee \otimes_{\bar{\mathbb{Q}}_\ell} V)^{G_{U'}},$$

see Example 4.1. The $\bar{\mathbb{Q}}_\ell$-dimension of the right-hand side is equal to the number of times the trivial representation appears in $P^\vee \otimes_{\bar{\mathbb{Q}}_\ell} V$. Let χ_P (resp. χ_V) be the character of P (resp. V); the character of $P^\vee$ is given by $\chi_{P^\vee}(\sigma) = \chi_P(\sigma^{-1})$. The character of $P^\vee \otimes_{\bar{\mathbb{Q}}_\ell} V$ is given by $\chi_{P^\vee} \cdot \chi_V$. Hence by Corollary 4.26 we obtain

$$\dim_{\bar{\mathbb{Q}}_\ell} \mathrm{Hom}_{\bar{\mathbb{Q}}_\ell[G_{U'}]}(P, V) = \frac{1}{|G_{U'}|} \sum_{\sigma \in G_{U'}} \chi_P(\sigma^{-1}) \cdot \chi_V(\sigma).$$

Therefore the statement for (d) follows from the formula (cf. (c))

$$\mathrm{Swan}_x(\mathcal{F}) = \dim_{\bar{\mathbb{Q}}_\ell} \mathrm{Hom}_{\bar{\mathbb{Q}}_\ell[G_{U'}]}(\mathrm{Sw}_{G_{U'},x} \otimes_{\mathbb{Z}_\ell} \bar{\mathbb{Q}}_\ell, \mathcal{F}_{\bar{\eta}}),$$

the discussion above and $\mathrm{sw}_{G_{U'}}(\sigma) = \mathrm{sw}_{G_{U'}}(\sigma^{-1})$, see (b). □

LEMMA 10.6. *In the situation of* 10.1.3 *denote by* $\Delta_{C'} \subseteq C' \times_k C'$ *the diagonal and by* Γ_σ *the graph of* σ *acting on* C'. *Then* Γ_σ *and* $\Delta_{C'}$ *intersect properly and*

$$(\Gamma_\sigma \cdot \Delta_{C'}) = \sum_{x \in C \setminus U} \sum_{\substack{x'/x \\ \sigma(x')=x'}} i_{G_{U',x'}}(\sigma), \quad \textit{for all } \sigma \neq 1,$$

where the left-hand side is defined as in (9.1).

PROOF. Take $\sigma \in G_{U'} \setminus \{1\}$ and $x' \in C'$ with $\sigma(x') = x'$. (Notice that this implies $x' \in C' \setminus U'$ since $G_{U'}$ acts transitively and freely on the fibers over U.) We have to show

$$i_{G_{U',x'}}(\sigma) = \mathrm{length}(\mathcal{O}_{C',x'}/I_\sigma),$$

where $I_\sigma \subseteq \mathcal{O}_{C',x'}$ is the ideal generated by the elements $a - \sigma^*(a)$, $a \in \mathcal{O}_{C',x'}$. Denote by x the image of x' in C. Denote by A the completion of $\mathcal{O}_{C,x}$ and by A' the completion of $\mathcal{O}_{C',x'}$. Then the right-hand side of the equality above is equal to $\mathrm{length}(A'/I_\sigma \cdot A')$. Further by Theorem 3.24 there exists a generator of A'/A, i.e. an element $\alpha \in A'$ such that $A' = A[\alpha]$. It follows that I_σ is the ideal generated by $\alpha - \sigma^*(\alpha)$. Thus by definition: $\mathrm{length}(A'/I_\sigma \cdot A') = v_{x'}(\alpha - \sigma^*(\alpha)) = i_{G_{U',x'}}(\sigma)$. □

NOTATION 10.7. Let D be a smooth projective curve over an algebraically closed field and $V \subsetneq D$ a strict non-empty open subset. In Corollary 10.4, we defined $\chi_c(V, A)$ and in part (b) of this corollary we saw that it is independent of the choice of A. We set

$$\chi_c(V) := \chi_c(V, A), \quad \chi(D) := \chi_c(V) + \mathrm{card}(D \setminus V)$$

where A is some ℓ-adic coefficient ring. We have an exact sequence $\cdots \to H^i(D, A) \to H^i(D \setminus V, A) \to H^{i+1}_c(V, A) \to \cdots$. This shows that if A is a field then $\chi(D) = \sum_i (-1)^i \dim_A H^i(D, A)$.

LEMMA 10.8. *In the situation of* 10.1.3 *we have*

$$\chi(C') = 2 - 2g(C') = |G_{U'}| \cdot \chi(C) - \sum_{x' \in C' \setminus U'} v_{x'}(\mathfrak{D}_{C'/C}),$$

where $v_{x'}(\mathfrak{D}_{C'/C})$ *is as in Proposition* 10.5, (b) *and* $g(C')$ *is the genus of the connected smooth projective curve* C'. *In particular,* $\chi_c(U') = 2 - 2g(C') - \mathrm{card}(C' \setminus U')$.

PROOF. Since we are over an algebraically closed field we have $\mu_\ell \cong \mathbb{Z}/\ell\mathbb{Z}$. Hence $\chi(C) = \sum_{i=0}^2 (-1)^i \dim_{\mathbb{F}_\ell} H^i(C, \mu_\ell)$ and [**12**, Arcata III, Cor. (3.5)] yields $\chi(C) = 2 - 2g(C)$. (This uses that if J is the Jacobian of C then $H^1(C, \mu_\ell) \cong J(k)[\ell] \cong (\mathbb{Z}/\ell\mathbb{Z})^{2g}$.) Now the lemma follows directly from the Hurwitz genus formula, see e.g. [**23**, IV, Cor 2.4]. □

Proof of Theorem 10.1. Given a closed point $x \in C$ the number of closed points $\bar{x} \in C \otimes_k \bar{k}$ lying over x is $[k(x) : k]$. Hence by Lemma 10.2 it suffices to consider the case $k = \bar{k}$. Let $\mathcal{F}$ be a lisse $\bar{\mathbb{Q}}_\ell$-sheaf on U. By Theorem 8.5 and Lemma 4.67 we find a finite field extension $E/\mathbb{Q}_\ell$ with ring of integers $R/\mathbb{Z}_\ell$ and a free lisse R-sheaf $\mathcal{F}'$ such that $\mathcal{F} = (\mathcal{F}' \otimes_R E) \otimes_E \bar{\mathbb{Q}}_\ell$. Thus by 10.1.1, (a) and Corollary 10.4, (b) it suffices to prove the following: Let $\mathbb{F}_\lambda$ be a finite field of characteristic ℓ and $\mathcal{F}$ a lisse $\mathbb{F}_\lambda$-sheaf of rank r, then

$$\chi_c(U, \mathcal{F}) = r \cdot \chi_c(U) - \sum_{x \in C \setminus U} \mathrm{Swan}_x(\mathcal{F}). \tag{10.4}$$

The sheaf $\mathcal{F}$ corresponds to a homomorphism $\pi_1(U,\bar{\eta}) \to \mathrm{GL}_r(\mathbb{F}_\lambda)$; since its target is a finite group we find a connected finite étale Galois cover $\pi : U' \to U$ with Galois group $G_{U'}$ which trivializes $\mathcal{F}$.

Let A be an ℓ-adic coefficient ring. By Theorem 8.5 we can identify the category of free lisse A-sheaves on U which are trivial on U' with the category of finitely generated $A[G_{U'}]$-modules which are free as A-modules. We can restrict the functors C^i from Lemma (b) to this category and obtain functors

$$T^i : (\text{fin. generated } A\text{-free } A[G_{U'}]\text{-modules}) \to (\text{free } A\text{-modules})$$

with a natural transformation $T^1 \to T^2$, such that for lisse A-sheaves $\mathcal{G}$ which are trivialized by $\pi : U' \to U$ we get

$$H^i_c(U,\mathcal{G}) = H^i(T^1(\mathcal{G}_{\bar{\eta}}) \to T^2(\mathcal{G}_{\bar{\eta}})), \quad i \geq 0.$$

We define a function

$$\nu_A : (\text{fin. generated } A\text{-free } A[G_{U'}]\text{-modules}) \to \mathbb{Z} \tag{10.5}$$

by

$$\nu_A(M) := \mathrm{rk}_A(T^2(M)) - \mathrm{rk}_A(T^1(M)) - (\mathrm{rk}_A(M)\chi_c(U,\bar{\mathbb{Q}}_\ell) - \sum_{x \in C \smallsetminus U} \mathrm{Swan}_x(M)).$$

Denote by $R_A(G_{U'})$ the Grothendieck group of the category on the left-hand side of (10.5) (see Definition 4.6). Then it follows from 10.1.1, (d) and Corollary 10.4, (c), that ν_A induces a well-defined group homomorphism

$$\nu_A : R_A(G_{U'}) \to \mathbb{Z}.$$

Since we reduced to showing (10.4), we have to prove that $\nu_{\mathbb{F}_\lambda} = 0$. Let E be a finite extension of $\mathbb{Q}_\ell$ with ring of integers R and residue field $\mathbb{F}_\lambda$. By Proposition 4.62 there is a homomorphism $d : R_E(G_{U'}) \to R_{\mathbb{F}_\lambda}(G_{U'})$. Recall that d is constructed as follows: Let V be an $E[G_{U'}]$-module, which is finite dimensional as an E-vector space. Take $M \subseteq V$ an $R[G_{U'}]$-submodule which is an R-lattice in V. Then $d([V]) = [M \otimes_R \mathbb{F}_\lambda]$. It follows from 10.1.1, (a) and Corollary 10.4, (b) that we obtain a commutative diagram

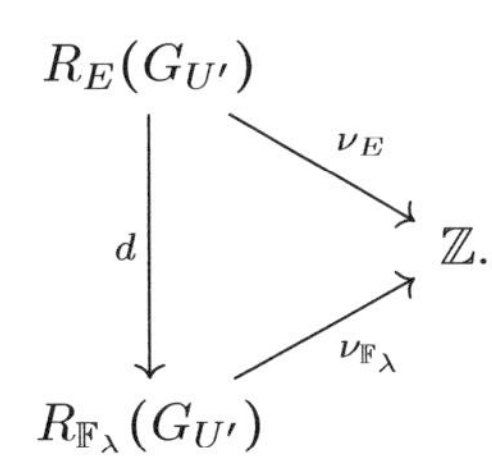

By Theorem 4.64, (a) the map d is surjective. Hence it suffices to show that $\nu_E = 0$. Retranslating this into lisse E-sheaves, we see that it suffices to prove equality (10.4) for $\mathcal{F}$ a lisse E-sheaf of rank r which is trivialized by $\pi : U' \to U$. Set $V = \mathcal{F}_{\bar{\eta}}$. It is an E-vector space of dimension $\dim_E V = r$ together with a $G_{U'}$-action. We have (see Theorem 8.5)

$$\pi^*\mathcal{F} \cong V_{U'}, \quad \mathcal{F} \cong (\pi_* V_{U'})^{G_{U'}},$$

where $V_{U'}$ is the constant lisse E-sheaf on U' defined by V. Denote by $\bar{\pi} : C' \to C$ the unique finite morphism between smooth proper curves over k, which restricts to $\pi : U' \to U$, and by $j : U \hookrightarrow C$, $j' : U' \hookrightarrow C'$ the open immersions. We obtain (via a direct computation)

$$j_!\mathcal{F} \cong j_!(\pi_* V_{U'})^{G_{U'}} \cong (j_!\pi_* V_{U'})^{G_{U'}} \cong (\bar{\pi}_* j'_! V_{U'})^{G_{U'}}.$$

By Lemma 9.8 and 9.1.2, (c) we get

$$H^i_c(U,\mathcal{F}) = H^i(C,(\bar{\pi}_* j'_! V_{U'})^{G_{U'}}) = H^i(C,\bar{\pi}_* j'_! V_{U'})^{G_{U'}} = H^i_c(U', V_{U'})^{G_{U'}}.$$

Next, if R is the ring of integers of E, then we find an R-lattice $M \subseteq V$ with $G_{U'}$-action such that $V \cong M \otimes_R E$ as $E[G_{U'}]$-modules. If $(R/\mathfrak{m}^n)_{U'} \to \mathcal{I}^\bullet$ is an injective resolution of the constant sheaf $R/\mathfrak{m}^n$ on U', then $\mathcal{I}^\bullet \otimes_{R/\mathfrak{m}^n} (M \otimes_R R/\mathfrak{m}^n)_{U'}$ is a resolution by injectives of the constant sheaf $(M \otimes_R R/\mathfrak{m}^n)$ which is compatible with the $G_{U'}$-action. This gives a $G_{U'}$-equivariant isomorphism $H^i_c(U', V_{U'}) \cong H^i_c(U', E_{U'}) \otimes_E V$. All together we obtain

$$H^i_c(U,\mathcal{F}) = (H^i_c(U', E_{U'}) \otimes_E V)^{G_{U'}}.$$

The dimension of the E-vector space $(H^i_c(U', E_{U'}) \otimes_E V)^{G_{U'}}$ is equal to the number of times the trivial rank one representation is contained in the $E[G_{U'}]$-module $H^i_c(U', E_{U'}) \otimes_E V$. Hence Corollary 4.26 yields

$$\begin{aligned}(10.6)\quad \chi_c(U,\mathcal{F}) &= -\dim_E(H^1_c(U',E)\otimes_E V)^{G_{U'}} + \dim_E(H^2_c(U',E)\otimes_E V)^{G_{U'}}\\ &= \frac{1}{|G_{U'}|}\sum_{\sigma\in G_{U'}} \mathrm{Tr}(\sigma|V)\big(\mathrm{Tr}(\sigma^*|H^2_c(U',E)) - \mathrm{Tr}(\sigma^*|H^1_c(U',E))\big).\end{aligned}$$

Set $Y' := C' \setminus U'$. The exact sequence $\cdots \to H^i(C',E) \to H^i(Y',E) \to H^{i+1}_c(U',E) \to \cdots$ from 9.1.2, (d) yields a short exact sequence

$$0 \to H^0(C',E) \to H^0(Y',E) \to H^1_c(U',E) \to H^1(C',E) \to 0$$

and an isomorphism

$$H^2_c(U',E) \cong H^2(C',E).$$

For $\sigma \in G_{U'}$ we obtain

$$\begin{aligned}(10.7)\quad &\mathrm{Tr}(\sigma^*|H^2_c(U',E)) - \mathrm{Tr}(\sigma^*|H^1_c(U',E))\\ &\qquad = -\mathrm{Tr}(\sigma^*|H^0(Y',E)) + \sum_{i=0}^{2}(-1)^i\,\mathrm{Tr}(\sigma^*|H^i(C',E)).\end{aligned}$$

We can write

$$H^0(Y',E) = \bigoplus_{x'\in Y'} E_{x'},$$

where $E_{x'} = E$ and $\sigma \in G_{U'}$ acts via:

$$\sigma^* : E_{x'} = E \xrightarrow{\mathrm{id}} E = E_{\sigma(x')}.$$

Thus

$$(10.8)\qquad \mathrm{Tr}(\sigma^*|H^0(\bar{Y}',E)) = \sum_{x\in Y}\sum_{\substack{x'/x\\ \sigma(x')=x'}} 1, \quad \sigma \in G_{U'}.$$

Further, for $\sigma \neq 1$ the graph Γ_σ and the diagonal $\Delta_{C'}$ intersect properly, see Lemma 10.6. Hence the Lefschetz trace formula (Theorem 9.7) yields

$$(10.9)\qquad \sum_{i=0}^{2}(-1)^i\,\mathrm{Tr}(\sigma^*|H^i(C',E)) = (\Gamma_\sigma \cdot \Delta_{C'}), \quad \text{if } \sigma \neq 1.$$

By Lemma 10.8, equation (10.8) and Proposition 10.5, (b) we get

$$
\begin{aligned}
-\operatorname{Tr}(1^*|H^0(Y',E)) &+ \sum_{i=0}^{2}(-1)^i \operatorname{Tr}(1^*|H^i(C',E)) \\
&= -\operatorname{card}(Y') + \chi(C') \\
&= -\operatorname{card}(Y') + |G_{U'}|\cdot(\chi_c(U) + \operatorname{card}(Y)) - \sum_{x'\in Y'} v_{x'}(\mathfrak{D}_{C'/C}) \\
&= |G_{U'}|\cdot\chi_c(U) - \sum_{x\in Y}\left(\left(\sum_{x'/x} 1 + v_{x'}(\mathfrak{D}_{C'/C})\right) - |G_{U'}|\right) \\
&= |G_{U'}|\cdot\chi_c(U) - \sum_{x\in Y}\operatorname{sw}_{G_{U',x}}(1).
\end{aligned}
$$

By (10.8), (10.9), Lemma 10.6 and Proposition 10.5, (b), we have for $\sigma \in G_{U'} \smallsetminus \{1\}$

$$
\begin{aligned}
-\operatorname{Tr}(\sigma^*|H^0(Y',E)) &+ \sum_{i=0}^{2}(-1)^i \operatorname{Tr}(\sigma^*|H^i(C',E)) \\
&= -\sum_{x\in Y}\sum_{\substack{x'/x\\ \sigma(x')=x'}} 1 + \sum_{x\in Y}\sum_{\substack{x'/x\\ \sigma(x')=x'}} i_{G_{U',x'}}(\sigma) \\
&= \sum_{x\in Y}\sum_{\substack{x'/x\\ \sigma(x')=x'}} (i_{G_{U',x'}}(\sigma) - 1) \\
&= -\sum_{x\in Y}\operatorname{sw}_{G_{U'},x}(\sigma).
\end{aligned}
$$

Hence by (10.6), (10.7) and Proposition 10.5, (d)

$$
\begin{aligned}
\chi_c(U,\mathcal{F}) &= \frac{1}{|G_{U'}|}\cdot\operatorname{Tr}(1^*|V)\cdot\left(|G_{U'}|\cdot\chi_c(U) - \sum_{x\in Y}\operatorname{sw}_{G_{U',x}}(1)\right) \\
&\quad - \frac{1}{|G_{U'}|}\cdot\sum_{\sigma\in G_{U'}\smallsetminus\{1\}}\left(\operatorname{Tr}(\sigma|V)\cdot\sum_{x\in Y}\operatorname{sw}_{G_{U',x}}(\sigma)\right) \\
&= r\cdot\chi_c(U) - \sum_{x\in Y}\operatorname{Swan}_x(\mathcal{F}).
\end{aligned}
$$

This finishes the proof. □

EXAMPLE 10.9. Let k be an algebraically closed field of characteristic $p > 0$, m a natural number with $(m,p) = 1$ and $\psi : \mathbb{F}_p \to \bar{\mathbb{Q}}_\ell^\times$ a group homomorphism. Let $\mathcal{L}_{m,\psi}$ be the lisse rank 1 sheaf on $\mathbb{A}_k^1$ from Example 8.8. Denote by ∞ the point in the complement of $\mathbb{A}_k^1 \subseteq \mathbb{P}_k^1$. Then $\operatorname{Swan}_\infty(\mathcal{L}_{m,\psi}) = m$, by Example 4.88, and $\chi_c(\mathbb{A}_k^1) = 1$ by Lemma 10.8. Thus the Grothendieck-Ogg-Shafarevich formula gives

$$\chi_c(\mathbb{A}_k^1, \mathcal{L}_\psi) = 1 - m.$$

10.2. Cohomological description of the Swan representation in equal characteristic. We fix the following situation:

- $k = \bar{k}$ is an algebraically closed field of positive characteristic p.
- We write $\mathbb{G}_m = \mathbb{G}_{m,k} = \operatorname{Spec} k[t, \frac{1}{t}]$ and denote by $\bar{\eta} \to \mathbb{G}_m$ a geometric point over the generic point. We embed $\mathbb{G}_m \subseteq \mathbb{P}^1$ and write $\mathbb{P}^1 \smallsetminus \mathbb{G}_m = \{0, \infty\}$.
- K is a complete discrete valuation field containing k with the property that k maps isomorphically onto the residue field of K. The choice of a

local parameter of K induces an isomorphism of K with the completion of $k(\mathbb{P}^1)$ at ∞, i.e. $K \cong k((t^{-1}))$. *We fix such an isomorphism.*

- Let L/K be a finite Galois extension with Galois group G.

In the following we explain Katz' cohomological description of Sw_G (see Theorem 4.53).

10.2.1. Let A be an ℓ-adic coefficient ring (see Convention 8.4). Let $\mathcal{F}$ be a lisse A-sheaf on $\mathbb{G}_m$. Following Katz, we define a *special* lisse A-sheaf on $\mathbb{G}_m$ as follows: If A is finite, we say $\mathcal{F}$ is special if it is tame at 0 and if its *monodromy group*, i.e., the image under the corresponding representation $\pi_1(\mathbb{G}_m, \bar{\eta}) \to \mathrm{Aut}(\mathcal{F}_{\bar{\eta}})$, has a unique p-Sylow group. In general we can write $\mathcal{F} = (\mathcal{F}_n) \otimes_R A$ as in Convention 8.4 and we say $\mathcal{F}$ is special if $\mathcal{F}_1$ is special. By [**26**, Cor. 1.5.7] the restriction functor induces an equivalence of categories

$$(\text{special lisse } A\text{-sheaves on } \mathbb{G}_m) \xrightarrow{\simeq} (\text{lisse } A\text{-sheaves on } \operatorname{Spec} K).$$

Composing the natural functor from the category of finitely generated $A[G]$- modules to the category of lisse A-sheaves on $\operatorname{Spec} K$, see Theorem 8.5, with the inverse of the above equivalence we obtain a fully faithful functor

$$(-)^{\mathrm{can}} : (\text{fin. gen. } A[G]\text{-modules}) \to (\text{special lisse } A\text{-sheaves on } \mathbb{G}_m). \tag{10.10}$$

If M is a finitely generated $A[G]$-module, then the sheaf M^{can} is called the *canonical extension of* M . The name is justified by the fact that restricting the representation corresponding to M^{can} to the decomposition group at ∞ gives back the original $A[G]$-module M. (Here we view G as a quotient of the absolute Galois group of $k((t^{-1}))$ via the isomorphism $K \cong k((t^{-1}))$ fixed above.) The functor $(-)^{\mathrm{can}}$ commutes with direct sums, tensor products, duals, and with changes of coefficients $A \to A'$.

THEOREM 10.10 ([**26**, Thm. 1.6.8]). *Denote by $j : \mathbb{G}_m \hookrightarrow \mathbb{A}^1$ the canonical open immersion. Then with the notation from Theorem* 4.53 *there is an isomorphism of $\mathbb{Z}_\ell[G]$-modules*

$$H^1(\mathbb{A}^1, j_!(\mathbb{Z}_\ell[G]^{\mathrm{can}})) \cong \mathrm{Sw}_G. \tag{10.11}$$

Here the G-module structure on the left-hand side is defined as follows: The left G-module $\mathbb{Z}_\ell[G]$ defines the constructible sheaf $j_!(\mathbb{Z}_\ell[G]^{\mathrm{can}})$ and the right G-module structure of $\mathbb{Z}_\ell[G]$ defines via functoriality a G-action on the sheaf $j_!(\mathbb{Z}_\ell[G]^{\mathrm{can}})$, which via functoriality induces a G-action on the left-hand side of (10.11).

PROOF. We sketch the proof following Katz. For an arbitrary ℓ-adic coefficient ring A we define the functor

$$T : (\text{fin. gen.} A[G] - \text{modules}) \to (\text{fin. gen. } A - \text{modules}) \tag{10.12}$$

by

$$T(M) := H^1(\mathbb{A}^1, j_!(M^{\mathrm{can}})).$$

We have $H^i(\mathbb{A}^1, j_!(M^{\mathrm{can}})) = 0$, for $i \neq 1$. (This is clear for $i = 0$ and follows for $i \geq 2$ from 9.1.2, (e).) Hence T is an exact functor. Furthermore, $A[G]$ acts left-$A[G]$-linearly on itself by right multiplication. Thus $T(A[G])$ becomes an $A[G]$-module by functoriality. Using an argument similar to the proof of Lemma 10.3 one can show:

(a) $T(M) = T(A[G]) \otimes_{A[G]} M$, for all finitely generated $A[G]$-modules M.
(b) $T(A[G])$ is a finitely generated projective $A[G]$-module.

(c) $T(A[G]) \otimes_A A' = T(A'[G])$, where $A \to A'$ is any map of ℓ-adic coefficient rings.

By (b), (c) and the uniqueness part of Theorem 4.53, (b), it suffices to show that $T(\bar{\mathbb{Q}}_\ell[G])$ is the Swan representation. By Proposition 4.25, (a) and Corollary 4.26, it suffices to show that for any irreducible $\bar{\mathbb{Q}}_\ell[G]$-module W we have

$$\dim_{\bar{\mathbb{Q}}_\ell}(T(\bar{\mathbb{Q}}_\ell[G]) \otimes_{\bar{\mathbb{Q}}_\ell} W^\vee)^G = \dim_{\bar{\mathbb{Q}}_\ell}(\mathrm{Sw}_G \otimes_{\mathbb{Z}_\ell} W^\vee)^G. \tag{10.13}$$

Since dualizing induces a bijection on the set of irreducible representations we may replace $W^\vee$ by W. The Swan representation is self dual (as one easily checks that $\mathrm{sw}_G(g) = \mathrm{sw}_G(g^{-1})$). Hence

$$\dim_{\bar{\mathbb{Q}}_\ell}(\mathrm{Sw}_G \otimes_{\mathbb{Z}_\ell} W)^G = \dim_{\bar{\mathbb{Q}}_\ell} \mathrm{Hom}_{\bar{\mathbb{Q}}_\ell[G]}(\mathrm{Sw}_G \otimes_{\mathbb{Z}_\ell} \bar{\mathbb{Q}}_\ell, W) = \mathrm{Swan}(W),$$

where the second equality follows from Theorem 4.86 and Remark 4.73. On the other hand by (a) above

$$\begin{aligned}\dim_{\bar{\mathbb{Q}}_\ell}(T(\bar{\mathbb{Q}}_\ell[G]) \otimes_{\bar{\mathbb{Q}}_\ell} W)^G &= \dim_{\bar{\mathbb{Q}}_\ell}(T(\bar{\mathbb{Q}}_\ell[G]) \otimes_{\bar{\mathbb{Q}}_\ell[G]} W)\\ &= \dim_{\bar{\mathbb{Q}}_\ell} H^1(\mathbb{A}^1, j_! W^{\mathrm{can}})\\ &= -\chi(\mathbb{A}^1, j_! W^{\mathrm{can}}).\end{aligned}$$

Furthermore, $\chi(\mathbb{A}^1, j_! W^{\mathrm{can}}) = \chi_c(\mathbb{A}^1, j_! W^{\mathrm{can}}) = \chi_c(\mathbb{G}_m, W^{\mathrm{can}})$, where the first equality follows e.g. from [**27**, 2.0.7] and the second equality holds by definition. We have $\chi_c(\mathbb{G}_m, \bar{\mathbb{Q}}_\ell) = 0$. Since W^{can} is lisse on $\mathbb{G}_m$ and tame in 0, the formula of Grothendieck-Ogg-Shafarevich yields $\chi_c(\mathbb{G}_m, W^{\mathrm{can}}) = -\mathrm{Swan}_\infty(W^{\mathrm{can}}) = -\mathrm{Swan}(W)$. Putting everything together yields the theorem. □

11. Higher dimensional ramification theory via restriction to curves

In this section we present the ramification theory of lisse sheaves on a higher dimensional smooth variety over a perfect field of positive characteristic, following [**16**, 3.].

11.0.1. Throughout this section we fix the following notation:

- k is a perfect field of characteristic $p > 0$ and $\bar{k}$ an algebraic closure.
- ℓ is a prime number different from p.
- A k-scheme is a scheme which is separated and of finite type over k.
- For a k-scheme X we denote by $\mathrm{Cu}(X)$ the set of normalizations of closed integral 1-dimensional subschemes of X. In other words, $C \in \mathrm{Cu}(X)$ is a smooth connected curve over k with a morphism $\nu : C \to X$, which is birational onto its image.
- If C is a smooth connected k-scheme of dimension 1, we denote by $\widehat{C}$ the (up to isomorphism) unique smooth projective curve over k, which admits an open dense embedding $C \hookrightarrow \widehat{C}$.

11.1. Lisse sheaves with bounded ramification.

DEFINITION 11.1. Let U be a normal k-scheme. A *divisorial compactification*[3] of U is a normal proper k-scheme X together with an open dense immersion $j : U \hookrightarrow X$ such that the complement $X \smallsetminus U$ is the support of an effective Cartier divisor.

REMARK 11.2. Let U be a normal k-scheme.

[3] This is non-standard terminology.

(a) A divisorial compactification of U exists. Indeed, by Nagata's compactification theorem (see e.g. [**11**, Thm. 4.1]), we find a dense open immersion of U into a proper k-scheme Y. Let $\tilde{Y}$ be the blow-up of Y in the closed subscheme $Y \smallsetminus U$ (with its reduced scheme structure). We obtain a dominant open immersion $U \hookrightarrow \tilde{Y}$ whose complement is the support of the exceptional divisor E of the blow-up, which is an effective Cartier divisor. Now let $\nu : X \to \tilde{Y}$ be the normalization of $\tilde{Y}$. Then ν is a finite morphism which is an isomorphism over U. We obtain an open dense embedding $U \hookrightarrow X$ whose complement is the support of the effective Cartier divisor $\nu^* E$. Then $U \hookrightarrow X$ is a divisorial compactification of U.

(b) If C is a smooth connected curve over k, then with the notation from 11.0.1 the inclusion $C \hookrightarrow \widehat{C}$ is the (up to isomorphism) unique divisorial compactification of C.

(c) Let X be a divisorial compactification of U and $C \in \mathrm{Cu}(U)$. Then by the valuative criterion for properness, the morphism $\nu : C \to U$ extends uniquely to a morphism $\nu_X : \widehat{C} \to X$. Furthermore $\widehat{C}$ is the normalization of its image in X, hence $\widehat{C} \in \mathrm{Cu}(X)$.

DEFINITION 11.3. Let C be a smooth connected curve over k, A an ℓ-adic coefficient ring in the sense of Convention 8.4 and $\mathcal{F}$ a lisse A-sheaf on C. We define the *Swan conductor of* $\mathcal{F}$ to be the following effective Cartier divisor on $\widehat{C}$:

$$\mathrm{Swan}(\mathcal{F}) := \sum_{x \in \widehat{C}} \mathrm{Swan}_x(\mathcal{F}) \cdot [x].$$

Here $\mathrm{Swan}_x(\mathcal{F})$ is the Swan conductor of $\mathcal{F}$ on C at x, see 10.1.1. Notice that $\mathrm{Swan}_x(\mathcal{F}) \in \mathbb{Z}_{\geq 0}$ by Theorem 4.86.

DEFINITION 11.4. Let U be a normal k-scheme, $U \hookrightarrow X$ a divisorial compactification and D an effective Cartier divisor supported in $X \smallsetminus U$. Let A be an ℓ-adic coefficient ring and $\mathcal{F}$ a lisse A-sheaf on U. Then we say that the *ramification of* $\mathcal{F}$ *is bounded by* D if the following condition is satisfied:

$$\mathrm{Swan}(\nu^*\mathcal{F}) \leq \nu_X^* D, \quad \text{for all } C \in \mathrm{Cu}(U). \tag{11.1}$$

Here $\nu : C \to U$ is the natural map and $\nu_X : \widehat{C} \to X$ its unique extension; the inequality takes place in the monoid of effective divisors on $\widehat{C}$.

EXAMPLE 11.5. Let $\mathcal{L}_{m,\psi}$ be the lisse sheaf on $\mathbb{A}^1 \subseteq \mathbb{P}^1$ from Example 8.8. Then the ramification of $\mathcal{L}_{m,\psi}$ is bounded by $m \cdot \{\infty\}$, by Example 4.88.

REMARK 11.6. In the situation above we find a DVR R which is finite over $\mathbb{Z}_\ell$ and a lisse R-sheaf $(\mathcal{F}_n)_n$ on U such that A is an R-algebra and $\mathcal{F} \cong (\mathcal{F}_n) \otimes_R A$. Then for any $\nu : C \to U$ in $\mathrm{Cu}(U)$ we have $\nu^*\mathcal{F} \cong (\nu^*\mathcal{F}_n) \otimes_R A$ and by 10.1.1, (a)

$$\mathrm{Swan}(\nu^*\mathcal{F}) = \mathrm{Swan}(\nu^*\mathcal{F}_1).$$

Thus whether the ramification of $\mathcal{F}$ is bounded or not by a given Cartier divisor D depends only on $\mathcal{F}_1$.

REMARK 11.7. In the situation of Definition 11.4 if $0 \to \mathcal{F}' \to \mathcal{F} \to \mathcal{F}'' \to 0$ is an exact sequence of lisse A-sheaves on U and the ramification of $\mathcal{F}'$ and $\mathcal{F}''$ is bounded by D' and D'', respectively, the ramification of $\mathcal{F}$ is bounded by $D' + D''$. This follows immediately from 10.1.1, (d).

PROPOSITION 11.8. *Let U be a normal k-scheme and $U \hookrightarrow X$ a divisorial compactification. Let A be an ℓ-adic coefficient ring and $\mathcal{F}$ a free lisse A-sheaf on U. Then there exists an effective Cartier divisor D on X supported in $X \smallsetminus U$ such that the ramification of $\mathcal{F}$ is bounded by D.*

Before we can prove the proposition we need a notion of discriminant in higher dimensions. The following definition is due to Alexander Schmidt.

DEFINITION 11.9. Let $\pi : Y' \to Y$ be a finite and generically étale morphism of degree n between connected normal k-schemes. Denote by $\mathrm{Tr}_{Y'/Y} : \pi_* \mathcal{O}_{Y'} \to \mathcal{O}_Y$ the trace map which is induced by $\mathrm{Tr}_{k(Y')/k(Y)}$. We define the discriminant $\mathcal{I}(D_{Y'/Y})$ of Y'/Y to be the sheaf associated to the following presheaf of ideals in $\mathcal{O}_Y$

$$Y \supseteq U \mapsto \sum_{\{x_1,\dots,x_n\}} \det((\mathrm{Tr}_{Y'/Y}(x_i x_j))_{i,j}) \cdot \mathcal{O}_Y(U) \quad \subseteq \mathcal{O}_Y(U),$$

where the sum is over all subset $\{x_1, \dots, x_n\} \subseteq \mathcal{O}_{Y'}(\pi^{-1}(U))$, which consist of $\mathcal{O}_Y(U)$-linearly independent elements.

LEMMA 11.10. *In the situation of Definition 11.9 the discriminant $\mathcal{I}(D_{Y'/Y})$ has the following properties:*

(a) *$\mathcal{I}(D_{Y'/Y}) \subseteq \mathcal{O}_Y$ is a coherent ideal sheaf.*
(b) *If $\pi : Y' \to Y$ is free and $e_1, \dots, e_n$ is a basis of the $\mathcal{O}_Y$-module $\pi_* \mathcal{O}_{Y'}$, then $\mathcal{I}(D_{Y'/Y}) = \det((\mathrm{Tr}_{Y'/Y}(e_i e_j))_{i,j}) \cdot \mathcal{O}_Y$.*
(c) *Assume $Y = \operatorname{Spec} A$ and $Y' = \operatorname{Spec} B$ are smooth affine curves. Then $\mathcal{I}(D_{Y'/Y})$ is the sheaf induced by the discriminant ideal $\mathfrak{d}_{B/A} \subseteq A$ from Definition 3.18.*
(d) *If $U \subseteq Y$ is an open subset such that $\pi^{-1}(U) \to U$ is étale, then $\mathcal{I}(D_{Y'/Y})_{|U} = \mathcal{O}_U$.*
(e) *Let $\nu : C \to Y$ be a morphism from a smooth connected curve C to Y and assume that $\nu(C)$ meets the locus over which π is étale. Let C' be the normalization of an irreducible component of $C \times_Y Y'$, which is generically étale over C. We obtain a commutative diagram*

$$\begin{array}{ccc} C' & \xrightarrow{\nu'} & Y' \\ {\scriptstyle \pi_{C'}}\downarrow & & \downarrow{\scriptstyle \pi} \\ C & \xrightarrow{\nu} & Y. \end{array}$$

Then the image of $\mathcal{I}(D_{Y'/Y})$ under $\nu^* : \nu^{-1}\mathcal{O}_Y \to \mathcal{O}_C$ is contained in $\mathcal{I}(D_{C'/C})$, i.e.

$$\nu^*(\mathcal{I}(D_{Y'/Y})) \subseteq \mathcal{I}(D_{C'/C}).$$

PROOF. By definition $\mathcal{I}(D_{Y'/Y})$ is an $\mathcal{O}_Y$-submodule of $\mathcal{O}_Y$ hence (a) is automatic. Now let $U = \operatorname{Spec} A$ be an affine open subset of Y and write $\pi^{-1}(U) = \operatorname{Spec} B$. Let $\{x_1, \dots, x_n\}$ and $\{y_1, \dots y_n\}$ be two sets of A-linearly independent elements of B and denote $B_x = \sum_i A x_i$, $B_y = \sum_i A y_i$. Assume $B_x \subseteq B_y$ then

$$\det((\mathrm{Tr}_{B/A}(x_i x_j))_{i,j}) \cdot A \subseteq \det((\mathrm{Tr}_{B/A}(y_i y_j))_{i,j}) \cdot A. \tag{11.2}$$

Indeed by assumption we find an $n \times n$-matrix M with coefficients in A such that (with the obvious vector notation) $x = My$ and we obtain

$$\det((\mathrm{Tr}_{B/A}(x_i x_j))_{i,j}) = \det(M)^2 \cdot \det((\mathrm{Tr}_{B/A}(y_i y_j))_{i,j}),$$

in particular (11.2) holds. This immediately implies (b) and using Lemma 3.21 also (c).

To prove (d) we can assume that B/A is étale and free with basis $e_1, \dots, e_n$ and we have to show $\delta := \det((\mathrm{Tr}_{B/A}(e_i e_j))_{i,j}) \in A^\times$. For this it suffices to show that for any prime $\mathfrak{p} \subseteq A$ the element δ maps to a unit in the residue field $k(\mathfrak{p})$. But since B/A is étale, $B \otimes_A k(\mathfrak{p})$ is étale over $k(\mathfrak{p})$ with basis $\bar{e}_i = e_i \otimes 1$, $i = 1, \dots, n$, and the image of δ in $k(\mathfrak{p})$ equals $\det((\mathrm{Tr}_{B\otimes_A k(\mathfrak{p})/k(\mathfrak{p})}(\bar{e}_i\bar{e}_j))_{i,j})$. It follows from Proposition 3.14, that this element is not zero in $k(\mathfrak{p})$. Hence (d) follows.

Finally, in the situation of (e), we can assume that $Y' \to Y$ is given by a finite ring extension B/A. If the image of C is a point, then by assumption it lies in the locus over which π is étale and hence $\mathcal{I}(D_{Y'/Y})$ is the unit ideal around this point. Thus we can assume that $\nu(C)$ is a curve and therefore $C' \to C$ corresponds to a finite and generically étale extension of Dedekind domains S/R. Let $x_1, \dots, x_n \in B$ be a sequence of A-linearly independent elements. We have to show that the image of $\det((\mathrm{Tr}_{B/A}(x_i x_j))_{i,j})$ under $A \to R$ is contained in $\mathfrak{d}_{S/R}$, which amounts to show that

$$\det((\mathrm{Tr}_{B\otimes_A R/R}(\bar{x}_i\bar{x}_j))_{i,j}) \in \mathfrak{d}_{S/R}, \tag{11.3}$$

where $\bar{x}_i$ denotes the image of x_i in $B \otimes_A R$. Observe that all irreducible components of $Y' \times_Y C$ map surjectively to C. Indeed, since $\pi : Y' \to Y$ is finite and surjective and Y is normal (hence universally unibranch) π is universally open, by [**19**, Cor. (14.4.4), (i)]. Let $S_1 = S, S_2, \dots, S_r$ be the affine coordinate rings of the normalizations of the irreducible components of $Y' \times_Y C$. We get a natural map $B \otimes_A R \to \prod_i S_i$ which becomes an isomorphism when tensored with $\otimes_A k(Y)$, since $Y' \times_Y C \to C$ is generically étale. It follows that we obtain a commutative diagram

$$\begin{array}{ccc} B\otimes_A R & \longrightarrow & \prod_i S_i \\ & \searrow^{\mathrm{Tr}} & \big\downarrow \scriptstyle\sum_i \mathrm{Tr}_{S_i/R} \\ & & R. \end{array}$$

The images of $\bar{x}_1, \dots, \bar{x}_n$ in $\prod_i S_i$ span an R-submodule of rank n. Further taking a basis $\{e_{ij}\}_{j=1,\dots,n_i}$ of S_i/R we get the following basis of $\prod_i S_i$

$$(e_{11}, 0 \dots, 0), \dots, (e_{1,n_1}, 0 \dots, 0), \dots, (0, \dots, 0, e_{r,1}), \dots, (0, \dots, 0, e_{r,n_r}).$$

Therefore the formula (11.2) (applied with $\{y_i\}$ the above basis) yields

$$\det((\mathrm{Tr}_{B\otimes_A R/R}(\bar{x}_i\bar{x}_j))_{i,j}) \cdot A \subseteq \mathfrak{d}_{S_1/R} \cdots \mathfrak{d}_{S_r/R} \subseteq \mathfrak{d}_{S_1/R}.$$

This proves (11.3) and finishes the proof of the lemma. □

REMARK 11.11. A more intrinsic way to define the discriminant ideal $\mathcal{I}(D_{Y'/Y})$ is the following: There is a well defined $\mathcal{O}_Y$-linear morphism

$$\varphi_{Y'/Y} : \bigwedge^n (\pi_* \mathcal{O}_{Y'}) \otimes_{\mathcal{O}_Y} \bigwedge^n (\pi_* \mathcal{O}_{Y'}) \to \mathcal{O}_Y,$$

which sends $(x_1 \wedge \dots \wedge x_n) \otimes (y_1 \wedge \dots \wedge y_n)$ to $\det((\mathrm{Tr}_{Y'/Y}(x_i y_j))_{i,j})$, see e.g. [**7**, §1, 9]. For an $\mathcal{O}_Y$-module $\mathcal{M}$ denote by $\Gamma(\mathcal{M})$ the divided power algebra of $\mathcal{M}$, see [**44**, III, 1. Def.]. Recall that it is a graded $\mathcal{O}_Y$-algebra which is locally generated by elements of the form $x^{[n]}$, $x \in \mathcal{M}$, $n \geq 1$, which have degree n and satisfy $n! x^{[n]} = (x^{[1]})^n$. Denote by $\Gamma^2(\mathcal{M})$ the $\mathcal{O}_Y$-module of homogeneous elements of degree 2. There is a canonical $\mathcal{O}_Y$-linear map $\gamma_2 : \Gamma^2(\mathcal{M}) \to \mathcal{M} \otimes_{\mathcal{O}_Y} \mathcal{M}$ satisfying $\gamma_2(x^{[2]}) = x \otimes x$ and if 2 is invertible in $\mathcal{O}_Y$ it induces an isomorphism with the symmetric tensors $\Gamma^2(\mathcal{M}) \xrightarrow{\simeq} (\mathcal{M} \otimes_{\mathcal{O}_Y} \mathcal{M})^{\Sigma_2}$, see [**44**, Prop. III.1, Prop. III.3]. Define $\varphi_{Y'/Y} := \varphi_{Y'/Y} \circ \gamma_2 : \Gamma^2(\bigwedge^n \pi_* \mathcal{O}_{Y'}) \to \mathcal{O}_Y$. Then we have $\mathcal{I}(D_{Y'/Y}) = \operatorname{im}(\varphi_{Y'/Y})$, as follows directly from the definition.

Proof of Proposition 11.8. We can assume that U is connected; let $\bar\eta \to U$ be a geometric point over the generic point of U. By Remark 11.6 it suffices to consider the case in which A is a finite field of characteristic ℓ. In particular $\mathcal{F}_{\bar\eta}$ is finite (as a set). Hence the representation $\pi_1(U, \bar\eta) \to \operatorname{Aut}(\mathcal{F}_{\bar\eta})$ defined by $\mathcal{F}$ factors over a finite quotient, i.e. there exists a finite connected étale Galois cover $\pi : U' \to U$ which trivializes $\mathcal{F}$. Let $\pi_X : X' \to X$ be the normalization of X inside the function field of U'. Then π_X is finite, $\pi_X^{-1}(U) = U'$ and X' is a divisorial compactification of U'. Let $\mathcal{I}(D_{X'/X})$ be the discriminant of X'/X (Definition 11.9). Since $\mathcal{I}(D_{X'/X})_{|U} = \mathcal{O}_U$ we find an effective Cartier divisor D on X with $\mathcal{O}_X(-D) \subseteq \mathcal{I}(D_{X'/X})$. We claim:

(11.4) The ramification of $\mathcal{F}$ is bounded by $\operatorname{rk}(\mathcal{F}) \cdot D$.

Take $\nu : C \to U$ in $\operatorname{Cu}(U)$. Let C' be a connected component of $C \times_U U'$. The projection maps induce maps $C' \to C$ and $C' \to U'$, which after compactification yield a commutative diagram

$$\begin{array}{ccc} \widehat{C}' & \xrightarrow{\nu_{X'}} & X' \\ {\scriptstyle \pi_{\widehat{C}}}\downarrow & & \downarrow{\scriptstyle \pi_X} \\ \widehat{C} & \xrightarrow{\nu_X} & X. \end{array}$$

By Lemma 11.10, (e), we have

$$\mathcal{O}_{\widehat{C}}(-D_{|\widehat{C}}) \subseteq \nu_X^* \mathcal{I}(D_{X'/X}) \subseteq \mathcal{I}(D_{\widehat{C}'/\widehat{C}}).$$

Denote by $D_{\widehat{C}'/\widehat{C}}$ the effective divisor on $\widehat{C}$ given by $\mathcal{I}(D_{\widehat{C}'/\widehat{C}})$; it is the discriminant divisor. We obtain

$$\sum_{x'/x} [x' : x] \cdot v_{x'}(\mathfrak{D}_{\widehat{C}'/\widehat{C}}) = \operatorname{mult}_x(D_{\widehat{C}'/\widehat{C}}) \leq \operatorname{mult}_x(D_{|\widehat{C}}), \tag{11.5}$$

where we use the notation from Proposition 10.5.

On the other hand $C' \to C$ is a connected finite étale Galois covering, say with Galois group $G_{C'}$, trivializing $\mathcal{E} := \nu^* \mathcal{F}$. Let $\bar\xi \to C$ be a geometric point over the generic point of C and $x \in \widehat{C} \setminus C$ a closed point. Then

$$\begin{aligned}
\operatorname{Swan}_x(\mathcal{E}) &= \dim_A \operatorname{Hom}_{A[G_{C'}]}(\operatorname{Sw}_{G_{C'},x} \otimes_{\mathbb{Z}_\ell} A, \mathcal{E}_{\bar\xi}) && \text{by 10.5(c), 4.73} \\
&= \dim_A \left(\underline{\operatorname{Hom}}_A(\operatorname{Sw}_{G_{C'},x} \otimes_{\mathbb{Z}_\ell} A, \mathcal{E}_{\bar\xi})^{G_{C'}}\right) && \text{by 4.1(e)} \\
&\leq \dim_A \underline{\operatorname{Hom}}_A(\operatorname{Sw}_{G_{C'},x} \otimes_{\mathbb{Z}_\ell} A, \mathcal{E}_{\bar\xi}) && \\
&= \operatorname{rk}(\mathcal{E}) \cdot \dim_A(\operatorname{Sw}_{G_{C'},x} \otimes_{\mathbb{Z}_\ell} A) && \\
&= \operatorname{rk}(\mathcal{E}) \cdot \operatorname{sw}_{G_{C'},x}(1) && \text{by (10.3)} \\
&= \operatorname{rk}(\mathcal{F}) \cdot \left(\left(\sum_{x'/x} [x' : x] \left(1 + v_{x'}\left(\mathfrak{D}_{\widehat{C}'/\widehat{C}}\right)\right)\right) - |G_{C'}|\right) && \text{by 10.5(b)} \\
&\leq \operatorname{rk}(\mathcal{F}) \cdot \sum_{x'/x} [x' : x] \cdot v_{x'}(\mathfrak{D}_{\widehat{C}'/\widehat{C}}) && \text{by (3.1), 5.3(c)} \\
&\leq \operatorname{rk}(\mathcal{F}) \cdot \operatorname{mult}_x(D_{|\widehat{C}}) && \text{by (11.5).}
\end{aligned}$$

Hence $\mathrm{Swan}(\nu^*\mathcal{F}) \leq \mathrm{rk}(\mathcal{F}) \cdot D_{|\widehat{C}}$ which proves the claim (11.4) and the proposition. □

As a consequence of the Grothendieck-Ogg-Shafarevich theorem we obtain a universal bound for the cohomolgy with compact support of a lisse $\bar{\mathbb{Q}}_\ell$-sheaf on a smooth curve over k with fixed rank and ramification bounded by a divisor of fixed degree. More precisely:

DEFINITION 11.12. Let C be a smooth and geometrically connected curve over k. Let $d \geq 0$ be a natural number. Then we define

$$\mathcal{C}_d := 2g(\widehat{C}) + 2d + 1,$$

where $g(\widehat{C})$ is the genus of $\widehat{C}$. In case D is an effective divisor of degree $\deg(D) \geq 0$ on $\widehat{C}$ we set

$$\mathcal{C}_D := \mathcal{C}_{\deg D}$$

and call it the *complexity of* D.

PROPOSITION 11.13. *Let C be a smooth geometrically connected curve over k. Let $r \geq 1$ and $d \geq 0$ be natural numbers. Then*

$$\dim_{\bar{\mathbb{Q}}_\ell} H^0_c(C \otimes_k \bar{k}, \mathcal{F}) + \dim_{\bar{\mathbb{Q}}_\ell} H^1_c(C \otimes_k \bar{k}, \mathcal{F}) \leq r \cdot \mathcal{C}_d$$

for all lisse $\bar{\mathbb{Q}}_\ell$-sheaves $\mathcal{F}$ on $C \otimes_k \bar{k}$ of rank r whose ramification is bounded by $r \cdot D$, where D is an effective divisor of degree $\deg(D) = d$ and with support $\mathrm{supp}(D) = \widehat{C} \otimes_k \bar{k} \smallsetminus C \otimes_k \bar{k}$.

PROOF. Observe that

$$\dim_{\bar{\mathbb{Q}}_\ell} H^0(C \otimes_k \bar{k}, \mathcal{F}) = \dim_{\bar{\mathbb{Q}}_\ell} \mathcal{F}_{\bar{\eta}}^{\pi_1(\widehat{C} \otimes_k \bar{k}, \bar{\eta})} \leq r, \tag{11.6}$$

where $\mathcal{F}_{\bar{\eta}}$ denotes the $\pi_1(\widehat{C} \otimes_k \bar{k}, \bar{\eta})$-representation corresponding to $\mathcal{F}$, see Theorem 8.5.

First assume $C = \widehat{C}$ is projective and hence $D = 0$. Let $x \in \widehat{C} \otimes_k \bar{k}$ be a closed point and denote by U the complement. Notice that $\mathrm{Swan}_x(\mathcal{F}) = 0$. The exact sequence

$$\cdots \to H^i(\widehat{C} \otimes_k \bar{k}, \mathcal{F}) \to H^i(x, \mathcal{F}) \to H^{i+1}_c(U, \mathcal{F}) \to \cdots,$$

Theorem 10.1 and Lemma 10.8 give

$$\begin{aligned} \sum_i (-1)^i \dim_{\bar{\mathbb{Q}}_\ell} H^i(\widehat{C} \otimes_k \bar{k}, \mathcal{F}) &= \chi_c(U, \mathcal{F}) + \dim_{\bar{\mathbb{Q}}_\ell} H^0(x, \mathcal{F}) \\ &= r\chi_c(U) + r \\ &= r\chi(\widehat{C} \otimes_k \bar{k}) \\ &= r(2 - 2g(\widehat{C})). \end{aligned} \tag{11.7}$$

Furthermore by Poincaré duality (Theorem 9.4) and (11.6)

$$\dim_{\bar{\mathbb{Q}}_\ell} H^2(\widehat{C} \otimes_k \bar{k}, \mathcal{F}) = \dim_{\bar{\mathbb{Q}}_\ell} H^0(\widehat{C} \otimes_k \bar{k}, \mathcal{F}^\vee(1)) \leq r.$$

Thus (11.6) and (11.7) yield

$$\dim_{\bar{\mathbb{Q}}_\ell} H^0_c(C \otimes_k \bar{k}, \mathcal{F}) + \dim_{\bar{\mathbb{Q}}_\ell} H^1_c(C \otimes_k \bar{k}, \mathcal{F}) \leq r \cdot \mathcal{C}_0.$$

Now assume the inclusion $C \subseteq \widehat{C}$ is strict, i.e. C is affine and the ramification of $\mathcal{F}$ is bounded by D with $\mathrm{supp}(D) = \widehat{C} \otimes_k \bar{k} \smallsetminus C \otimes_k \bar{k}$. Then by Corollary 9.5, Poincaré duality 9.4 and (11.6)

$$\dim_{\bar{\mathbb{Q}}_\ell} H^0_c(C \otimes_k \bar{k}, \mathcal{F}) = 0, \quad \dim_{\bar{\mathbb{Q}}_\ell} H^2_c(C \otimes_k \bar{k}, \mathcal{F}) = \dim_{\bar{\mathbb{Q}}_\ell} H^0(C \otimes_k \bar{k}, \mathcal{F}^\vee(1)) \le r.$$

Thus by Grothendieck-Ogg-Shafarevich 10.1 and Lemma 10.8

$$\begin{aligned}
&\dim_{\bar{\mathbb{Q}}_\ell} H^1_c(C \otimes_k \bar{k}, \mathcal{F})\\
&\quad = r \cdot (2g(\widehat{C}) - 2) + r \cdot \mathrm{card}(\mathrm{supp}(D)) + \dim_{\bar{\mathbb{Q}}_\ell} H^2_c(C \otimes_k \bar{k}, \mathcal{F})\\
&\qquad + \sum_{x \in \mathrm{supp}(D)} \mathrm{Swan}_x(\mathcal{F})\\
&\quad \le r \cdot (2g(\widehat{C}) - 2) + rd + r + rd\\
&\quad = r(2g(\widehat{C}) + 2d - 1)\\
&\quad < r \cdot \mathcal{C}_d.
\end{aligned}$$

□

DEFINITION 11.14. Let U be a normal k-scheme, A an ℓ-adic coefficient ring and $\mathcal{F}$ a lisse A-sheaf on U. We say that $\mathcal{F}$ is *tame* if for any $\nu : C \to U$ in $\mathrm{Cu}(U)$ we have $\mathrm{Swan}(\nu^*\mathcal{F}) = 0$.

REMARK 11.15. We immediately see that $\mathcal{F}$ is tame if and only if there exists a divisorial compactification $U \hookrightarrow X$ with respect to which the ramification of $\mathcal{F}$ is bounded by the zero divisor. Moreover, this is equivalent to the ramification of $\mathcal{F}$ being bounded by the zero divisor with respect to all divisorial compactifications.

The following proposition is a direct consequence of Theorem 6.18 due to Kerz and Schmidt.

PROPOSITION 11.16 (Kerz-Schmidt). *Let U be a normal connected scheme over k, $\bar{x} \to U$ a geometric point and A an ℓ-adic coefficient ring. Let $\pi_1^{\mathrm{tame}}(U, \bar{x})$ be the tame fundamental group of U at $\bar{x}$, see Definition 6.20. Then the equivalence of categories from Theorem 8.5 restricts to an equivalence between the following full subcategories*

$$(\textit{tame } A\textit{-sheaves on } U) \xrightarrow{\simeq} (A\textit{-representations of } \pi_1^{\mathrm{tame}}(U, \bar{x})).$$

PROOF. In Theorem 8.5 we constructed two quasi-inverse functors

$$(\text{lisse } A\text{-sheaves on } U) \underset{G}{\overset{F}{\rightleftarrows}} (A\text{-representations of } \pi_1(U, \bar{x}))$$

It suffices to show F and G induce functors between the full subcategories from the statement. In view of the construction of F and G it suffices to consider the case where A is a finite ring. Let $\mathcal{F}$ be a tame A-sheaf on U. Then we have to show that the $\pi_1(U, \bar{x})$-representation $F(\mathcal{F}) = \mathcal{F}_{\bar{x}}$ factors over the continuous quotient map $\pi_1(U, \bar{x}) \twoheadrightarrow \pi_1^{\mathrm{tame}}(U, \bar{x})$, see Proposition 6.21, (a). By Proposition 7.6 the sheaf $\mathcal{F}$ is represented by a finite étale U-group scheme $U_{\mathcal{F}}$. The stalk $\mathcal{F}_{\bar{x}}$ is equal to the fiber of $U_{\mathcal{F}} \to U$ at $\bar{x}$. By the definition of $\pi_1^{\mathrm{tame}}(U, \bar{x})$ it suffices to prove that $U_{\mathcal{F}} \to U$ is tame, see Definition 6.15, (b). By Theorem 6.18 it suffices to check that for any smooth connected k-curve C and any map $\nu : C \to U$ the base changed morphism $U_{\mathcal{F}} \times_U C \to C$ is tamely ramified along $\widehat{C} \smallsetminus C$. If $\nu(C)$ is a point there is nothing to show. Otherwise ν factors over the normalization $\widetilde{\nu(C)} \to \nu(C) \subseteq U$.

If the base change of $U_{\mathcal{F}} \to U$ over $\widetilde{\nu(C)}$ is tame, then so is the base change over C, see e.g. [**41**, II, Cor. 7.8]. Thus we can assume that $\nu : C \to U$ is in $\mathrm{Cu}(U)$ (see 11.0.1 for the Definition of $\mathrm{Cu}(U)$). Notice that $U_{\nu^*\mathcal{F}} := U_{\mathcal{F}} \times_U C \to C$ is a finite étale morphism representing $\nu^*\mathcal{F}$. Since $\mathcal{F}$ is tame we have $\mathrm{Swan}_y(\nu^*\mathcal{F}) = 0$ for all $y \in \widehat{C}$, i.e. for $\bar{\eta} \to C$ a geometric point over the generic point of C and P_y the wild inertia group at y we have $U_{\mathcal{F}} \times_U \bar{\eta} = \mathcal{F}_{\bar{\eta}} = \mathcal{F}_{\bar{\eta}}^{P_y} = (U_{\mathcal{F}} \times_U \bar{\eta})^{P_y}$ (see 10.1.1, (c)). This means that $U_{\mathcal{F}} \times_U C \to C$ is tamely ramified.

It remains to show that if M is an A-representation of $\pi_1^{\mathrm{tame}}(U, \bar{x})$ then the lisse sheaf $G(M)$ is a tame sheaf on U. But by construction of $\pi_1^{\mathrm{tame}}(U, \bar{x})$ there is a finite étale Galois cover $P \to U$ which is tame such that $G(M)_{|P}$ is trivial. It follows that $G(M)$ is represented by a finite étale U-group-scheme $U_{G(M)}$ sitting inside $P \to U_{G(M)} \to U$. It follows that $U_{G(M)} \to U$ is tame. Hence for $\nu : C \to U$ in $\mathrm{Cu}(U)$, the base change $U_{G(M)} \times_U C \to C$ is tame (by Theorem 6.18). This means $\mathrm{Swan}(\nu^* G(M)) = 0$, so $G(M)$ is tame. □

REMARK 11.17.
(a) Let A and U be as above and $\mathcal{F}$ a lisse A-sheaf on U and $\mathcal{L}$ a tame A-sheaf on U, which is free of rank 1. Let $U \hookrightarrow X$ be a divisorial compactification and D an effective Cartier divisor supported in $X \smallsetminus U$. Then the ramification of $\mathcal{F}$ is bounded by D if and only if the ramification of $\mathcal{F} \otimes \mathcal{L}$ is bounded by D. Indeed if U is a curve and $x \in X \smallsetminus U$ a point at infinity, then the restriction of the representation corresponding to $\mathcal{L}$ to the wild inertia group at x is trivial; in particular $\mathrm{Swan}_x(\mathcal{F}) = \mathrm{Swan}_x(\mathcal{F} \otimes_A \mathcal{L})$. The general case follows from the curve case.
(b) If $\mathcal{L}$ is a lisse A-sheaf on $\mathrm{Spec}\, k$, then the pullback $\mathcal{L}_{|U}$ to U is tame.

11.1.1. *Weil sheaves.* (See [**13**, (1.1.7)-(1.1.12)].)

(a) Let $\mathbb{F}_q$ be the finite field with $q = p^n$ elements and fix an algebraic closure $\mathbb{F}$. Denote by $F \in \mathrm{Gal}(\mathbb{F}/\mathbb{F}_q)$ the *geometric Frobenius* element given by $F(a) = a^{-q}$. We define the *Weil group* $W(\mathbb{F}/\mathbb{F}_q)$ *of* $\mathbb{F}_q$ to be the subgroup of $\mathrm{Gal}(\mathbb{F}/\mathbb{F}_q)$ generated by F. Notice that under the isomorphism $\mathrm{Gal}(\mathbb{F}/\mathbb{F}_q) \xrightarrow{\simeq} \hat{\mathbb{Z}}$ which sends F to 1, the group $W(\mathbb{F}/\mathbb{F}_q)$ is identified with $\mathbb{Z}$. It is equipped with the discrete topology.
(b) Let X be a geometrically connected $\mathbb{F}_q$-scheme and $\bar{x} \to X \otimes_{\mathbb{F}_q} \mathbb{F}$ a geometric point. Then there is a short exact sequence (see [**22**, IX, Thm. 6.1])

$$0 \to \pi_1(X \otimes_{\mathbb{F}_q} \mathbb{F}, \bar{x}) \to \pi_1(X, \bar{x}) \to \mathrm{Gal}(\mathbb{F}/\mathbb{F}_q) \to 0.$$

We define the *Weil group of* X *at* $\bar{x}$ to be the topological group

$$W(X, \bar{x}) = \pi_1(X, \bar{x}) \times_{\mathrm{Gal}(\mathbb{F}/\mathbb{F}_q)} W(\mathbb{F}/\mathbb{F}_q).$$

Notice that since $W(\mathbb{F}/\mathbb{F}_q)$ has the discrete topology the subgroup

$$\pi_1(X \otimes_{\mathbb{F}_q} \mathbb{F}, \bar{x}) = \pi_1(X, \bar{x}) \times_{\mathrm{Gal}(\mathbb{F}/\mathbb{F}_q)} \{0\} \subseteq W(X, \bar{x})$$

is open and closed. (In particular, $W(X, \bar{x})$ is *not* equipped with the induced topology from its inclusion into $\pi_1(X, \bar{x})$, since $\pi_1(X \otimes_{\mathbb{F}_q} \mathbb{F}, \bar{x}) \subseteq \pi_1(X, \bar{x})$ is not open, the quotient group being infinite.) We obtain a short exact sequence of topological groups

$$0 \to \pi_1(X \otimes_{\mathbb{F}_q} \mathbb{F}, \bar{x}) \to W(X, \bar{x}) \to W(\mathbb{F}/\mathbb{F}_q) \to 0.$$

(c) Let X and $\bar{x} \to X \otimes_{\mathbb{F}_q} \mathbb{F}$ be as above. The action of $W(\mathbb{F}/\mathbb{F}_q)$ on $\mathbb{F}$ induces an action of $W(\mathbb{F}/\mathbb{F}_q)$ on $X \otimes_{\mathbb{F}_q} \mathbb{F}$. By definition a *Weil sheaf on* X is a constructible $\bar{\mathbb{Q}}_\ell$-sheaf $\mathcal{F}$ on $X \otimes_{\mathbb{F}_q} \mathbb{F}$ together with an action of $W(\mathbb{F}/\mathbb{F}_q)$ on $\mathcal{F}$, i.e. morphisms $\mathcal{F}(\sigma) : \mathcal{F} \to \sigma^*\mathcal{F}$, such that $\mathcal{F}(\mathrm{id}) = \mathrm{id}$ and $\tau^*\mathcal{F}(\sigma) \circ \mathcal{F}(\tau) = \mathcal{F}(\tau \circ \sigma)$. We say that $\mathcal{F}$ is a *lisse Weil sheaf* if the underlying $\bar{\mathbb{Q}}_\ell$-sheaf on $X \otimes_{\mathbb{F}_q} \mathbb{F}$ is lisse. Morphisms between Weil sheaves are morphisms between the underlying $\bar{\mathbb{Q}}_\ell$-sheaves, which are compatible with the $W(\mathbb{F}/\mathbb{F}_q)$-action in the obvious sense.

We record the following properties:

(i) There are natural functors

$$(\text{constructible } \bar{\mathbb{Q}}_\ell\text{-sheaves on } X) \hookrightarrow (\text{Weil sheaves on } X)$$
$$\to (\text{constructible } \bar{\mathbb{Q}}_\ell\text{-sheaves on } X \otimes_{\mathbb{F}_q} \mathbb{F}).$$

Here the first functor is given by pulling back a constructible sheaf on X to $X \otimes_{\mathbb{F}_q} \mathbb{F}$ with $W(\mathbb{F}/\mathbb{F}_q)$ acting via restriction of the natural $\mathrm{Gal}(\mathbb{F}/\mathbb{F}_q)$-action; this functor is fully faithful. The second functor is given by forgetting the $W(\mathbb{F}/\mathbb{F}_q)$-action.

(ii) Furthermore one can show that there is an equivalence of categories

$$(\text{lisse Weil sheaves on } X) \xrightarrow{\simeq} (\bar{\mathbb{Q}}_\ell\text{-representations of } W(X,\bar{x})), \quad \mathcal{F} \mapsto \mathcal{F}_{\bar{x}}.$$

Here we define a $\bar{\mathbb{Q}}_\ell$-representations of $W(X,\bar{x})$ as a group homomorphism $W(X,\bar{x}) \to \mathrm{Aut}_{\bar{\mathbb{Q}}_\ell}(V)$ with V a finite dimensional $\bar{\mathbb{Q}}_\ell$-vector space, which factors over a continuous homomorphism $W(X,\bar{x}) \to \mathrm{Aut}_E(V_E)$, where E is a finite field extension of $\mathbb{Q}_\ell$ and V_E is a finite dimensional E-vector space with an isomorphism $V_E \otimes_E \bar{\mathbb{Q}}_\ell \cong V$.

(d) Let U be a normal $\mathbb{F}_q$-scheme, $U \hookrightarrow X$ a divisorial compactification and D an effective Cartier divisor on X supported in $X \smallsetminus D$. Let $\mathcal{F}$ be a lisse Weil sheaf on U. Then we say that the *ramification of $\mathcal{F}$ is bounded by D* if the ramification of the underlying lisse $\bar{\mathbb{Q}}_\ell$-sheaf on $U \otimes_{\mathbb{F}_q} \mathbb{F}$ is bounded by $D \otimes_{\mathbb{F}_q} \mathbb{F}$.

REMARK 11.18. Notice that (up to isomorphism) a lisse Weil sheaf of rank 1 on $\mathrm{Spec}\,\mathbb{F}_q$ is just a group homomorphism $W(\mathbb{F}/\mathbb{F}_q) = \mathbb{Z} \to \bar{\mathbb{Q}}_\ell^\times$, i.e. it corresponds uniquely to an element in $\bar{\mathbb{Q}}_\ell^\times$.

THEOREM 11.19 (Deligne, see [**16**]). *Let U be a* smooth *and connected $\mathbb{F}_q$-scheme, $U \hookrightarrow X$ a divisorial compactification and D an effective Cartier divisor supported in $X \smallsetminus U$. Let $r \geq 1$ be a natural number. Then, up to twist with a lisse rank 1 Weil sheaf coming from $\mathrm{Spec}\,\mathbb{F}_q$, there are only finitely many irreducible lisse Weil sheaves of rank r on U whose ramification is bounded by D.*

In fact Deligne proves a more general finiteness result (see Theorem 11.32). We need some preparations to state it.

11.2. Skeleton sheaves.

11.2.1. *Semi-simplification.* Let G be a group (possibly infinite), E a field and V a finite dimensional E-vector space with G action, i.e. a representation of G. We can view V as a left $E[G]$-module and $\dim_E V < \infty$ implies that it is artinian and noetherian. Therefore V has finite length as $E[G]$-module, i.e. there is a chain of $E[G]$-submodules

$$0 = V_0 \subseteq V_1 \subseteq \ldots \subseteq V_r = V$$

such that the successive subquotients V_i/V_{i-1} are simple (or irreducible) $E[G]$-modules. Such a chain is called a decomposition series of V. By the Jordan-Hölder Theorem (see e.g. [**8**, I, §4.7, Thm. 6]) the length r is independent of the chosen decomposition series as is the sequence $V_r/V_{r-1}, \dots, V_1/V_0$, at least up to permutation and isomorphism. Therefore we can associate to V the $E[G]$-module

$$V^{\mathrm{ss}} = \oplus_{i=1}^{r} V_i/V_{i-1}.$$

The isomorphism class of this module depends only on V and it is called the *semi-simplification of* V. Note that if E is a finite extension of $\mathbb{Q}_\ell$ and if G is a topological group which acts continuously on V equipped with the ℓ-adic topology (cf. 8.0.5), then the action of G on the subspaces V_i is automatically continuous. The semi-simplification V^{ss} therefore is a continuous semi-simple G-representation.

In particular if $\mathcal{F}$ is a lisse $\bar{\mathbb{Q}}_\ell$-sheaf on a connected k-scheme X (resp. a lisse Weil sheaf), then we can view it as an $\bar{\mathbb{Q}}_\ell[\pi_1(X)]$-module (resp. $\bar{\mathbb{Q}}_\ell[W(X)]$-module) and can talk about its semi-simplification $\mathcal{F}^{\mathrm{ss}}$. Notice that $\mathcal{F}^{\mathrm{ss}}$ is a direct sum of irreducible lisse $\bar{\mathbb{Q}}_\ell$-sheaves (resp. lisse Weil sheaves).

NOTATION 11.20. Let U be a smooth and connected $\mathbb{F}_q$-scheme and $r \geq 1$ a natural number. We set

$$\mathcal{R}_r(U) = \{\text{isomorphism classes of lisse rank } r \text{ Weil sheaves on } U\}/\sim_{\mathrm{ss}},$$

where $\sim_{\mathrm{ss}}$ is the equivalence relation defined by

$$\mathcal{F} \sim_{\mathrm{ss}} \mathcal{G} :\Longleftrightarrow \mathcal{F}^{\mathrm{ss}} \cong \mathcal{G}^{\mathrm{ss}}.$$

Let $U \hookrightarrow X$ be a divisorial compactification and D an effective Cartier divisor supported on the complement. We set

$$\begin{aligned}\mathcal{R}_r(U,D) &:= \mathcal{R}_r(X,D)\\ &:= \{[\mathcal{F}] \in \mathcal{R}_r(U) \,|\, \text{the ramification of } \mathcal{F}^{\mathrm{ss}} \text{ is bounded by } D\},\end{aligned}$$

here $[\mathcal{F}]$ denotes the class of a lisse Weil sheaf $\mathcal{F}$ of rank r in $\mathcal{R}_r(U)$.

DEFINITION 11.21. Let U be a smooth connected $\mathbb{F}_q$-scheme.

(a) A *skeleton sheaf* $\mathcal{F}$ on U of rank r is a collection $\mathcal{F} = (\mathcal{F}_C)_{C \in \mathrm{Cu}(U)}$, with $\mathcal{F}_C \in \mathcal{R}_r(C)$, which is compatible in the sense

$$\mathcal{F}_{C|(C\times_U C')_{\mathrm{red}}} = \mathcal{F}_{C'|(C\times_U C')_{\mathrm{red}}}, \quad \text{for all } C, C' \in \mathrm{Cu}(U).$$

We denote by $\mathcal{V}_r(U)$ the set of rank r skeleton sheaves on U.

(b) We say that a skeleton sheaf $\mathcal{F}$ on U is *irreducible* if it cannot be written as a direct sum $\mathcal{F}_1 \oplus \mathcal{F}_2$, with $\mathcal{F}_i \in \mathcal{V}_{r_i}(U)$, $r_i \geq 1$. (For $C \in \mathrm{Cu}(U)$ we have a well-defined notion of direct sum in $\bigcup_{r\geq 0} \mathcal{R}_r C$ and for $\mathcal{F}_i = (\mathcal{F}_{i,C})_{C\in\mathrm{Cu}(U)} \in \mathcal{V}_{r_i}(U)$, $r_i \geq 1$, we define $\mathcal{F}_1 \oplus \mathcal{F}_2 = (\mathcal{F}_{1,C} \oplus \mathcal{F}_{2,C})_{C\in\mathrm{Cu}(U)}$; clearly $\mathcal{F}_1 \oplus \mathcal{F}_2$ lies in $\mathcal{V}_{r_1+r_2}(U)$.)

(c) Let $U \hookrightarrow X$ be a divisorial compactification and D an effective Cartier divisor supported in $X \setminus U$. We say that a skeleton sheaf $\mathcal{F}$ has *ramification bounded by* D if

$$\mathcal{F}_C \in \mathcal{R}_r(\widehat{C}, \nu_X^* D), \quad \text{for all } \nu : C \to U \text{ in } \mathrm{Cu}(U).$$

We denote by $\mathcal{V}_r(U,D)$ or $\mathcal{V}_r(X,D)$ the set of rank r skeleton sheaves on U whose ramification is bounded by D.

REMARK 11.22. Skeleton sheaves were introduced by Deligne and Drinfeld motivated by work of Wiesend, Kerz and Schmidt. See [**16**, 2.2] for a discussion where the name is coming from.

PROPOSITION 11.23. *Let U be a smooth connected $\mathbb{F}_q$-scheme and $r \geq 1$. There is a well-defined map*

$$\mathrm{sk} : \mathcal{R}_r(U) \to \mathcal{V}_r(U), \quad [\mathcal{F}] \mapsto \mathrm{sk}([\mathcal{F}]) := ([\mathcal{F}_{|C}])_{C \in \mathrm{Cu}(U)}.$$

It has the following properties:

(a) *If $U \hookrightarrow X$ is a divisorial compactification and D an effective Cartier divisor supported on $X \smallsetminus U$, then* sk *restricts to a map $\mathcal{R}_r(X,D) \to \mathcal{V}_r(X,D)$.*
(b) *The map* sk *is injective.*
(c) *A lisse Weil sheaf $\mathcal{F}$ on U is irreducible if and only if* $\mathrm{sk}([\mathcal{F}])$ *is irreducible in the sense of Definition* 11.21, (b).

PROOF. For the well-definedness of sk we have to show that when we have two lisse Weil sheaves $\mathcal{F}$, $\mathcal{G}$ of rank r on U with $\mathcal{F}^{\mathrm{ss}} = \mathcal{G}^{\mathrm{ss}}$, then $(\mathcal{F}_{|C})^{\mathrm{ss}} = (\mathcal{G}_{|C})^{\mathrm{ss}}$ for all $C \to U$ in $\mathrm{Cu}(U)$. But if we identify $\mathcal{F}$ with a representation of $\pi_1(U)$ and $0 = \mathcal{F}_0 \subseteq \mathcal{F}_1 \subseteq \ldots \subseteq \mathcal{F}_r = \mathcal{F}$ is a decomposition series of $\mathcal{F}$, then the chain $0 = \mathcal{F}_{0|C} \subseteq \mathcal{F}_{1|C} \subseteq \ldots \subseteq \mathcal{F}_{r|C} = \mathcal{F}_{|C}$ can be refined to a composition series of $\mathcal{F}_{|C}$. Hence

$$(\mathcal{F}_{|C})^{\mathrm{ss}} \cong ((\mathcal{F}^{\mathrm{ss}})_{|C})^{\mathrm{ss}},$$

which immediately gives the well-definedness.

Part (a) of the Proposition follows directly from the definitions. The parts (b) and (c) require some preliminary interludes. We start with part (b), for which we recall some well-known facts:

11.2.2. *Dirichlet density.* We follow [**48**]. Let Y be an irreducible $\mathbb{F}_q$-scheme of dimension $n \geq 1$. Let $||Y||$ be the set of closed points in Y. Then the sum $\sum_{y \in ||Y||} \frac{1}{q^{\deg(y)\cdot s}}$ converges absolutely for $s \in \mathbb{C}$ with $\mathrm{Re}(s) > n$, where $\deg(y) = [\mathbb{F}_q(y) : \mathbb{F}_q]$. (This is a consequence of the fact that the zeta function $\zeta(Y,s)$ of Y converges in this domain.) Let $M \subseteq ||Y||$ be a subset. We say that M has *Dirichlet density* δ if

$$\lim_{s \to n} \left(\frac{\sum_{y \in M} \frac{1}{q^{\deg(y)\cdot s}}}{\log(\frac{1}{s-n})} \right) = \delta.$$

It follows from the fact that the zeta function of Y has a simple pole at $s = n$, that the Dirichlet density of $||Y||$ equals 1.

11.2.3. Let Y be an irreducible $\mathbb{F}_q$-scheme and $\bar{\eta} \to Y$ a geometric point over the generic point of Y. Let K be the residue field of the generic point of Y and $K \hookrightarrow \bar{K}$ the inclusion corresponding to $\bar{\eta} \to \eta$. Let $\mathbb{F}$ be an algebraic closure of $\mathbb{F}_q$ and denote by $F \in \mathrm{Gal}(\mathbb{F}/\mathbb{F}_q)$ the geometric Frobenius ($F(a) = a^{-q}$). For $y \in ||Y||$ we denote by $F_y \in \pi_1(Y,\bar{\eta})$ the following conjugation class, called the *Frobenius class at y*: Denote by K_y the completion of K at y and choose an algebraic closure $\bar{K}_y$ with an embedding $\bar{K} \hookrightarrow \bar{K}_y$. Denote by $I_y \subseteq \mathrm{Gal}(\bar{K}_y/K_y)$ the inertia group. We have natural maps

$$\mathrm{Gal}(\mathbb{F}/k(y)) \cong \mathrm{Gal}(\bar{K}_y/K_y)/I_y \to \pi_1(Y,\bar{\eta})$$

and we define F_y to be the conjugacy class of the image of $F^{\deg(y)} \in \mathrm{Gal}(\mathbb{F}/k(y))$ in $\pi_1(Y,\bar{\eta})$. If we choose a different embedding $\bar{K} \hookrightarrow \bar{K}_y$ then the image of $F^{\deg(y)}$

in $\pi_1(U,\bar{\eta})$ will differ by conjugation, so that the class F_y is independent of all the choices.

Notice that if we have a representation $\rho : \pi(Y,\bar{\eta}) \to \mathrm{Aut}_A(M)$ where M is a finite free module over some ring A, then we can talk about the trace and the characteristic polynomial of $\rho(F_y)$.

THEOREM 11.24 (Artin-Chebotarev density, see [**48**, Thm. 7]). *Let U be an irreducible $\mathbb{F}_q$-scheme, $\bar{\eta} \to U$ a geometric point over its generic point and $V \to U$ a pointed finite étale Galois covering with Galois group G. Let $S \subseteq G$ be a subset of G which is stable under conjugation. Then the set*

$$\{u \in \|U\| \mid F_u \text{ is mapped into } S \text{ under } \pi_1(U,\bar{\eta}) \twoheadrightarrow G\}$$

has Dirichlet density equal to $|S|/|G|$.

COROLLARY 11.25. *Let $\bar{\eta} \to U$ be as in Theorem 11.24 above. Then the subset $\bigcup_{u\in\|U\|} F_u$ is dense in $\pi_1(U,\bar{\eta})$.*

PROOF. Since $\|U\|$ has Dirichlet density 1, Theorem 11.24 yields that $S := \bigcup_{u\in\|U\|} F_u$ maps surjectively to any finite quotient of $\pi_1(U,\bar{\eta})$. Since $\pi_1(U,\bar{\eta})$ is profinite the statement follows. □

The following Proposition is a generalization of Corollary 4.26.

PROPOSITION 11.26 ([**9**, §12, No. 1, Prop. 3]). *Let E be a field of characteristic zero and A an E-algebra with 1 (not necessarily commutative). Let M and M' be two semi-simple (left) A-modules which are finite dimensional as E-vector spaces. Then M and M' are isomorphic as A-modules if and only if $\mathrm{Tr}_{M/E}(a) = \mathrm{Tr}_{M'/E}(a)$, for all $a \in A$. Here $\mathrm{Tr}_{M/E}(a)$ denotes the trace of the E-linear map $M \to M$, $m \mapsto a\cdot m$.*

Proof of Proposition 11.23, (b). First of all notice that for every $u \in \|U\|$ there is a curve $\nu : C \to U$ in $\mathrm{Cu}(U)$ and a point $u_1 \in C$ such that ν induces an isomorphism $k(u) \xrightarrow{\simeq} k(u_1)$. Indeed we can take a regular sequence $t_1,\dots,t_d \in \mathcal{O}_{U,u}$ which generates the maximal ideal. Then the vanishing locus of $t_1,\dots,t_{d-1}$ defines a smooth curve C_0 in an open neighborhood of $u \in U$ and we can take $\nu : C \to U$ as the normalization of the closure of C_0 in U.

Now let $\mathcal{F}$ and $\mathcal{G}$ be two semi-simple lisse Weil sheaves on U of rank r and assume $\mathcal{F}_{|C} \cong \mathcal{G}_{|C}$, for all $\nu : C \to U$. By the above we get $\mathrm{Tr}_{\mathcal{F}/\bar{\mathbb{Q}}_\ell}(F_u) = \mathrm{Tr}_{\mathcal{G}/\bar{\mathbb{Q}}_\ell}(F_u)$ for all $u \in \|U\|$. (Here we identify $\mathcal{F}$ and $\mathcal{G}$ with their representations and use the notation from 11.2.3.) Therefore, there exists a finite field extension $E/\mathbb{Q}_\ell$ such that $\mathrm{Tr}_{\mathcal{F}/\bar{\mathbb{Q}}_\ell}, \mathrm{Tr}_{\mathcal{G}/\bar{\mathbb{Q}}_\ell} : \pi_1(U) \to E$ are continuous class functions which by Corollary 11.25 coincide on a dense open subset. Since the ℓ-adic topology on E is Hausdorff this implies $\mathrm{Tr}_{\mathcal{F}/\bar{\mathbb{Q}}_\ell} = \mathrm{Tr}_{\mathcal{G}/\bar{\mathbb{Q}}_\ell}$. Now Proposition 11.26 with $A = E[\pi_1(U)]$ gives $\mathcal{F} = \mathcal{G}$. □

Part (c) of Proposition 11.23 will be a consequence of a series of lemmas and propositions. We follow closely the arguments in [**16**, Prop. B.1]:

LEMMA 11.27. *Let ℓ be a prime number and $K \to H$ a morphism of profinite groups with H a pro-ℓ group. Denote by H^{ab} the abelianization of H in the category of profinite groups; it is the quotient of H by the closure of the commutator subgroup. If the induced map $K \to H^{\mathrm{ab}}/\ell H^{\mathrm{ab}}$ is surjective, then so is $K \to H$.*

PROOF (CF. [**51**, I.4.2]). We may assume without loss of generality that we have an inverse system of maps between finite groups $K_i \to H_i$, $i \in I$, which in the limit gives the map $K \to H$. (Indeed we know $H = \varprojlim_U H/U$, where the limit is over all normal open subgroups $U \subseteq H$. We denote by K_U the preimage of U in K and can replace K by $\varprojlim_U K/K_U$.) Since $H^{\mathrm{ab}} = \varprojlim_i H_i^{\mathrm{ab}}$, we are reduced to the case in which K is a finite group and H a finite ℓ-group. Now assume the map $K \to H^{\mathrm{ab}}/\ell$ is surjective. Since H^{ab} is a finitely generated $\mathbb{Z}/\ell^N\mathbb{Z}$-module (for some $N >> 0$), Nakayama's Lemma implies that $K \to H^{\mathrm{ab}}$ is surjective.

Let ℓ^r be the order of H, then by [**8**, I, §6.5, Thm. 1] there exists a sequence

$$H = H^1 \supseteq H^2 \supseteq \ldots \supseteq H^{r+1} = \{1\} \tag{11.8}$$

such that $[H, H^i] \subseteq H^{i+1}$, $1 \le i \le r$, and H^i/H^{i+1} is cyclic of order ℓ. Denote by $K' \subseteq H$ the image of K in H. Assume K' is strictly contained in H. Then there exists a minimal $i \le r$ such that $N := K' \cdot H^{i+1} \subsetneq K' \cdot H^i = H$. Then $N \subseteq H$ is a normal subgroup. (Indeed it suffices to see that H^i normalizes N. It clearly normalizes H^{i+1} and for $k \in K'$ and $h \in H^i$ we have $h \cdot k \cdot h^{-1} = k \cdot [k^{-1}, h] \in N$.) Further H/N is a non-zero quotient of H^i/H^{i+1} hence is cyclic of order ℓ. Thus the quotient map $H \to H/N$ factors over $H^{\mathrm{ab}} \to H/N$. But K maps to 1 in H/N, a contradiction to the surjectivity of $K \to H^{\mathrm{ab}}$. Thus $K' = H$. □

LEMMA 11.28 ([**16**, Lem. B.2]). *Let U be a smooth connected $\mathbb{F}_q$-scheme and $\mathcal{F}$ an irreducible lisse $\bar{\mathbb{Q}}_\ell$-sheaf on U. Then there exists a connected finite étale Galois cover $U' \to U$ with the following property: For any $\nu : C \to U$ in* Cu(U) *such that $C \times_U U'$ is connected the pullback $\nu^*\mathcal{F}$ is irreducible.*

PROOF. Take a DVR R finite over $\mathbb{Z}_\ell$ and a free lisse R-sheaf $(\mathcal{F}_n)_n$ such that $\mathcal{F} = (\mathcal{F}_n) \otimes_R \bar{\mathbb{Q}}_\ell$. Let $\mathbb{F}_\lambda = R/\mathfrak{m}$ be the residue field of R and denote by $\rho : \pi_1^{\text{ét}}(U) \to \mathrm{GL}_r(R)$ the R-representation corresponding to $(\mathcal{F}_n)$, where $r = \operatorname{rank}\mathcal{F}$. Let $G = \rho(\pi_1^{\text{ét}}(U))$ be the image of ρ and set

$$H_1 := \ker\left(\pi_1^{\text{ét}}(U) \xrightarrow{\rho} \mathrm{GL}_r(R) \to \mathrm{GL}_r(\mathbb{F}_\lambda)\right), \quad H_2 := \bigcap_{\varphi \in \mathrm{Hom}_{\mathrm{cont}}(H_1, \mathbb{Z}/\ell)} \ker(\varphi).$$

Then H_1 is an open normal subgroup of $\pi_1^{\text{ét}}(U)$, hence there exists a finite étale Galois cover $U_{H_1} \to U$ with $\pi_1^{\text{ét}}(U_{H_1}) = H_1$. Now the group H_1^{ab}/ℓ is profinite, in particular compact, and its Pontryagin dual is equal to

$$\mathrm{Hom}_{\mathrm{cont}}(H_1^{\mathrm{ab}}/\ell, \mathbb{R}/\mathbb{Z}) = \mathrm{Hom}_{\mathrm{cont}}(\pi_1^{\text{ét}}(U_{H_1}), \mathbb{Z}/\ell) = H^1(U_{H_1,\text{ét}}, \mathbb{Z}/\ell).$$

Here the second equality follows from the fact that for a finite abelian group A the first cohomology group $H^1(U_{\mathrm{et}}, A)$ is in bijection to the set of A-torsors over U (see e.g. [**39**, III, §4]). In particular H_1^{ab}/ℓ is a finite group, cf. 9.1.2, (a). Further, by definition $H_2 \subseteq H_1$ is a closed normal subgroup. We have $H_2 = \ker(H_1 \to H_1^{\mathrm{ab}}/\ell)$. (Clearly $\supseteq$; for the other inclusion it suffices to see that for any non-zero element $\sigma \in H_1^{\mathrm{ab}}/\ell$ there exists a continuous map $H_1^{\mathrm{ab}}/\ell \to \mathbb{Z}/\ell$ sending σ to a non-zero element. The construction of such a map is easily achieved by considering a filtration as in (11.8).) Hence $H_1/H_2 = H_1^{\mathrm{ab}}/\ell$ and therefore H_2 is an open normal subgroup of H_1. Denote by $U' \to U$ the associated Galois cover.

Now assume $\nu : C \to U$ is in Cu(U) and $C' := C \times_U U'$ is connected. Then $C' \to C$ is a Galois cover with Galois group $\pi^{\text{ét}}(U)/H_2$. Denote by $H_2' := \rho(H_2) \subseteq H_1' := \rho(H_1)$ the images of H_i in G. Notice that H_1' is a normal subgroup and is contained in $1 + \mathfrak{m}\operatorname{End}_R(R^r)$, hence is pro-$\ell$. Denote by ρ_C the composition (well-defined up to conjugation) $\rho_C : \pi_1^{\text{ét}}(C) \to \pi_1^{\text{ét}}(U) \to G$. By the above $\pi_1^{\text{ét}}(U)/H_2$ is a quotient of $\pi_1^{\text{et}}(C)$. Hence ρ_C maps onto G/H_2' and a fortiori onto G/H_1'. Let $K \subseteq \pi_1^{\text{ét}}(C)$ be the preimage of $H_1'/H_2' = H_1'^{\text{ab}}/\ell$. Then $K \to H_1'$ is a map of profinite groups which is surjective by Lemma 11.27. All together we see that ρ_C is surjective. Since ρ is irreducible so is ρ_C and this finishes the proof. □

We will need the following version of the Hilbert irreducibility theorem.

THEOREM 11.29 (see e.g. [**14**, Cor. A.2]). *Let C be a smooth projective curve over $\mathbb{F}_q$ with function field K. Let $x \in C$ be a closed point and denote by K_x the completion of K along x. Let $V \subseteq \mathbb{A}_K^n$ be a non-empty open subset and $V' \to V$ a connected finite étale covering. Then the set of K-rational points $y \in V(K)$ which do not split in V' (i.e. $y \times_V V'$ is a single point) is dense in $V(K_x)$. Here we equip $V(K_x) \subseteq K_x^n$ with the subspace topology.*

LEMMA 11.30. *Let k be a perfect field and $y, y' \in \mathbb{A}_k^d$ two closed points. Then there exists an integral curve $\Gamma \subseteq \mathbb{A}_k^d$ with y, y' contained in the smooth locus of Γ.*

PROOF. In case y, y' are k-rational points we can simply take Γ to be the line L connecting y and y'. In general we find a finite Galois extension k'/k such that y, y' split completely in $\mathbb{A}_{k'}^d$, i.e.

$$y \times_{\operatorname{Spec} k} \operatorname{Spec} k' = \{y_1, \dots, y_n\}, \quad y' \times_{\operatorname{Spec} k} \operatorname{Spec} k' = \{y_1', \dots, y_{n'}'\}$$

with k'-rational points $y_i, y_j' \in \mathbb{A}_{k'}^d$. Denote by $L_{i,j}$ the line in $\mathbb{A}_{k'}^d$ connecting y_i with y_j'. Set and let $\Gamma' = \cup_{i,j} L_{i,j}$; it is a curve in $\mathbb{A}_{k'}^d$, which is smooth in the points y_i, y_j'. Since $\operatorname{Gal}(k'/k)$ acts via permutation on $\{y_i\}$ and on $\{y_j'\}$ it also permutes the $L_{i,j}$. In particular Γ' is stable under the $\operatorname{Gal}(k'/k)$ action and by Galois descent (a special case of Theorem 7.7) there exists a curve $\Gamma_0 \subseteq \mathbb{A}_k^d$ with $\Gamma_0 \times_k k' = \Gamma'$. Let $\Gamma \subseteq \Gamma_0$ be an irreducible component (with its reduced scheme structure.) Then $\Gamma \times_k k'$ is a closed subcurve of Γ', hence is the union of certain $L_{i,j}$ and therefore contains certain y_i, y_j'. It follows that Γ contains y, y' and that these lie in the smooth locus of Γ. □

PROPOSITION 11.31 ([**16**, Prop. B.1]). *Let U be a smooth connected $\mathbb{F}_q$-scheme, $x \in U$ a closed point and $\mathcal{F} \in \mathcal{R}_r(U)$ irreducible. Then there exists a curve $\nu : C \to U$ in $\operatorname{Cu}(U)$ such that $x \in \nu(C)$ and $\nu^*\mathcal{F}$ is irreducible.*

PROOF. By [**13**, Prop. (1.3.14)] there exists a character $\chi \in \mathcal{R}_1(\mathbb{F}_q)$ such that $\mathcal{F} \otimes \chi$ is a lisse $\bar{\mathbb{Q}}_\ell$-sheaf. Thus we may assume from the beginning that $\mathcal{F}$ is an irreducible lisse $\bar{\mathbb{Q}}_\ell$-sheaf on U. Let $U' \to U$ be the finite étale Galois cover from Lemma 11.28. Then it suffices to construct a curve $\nu : C \to U$ in $\operatorname{Cu}(U)$ such that $U' \times_U C$ is connected and $x \in \nu(C)$.

To this end we proceed as follows. By Noether normalization (see e.g. [**15**, Cor. 16.18]) there exists a finite and generically étale $\mathbb{F}_q$-morphism $f : U \to \mathbb{A}_{\mathbb{F}_q}^d$. Set $y := f(x) \in \mathbb{A}_{\mathbb{F}_q}^d$. Denote by $V \subseteq \mathbb{A}_{\mathbb{F}_q}^d$ a non-empty open subset such that $f^{-1}(V) \to V$ is finite étale. Let $y' \in V$ be a closed point. Then by Lemma 11.30 we find $\mu : \Gamma \to \mathbb{A}_{\mathbb{F}_q}^n$ in $\operatorname{Cu}(\mathbb{A}_{\mathbb{F}_q}^n)$, such that μ is an isomorphism over y, y'. Let K be the function field of Γ and K_y the completion along y. Since Γ is smooth the choice of a local parameter at y yields a unique isomorphism of $\mathbb{F}_q$-algebras $K_y \cong k(y)((t))$ (see e.g. [**50**, II, §4]). On the other hand $k(y)$ is a finite extension of $\mathbb{F}_q$ hence of the form $k(y) = \mathbb{F}_q[x]/f$ for some irreducible polynomial f. Thus there is a closed point $z \in \mathbb{A}^1_k$ with $k(z) = k(y)$ and the completion of $k(\mathbb{A}^1)$ at z is isomorphic to K_y, i.e. $k(\mathbb{A}^1)_z \cong K_y = k(z)((t))$.

Let $\mathbb{A}^d_{\mathbb{F}_q} \to \mathbb{A}^1_{\mathbb{F}_q}$ be a linear projection such that the composition $\Gamma \to \mathbb{A}^d_{\mathbb{F}_q} \to \mathbb{A}^1_{\mathbb{F}_q}$ is finite. Denote by $\mathbb{A}^{d-1}_{k(\mathbb{A}^1)}$ the base change of this projection along $\operatorname{Spec} k(\mathbb{A}^1) \to \mathbb{A}^1_{\mathbb{F}_q}$ (inclusion of the generic point) and by $V_{k(\mathbb{A}^1)} \subseteq \mathbb{A}^{d-1}_{k(\mathbb{A}^1)}$ the corresponding base change of V. Since Γ is finite over $\mathbb{A}^1$ and $\mu(\Gamma) \cap V \ni y'$ the map μ induces a morphism $\operatorname{Spec} k(\mathbb{A}^1)_z = \operatorname{Spec} K_y \to V_{k(\mathbb{A}^1)}$, i.e. an element $\mu_z \in V_{k(\mathbb{A}^1)}(k(\mathbb{A}^1)_z)$. By Theorem 11.29 we find a point $v \in V_{k(\mathbb{A}^1)}(k(\mathbb{A}^1))$ which is t-adically arbitrarily close to μ_z and which does not split in $U'_{k(\mathbb{A}^1)}$. Since μ_z spreads out to $\operatorname{Spec}(k(z)[\![t]\!]) = \operatorname{Spec} \mathcal{O}_{K_y} \to \operatorname{Spec} \mathcal{O}_{\mathbb{A}^d_{\mathbb{F}_q},y}$, where the closed point maps to y, we can achieve that v spreads out to a map $\operatorname{Spec} \mathcal{O}_{\mathbb{A}^1,z} \to \operatorname{Spec} \mathcal{O}_{\mathbb{A}^d_{\mathbb{F}_q},y}$, which sends the closed point to y. Then v defines a point in V whose closure in $\mathbb{A}^d_{\mathbb{F}_q}$ is a curve containing y and over which in U' – a fortiori in U – lies exactly one point. Let $u \in U$ be the unique point lying over v. By the going-down theorem (see e.g. [**38**, Thm. 9.4, (ii)]) there exists a curve in U whose generic point lies over v and which contains x; since u is the only point in U mapping to v we get that the closure of u in U contains x. Let $\nu : C \to U$ be the normalization of $\overline{\{u\}}$. Then $x \in \nu(C)$ and $U' \times_U C$ is irreducible by the choice of v. This finishes the proof. □

Proof of Proposition 11.23, (c). Let $\mathcal{F}$ be a lisse Weil sheaf of rank r on U. If $\mathcal{F}$ is not irreducible, then the class $[\mathcal{F}]$ in $\mathcal{R}_r(U)$ can be written as a non-trivial sum $[\mathcal{F}] = [\mathcal{F}_1] \oplus [\mathcal{F}_2]$. Hence $\operatorname{sk}([\mathcal{F}]) = \operatorname{sk}([\mathcal{F}_1]) \oplus \operatorname{sk}([\mathcal{F}_2])$ is not irreducible in $\mathcal{V}_r(U)$. If $\mathcal{F}$ is irreducible, then for all closed points $x \in U$ there exists a curve $\nu_x : C_x \to U$ such that $x \in \nu_x(C_x)$ and $\nu_x^*\mathcal{F}$ is irreducible, by Proposition 11.31. Assume $\operatorname{sk}([\mathcal{F}]) = \mathcal{F}_1 \oplus \mathcal{F}_2$, with $\mathcal{F}_i \in \mathcal{V}_{r_i}(U)$. Then it follows that either $\nu_x^*\mathcal{F}_1 = 0$ for all $x \in U$ or $\nu_x^*\mathcal{F}_2 = 0$ for all $x \in U$. (Indeed the set Z_i of all $x \in U$ such that $\nu_x^*\mathcal{F}_i = 0$ is closed; since $U = Z_1 \sqcup Z_2$ and U is connected we have either $Z_1 = U$ or $Z_2 = U$.) Since the $\mathcal{F}_i$ are lisse we obtain that either $\mathcal{F}_1 = 0$ or $\mathcal{F}_2 = 0$. Hence $\operatorname{sk}([\mathcal{F}])$ is irreducible. This finishes the proof of Proposition 11.23. □

In view of Proposition 11.23, Theorem 11.19 is a consequence of the following theorem.

THEOREM 11.32 (Deligne). *Let U be a smooth connected $\mathbb{F}_q$-scheme, $U \hookrightarrow X$ a divisorial compactification and D an effective Cartier divisor supported in $X \setminus U$. Then the set of irreducible skeleton sheaves $\mathcal{F} \in \mathcal{V}_r(X,D)$ is finite up to twists by elements from $\mathcal{R}_1(\operatorname{Spec}\mathbb{F}_q)$ and its cardinality does not depend on the choices of $\ell \neq p$.*

For a proof of this Theorem, which relies on the Langlands correspondence [**31**] and Weil II [**13**], we refer the reader to [**16**].

QUESTION 11.33 (Deligne, see [**16**, Question 1.2]). Let U be a smooth connected $\mathbb{F}_q$-scheme, $U \hookrightarrow X$ a divisorial compactification and D an effective Cartier divisor supported in $X \setminus U$. Is the injection $\operatorname{sk} : \mathcal{R}_r(X,D) \hookrightarrow \mathcal{V}_r(X,D)$ actually a bijection?

REMARK 11.34. The above question has a positive answer for $r = 1$, see [**28**, Cor. V].

12. Grothendieck-Ogg-Shafarevich in higher dimensions following Kato-Saito

In this section we give a short account of the generalization of the Grothendieck-Ogg-Shafarevich formula to higher dimension due to Kato and Saito.

12.0.1. Throughout this section we fix the following notation:

- k is a perfect field of characteristic $p > 0$ and $\bar{k}$ an algebraic closure.
- ℓ is a prime number different from p.
- A k-scheme is a scheme which is separated and of finite type over k.
- If X is a k-scheme we denote by $CH_0(X)$ the Chow group of zero-cycles on X modulo rational equivalence, i.e. it is the free abelian group on the set of closed points of X modulo the subgroup generated by elements of the form $\operatorname{div}(f)_C$, where $C \subseteq X$ is a closed integral subscheme of dimension 1, $f \in k(C)^\times$ and $\operatorname{div}(f)_C$ denotes the divisor on C associated to f, which we view as a zero-cycle on X.

12.1. Higher dimensional version of Grothendieck-Ogg-Shafarevich.

12.1.1. *Intersection product with the log-diagonal.* Let U be a smooth connected k-scheme of dimension d. Let $j : U \hookrightarrow X$ and $j' : U \hookrightarrow X'$ be two open embeddings of U into integral proper k-schemes. If there exists a morphism $\pi : X' \to X$ such that $\pi \circ j' = j$, then π is unique and induces a proper morphism $\pi : X' \smallsetminus U \to X \smallsetminus U$ and hence also a pushforward

$$\pi_* : CH_0(X' \smallsetminus U) \to CH_0(X \smallsetminus U),$$

which sends the class of a closed point $x' \in X' \smallsetminus U$ to the class of $[x' : \pi(x')] \cdot \pi(x')$. We obtain a projective system indexed by open embeddings $j : U \hookrightarrow X$ as above. We set

$$CH_0(\partial U) := \varprojlim_j CH_0(X \smallsetminus U), \quad CH_0(\partial U)_{\mathbb{Q}} := \varprojlim_j (CH_0(X \smallsetminus U) \otimes_{\mathbb{Z}} \mathbb{Q}).$$

The degree map $\deg_k : CH_0(X \smallsetminus U) \to CH_0(\operatorname{Spec} k) = \mathbb{Z}$ is induced by pushforward along the structure map $X \to \operatorname{Spec} k$ and hence induces well-defined degree maps

$$\deg_k : CH_0(\partial U) \to \mathbb{Z}, \quad \deg_k : CH_0(\partial U)_{\mathbb{Q}} \to \mathbb{Q}.$$

Notice that if U is a curve with unique smooth compactification $j : U \hookrightarrow C$, then $CH_0(\partial U) = CH_0(C \smallsetminus U)$ and $CH_0(\partial U)_{\mathbb{Q}} = CH_0(C \smallsetminus U) \otimes_{\mathbb{Z}} \mathbb{Q}$.

Let $V \to U$ be a finite étale Galois cover with Galois group G. Denote by $\Delta_V \subseteq V \times_U V$ the diagonal. Notice that

$$(V \times_U V) \smallsetminus \Delta_V = \bigsqcup_{\sigma \in G \smallsetminus \{1\}} \Gamma_\sigma,$$

where Γ_σ is the graph of $\sigma : V \to V$. In [**24**, Thm 3.2.3] Kato-Saito define for each $\sigma \neq 1$ an element

$$(\Gamma_\sigma, \Delta_{\bar{V}})^{\log} \in CH_0(\partial V)_{\mathbb{Q}},$$

which is called *the intersection product of Γ_σ with the log-diagonal.* It has the following properties:

(a) The subscheme $\Gamma_\sigma \subseteq V \times_U V$ defines a map

$$\Gamma_\sigma^* : H^i_c(V \otimes_k \bar{k}, \mathbb{Q}_\ell) \xrightarrow{p_1^*} H^i_c(\Gamma_\sigma \otimes_k \bar{k}, \mathbb{Q}_\ell) \xrightarrow{p_{2*}} H^i_c(V \otimes_k \bar{k}, \mathbb{Q}_\ell).$$

Here the $p_i : \Gamma_\sigma \to V$, $i = 1, 2$, are induced by the two projection maps $V \times_U V \to V$ and the pushforward p_{2*} is induced by

$$p_{2*}(\mathbb{Q}_{\ell,\Gamma_\sigma}) = (\mathbb{Q}_{\ell,V})^{|G|} \xrightarrow{\Sigma} \mathbb{Q}_{\ell,V}.$$

Then (see [**24**, Prop. 3.2.4])

$$\sum_{i=0}^{2d}(-1)^i \operatorname{Tr}(\Gamma_\sigma^* | H_c^i(V \otimes_k \bar{k}, \mathbb{Q}_\ell)) = \deg_k(\Gamma_\sigma, \Delta_{\bar{V}})^{\log}.$$

(b) Assume there exist compactifications $V \hookrightarrow Y$ and $U \hookrightarrow X$ and an *étale* morphism $Y \to X$ extending the covering $V \to U$. Then the image of $(\Gamma_\sigma, \Delta_V)^{\log}$ under the projection

(12.1) $$CH_0(\partial V)_{\mathbb{Q}} \to CH_0(Y \setminus V) \otimes_{\mathbb{Z}} \mathbb{Q}$$

is zero (see [**24**, Cor. 3.3.4]).

(c) Let $V \hookrightarrow Y$ and $U \hookrightarrow X$ be compactifications and $Y \to X$ a morphism extending $V \to U$. Assume X is smooth and $X \setminus U$ is a strict normal crossings divisor. If $Y \to X$ is tamely ramified along $X \setminus U$ (in the sense of Definition 6.15,(a)), then $(\Gamma_\sigma, \Delta_V)^{\log}$ maps to zero under the projection (12.1) (see [**24**, Prop. 3.3.5, 2.]).

12.1.2. *Swan character class and Swan class.* Let U be a smooth connected k-scheme.

(a) Let $\pi : V \to U$ be a finite étale Galois covering with Galois group G. Then in [**24**, Def. 4.1.1] the *Swan character* $s_{V/U}(\sigma) \in CH_0(\partial V)$, $\sigma \in G$, is defined by

$$s_{V/U}(\sigma) = \begin{cases} -(\Gamma_\sigma, \Delta_V)^{\log}, & \text{if } \sigma \neq 1 \\ \sum_{\sigma \neq 1}(\Gamma_\sigma, \Delta_V)^{\log}, & \text{if } \sigma = 1. \end{cases}$$

(b) Let $\mathcal{F}$ be a lisse $\mathbb{F}_\lambda$-sheaf on U, for some finite extension $\mathbb{F}_\lambda$ of $\mathbb{F}_\ell$. Assume $\mathcal{F}$ is trivialized by a Galois covering $\pi : V \to U$ as in (a). Then in [**24**, Def. 4.2.1] the *Swan class* $\mathrm{Sw}_{V/U}(\mathcal{F}) \in CH_0(\partial V)_{\mathbb{Q}}$ is defined by

$$\mathrm{Sw}_{V/U}(\mathcal{F}) := \sum_{\sigma \in G_{(p)}} \left(\dim_{\mathbb{F}_\ell} M^\sigma - \frac{\dim_{\mathbb{F}_\ell} M^{\sigma^p}/M^\sigma}{p-1} \right) \cdot s_{V/U}(\sigma).$$

Here M is the G-representation corresponding to $\mathcal{F}$, $M^\sigma = \{m \in M \,|\, \sigma(m) = m\}$ and $G_{(p)} = \{\sigma \in G \,|\, \sigma \text{ has } p\text{-power order}\}$.

(c) Let A be an ℓ-adic coefficient ring and $\mathcal{F}$ a lisse A-sheaf on U. Then in [**24**, Def. 4.2.8] the *Swan class* $\mathrm{Sw}_U(\mathcal{F}) \in CH_0(\partial U)_{\mathbb{Q}}$ is defined as follows: We find a DVR R finite over $\mathbb{Z}_\ell$ and a lisse R-sheaf $(\mathcal{F}_n)$ such that $\mathcal{F} = (\mathcal{F}_n) \otimes_R A$. We find a Galois cover $\pi : V \to U$ as in (a) trivializing $\mathcal{F}_1$ and we define

$$\mathrm{Sw}_U(\mathcal{F}) := \frac{1}{|G|} \cdot \pi_* \mathrm{Sw}_{V/U}(\mathcal{F}_1).$$

This definition is independent of the choice of R, $(\mathcal{F}_n)$, and π.

REMARK 12.1. (a) Considering Lemma 10.6, the definition of $s_{V/U}(\sigma)$ above should remind you of the definition of the character a_G in Theorem 4.44.

(b) If U is a smooth curve with smooth compactification X and $\mathcal{F}$ is a lisse A-sheaf on U, then $\mathrm{Sw}_U(\mathcal{F}) = \sum_{x\in X\setminus U} \mathrm{Swan}_x(\mathcal{F})[x]$ in $CH_0(\partial U)_{\mathbb{Q}} = CH_0(X\setminus U)\otimes_Z \mathbb{Q}$, see [**24**, Introduction].

(c) In [**24**, Conj. 4.3.7] it is conjectured that $\mathrm{Sw}_U(\mathcal{F})$ always lies in the image of $CH_0(\partial U) \to CH_0(\partial U)_{\mathbb{Q}}$. In dimension 1 this is a consequence of the Hasse-Arf theorem and the remark above. In dimension 2 this conjecture is proved in [**24**, Cor. 5.1.7].

The higher dimensional version of the Grothendieck-Ogg-Shafarevich formula now takes the following form:

THEOREM 12.2 ([**24**, 4.2.9]). *Let U be a smooth and connected k-scheme of dimension d and $\mathcal{F}$ a lisse $\bar{\mathbb{Q}}_\ell$-sheaf on U. Then*

$$\chi_c(U\otimes_k \bar{k}, \mathcal{F}) := \sum_{i=0}^{2d} (-1)^i \dim_{\bar{\mathbb{Q}}_\ell} H^i_c(U\otimes_k \bar{k}, \mathcal{F})$$
$$= \mathrm{rk}(\mathcal{F})\cdot \chi_c(U, \bar{\mathbb{Q}}_\ell) - \deg_k(\mathrm{Sw}_U(\mathcal{F})).$$

We conclude with some remarks on more recent developments. For a more extensive overview we refer to [**54**].

12.2. Outlook on ramification theory following Abbes-Saito.

(a) In [**3**], Abbes and Saito give a refinement of Theorem 12.2. If $\pi : X \to \mathrm{Spec}\, k$ is a k-scheme and A an ℓ-adic coefficient ring, then the dualizing complex $K_X := \pi^!(A_{\mathrm{Spec}\, k})$ is defined as an object in the derived category of complexes of A-modules of finite tor-dimension with constructible cohomology. (In case π is smooth of pure dimension d, $K_X \cong A(d)[2d]$.) Let $\mathcal{F}$ be a constructible sheaf of free A-modules. Then in [**3**, Def. 2.1.1] the *characteristic class of* $\mathcal{F}$ is defined as an element $C(\mathcal{F}) \in H^0(X, K_X)$. It has the property that if π is proper, then

$$\mathrm{Tr}_\pi(C(\mathcal{F})) = \chi(X_{\bar{k}}, \mathcal{F}) \in A$$

where $\mathrm{Tr}_\pi : H^0(X, K_X) \to H^0(\mathrm{Spec}\, k, A) = A$ is the trace map.

Now assume U is a smooth k-scheme and that $j : U \hookrightarrow X$ is a compactification. Let $\mathcal{F}$ be a free lisse A-sheaf on U which is of Kummer type with respect to X (see [**3**, Def. 3.1.1]). Then ([**3**, Thm. 3.3.1])

$$C(j_!\mathcal{F}) = \mathrm{rk}(\mathcal{F})\cdot C(j_!A_U) - [\mathrm{Sw}^{\mathrm{naive}}_U(\mathcal{F})] \quad \text{in } H^0(X, K_X), \tag{12.2}$$

where $\mathrm{Sw}^{\mathrm{naive}}_U(\mathcal{F})$ is the naive Swan class (it is conjectured to be equal to $\mathrm{Sw}_U(\mathcal{F})$ and it is known that they have the same degree) and $[\mathrm{Sw}^{\mathrm{naive}}_U(\mathcal{F})]$ denotes the cycle class of $\mathrm{Sw}^{\mathrm{naive}}_U(\mathcal{F})$ in $H^0(X, K_X)$. In particular applying Tr_π we get back Theorem 12.2 for $\mathcal{F}$ as above. Actually, Theorem 12.2 can be reproved in its full generality using the equality (12.2), see [**3**, Cor. 3.3.2].

(b) We comment on the relation between the ramification theory via curves which is described in Section 11 and the ramification theory developed by Abbes-Saito in [**1**], [**2**], [**45**], [**4**].

Let K be a complete discrete valuation field with arbitrary residue field, K^{sep} a separable closure of K and $G = \mathrm{Gal}(K^{\mathrm{sep}}/K)$ the absolute Galois group. Then in [**1**] a *logarithmic ramification filtration* $(G^r_{\log})_{r\in\mathbb{Q}_{\geq 0}}$ is defined, which has the following properties:

(i) $G^s_{\log} \subseteq G^r_{\log}$, for $s < r$.

(ii) $G^0_{\log} = I =$ inertia subgroup of G, $G^{0+}_{\log} := \overline{\bigcup_{r>0} G^r_{\log}} = P =$ wild inertia subgroup.

(iii) Set $G^{r+}_{\log} := \overline{\bigcup_{s>r} G^s_{\log}}$. If K has characteristic $p > 0$ and "comes from geometry", then the quotient $G^r_{\log}/G^{r+}_{\log}$ is abelian and is killed by p for all $r > 0$.

(iv) In particular, if ℓ is a prime different from p, then according to Lemma 4.70, every ℓ-adic representation V of $G^{0+}_{\log}$ factors through a finite quotient, and as in Proposition 4.78 we obtain a break decomposition $V = \bigoplus_{x\in\mathbb{Q}_{\geq 0}} V(x)$. One then defines the *logarithmic Swan conductor* $\mathrm{Sw}^{\log}(V) := \sum_{x\in\mathbb{Q}_{\geq 0}} x \dim V(x)$.

(v) If the residue field of K is perfect, the filtration $G^r_{\log}$ coincides with the classical ramification filtration from Definition 3.54, and for an ℓ-adic representation V of G we have $\mathrm{Sw}(V) = \mathrm{Sw}^{\log}(V)$.

Now let U be a smooth connected k-scheme and denote by $\bar{\eta} \to U$ a geometric point over the generic point of U. Assume that we have a dominant open embedding $j : U \hookrightarrow X$ into a smooth k-scheme such that the complement $X \setminus U$ with its reduced structure is a strict normal crossings divisor $D = \cup_i D_i$. Denote the generic points of D by $\xi_i \in D_i$ and by G_{ξ_i} the decomposition subgroups at ξ_i of the absolute Galois group of $k(X)$ (well-defined up to conjugation). Let $\mathcal{F}$ be a free lisse A-sheaf on U. Define the *logarithmic Swan conductor of $\mathcal{F}$ with respect to X as*

$$\mathrm{Sw}^{\log}_X(\mathcal{F}) := \sum_i \mathrm{Sw}^{\log}_{\xi_i}(\mathcal{F}) \cdot D_i,$$

where $\mathrm{Sw}^{\log}_{\xi_i}(\mathcal{F})$ is the logarithmic Swan conductor of the representation $\mathcal{F}|_{G_\xi}$ in the sense of (b), (iv) above. Note that this is well-defined, even though G_{ξ_i} is only well-defined up to conjugation.

In [**16**, after Remark 3.10] the following conjecture is formulated (see also [**6**, Conj. B]):

CONJECTURE 12.3. *Let U be a smooth connected k-scheme and $\mathcal{F}$ a lisse A-sheaf on U. Let $j : U \hookrightarrow X$ be a divisorial compactification and D an effective Cartier divisor supported in $X \setminus U$. Then the following two statements are equivalent:*

(a) *The ramification of $\mathcal{F}$ is bounded by D in the sense of Definition 11.4.*

(b) *For all k-morphisms $h : V \to X$ fitting into a commutative diagram*

$$\begin{array}{ccc} & & V \\ & {}^{j'}\nearrow & \downarrow{\scriptstyle h} \\ U & \xrightarrow[j]{} & X, \end{array}$$

such that V is a smooth k-scheme, j' is an open dominant embedding and the complement $V \setminus U$ is a strict normal crossings divisor, we have

$$\mathrm{Sw}^{\log}_V(\mathcal{F}) \leq h^* D.$$

For $D = 0$ this is Proposition 11.16; for X smooth and $\mathcal{F}$ of rank 1 this conjecture is proved in [**6**, Thm. 7.1].

Acknowledgments

The authors wish to thank the participants of the summer school for their attention and input. In particular they wish to thank Pedro Ángel Castillejo, Javier Fresán and Shahab Rajabi for pointing out numerous misprints. The authors are also grateful to the referees for making several valuable comments.

References

[1] Ahmed Abbes and Takeshi Saito, *Ramification of local fields with imperfect residue fields*, Amer. J. Math. **124** (2002), no. 5, 879–920. MR1925338

[2] Ahmed Abbes and Takeshi Saito, *Ramification of local fields with imperfect residue fields. II*, Doc. Math. **Extra Vol.** (2003), 5–72. Kazuya Kato's fiftieth birthday. MR2046594

[3] Ahmed Abbes and Takeshi Saito, *The characteristic class and ramification of an l-adic étale sheaf*, Invent. Math. **168** (2007), no. 3, 567–612, DOI 10.1007/s00222-007-0040-7. MR2299562

[4] Ahmed Abbes and Takeshi Saito, *Ramification and cleanliness*, Tohoku Math. J. (2) **63** (2011), no. 4, 775–853, DOI 10.2748/tmj/1325886290. MR2872965

[5] M. Artin, A. Grothendieck, and J.-L. Verdier, *Théorie des topos et cohomologie étale des schémas: Séminaire de Géométrie Algébrique de Bois-Marie 1963/1964.*, Lecture Notes in Mathematics, vol. 269, 270, 305, Springer-Verlag, 1972/1973. MR0354652

[6] Ivan Barrientos, *Log ramification via curves in rank 1*, Int. Math. Res. Not. IMRN **19** (2017), 5769–5799. MR3741881

[7] N. Bourbaki, *Éléments de mathématique. Algèbre. Chapitre 9* (French), Springer-Verlag, Berlin, 2007. Reprint of the 1959 original. MR2325344

[8] ———, *Éléments de mathématique. Algèbre. Chapitres 1 à 3*, Springer, Berlin, 2007, reprint of the 1970 original.

[9] N. Bourbaki, *Éléments de mathématique. Algèbre. Chapitre 8. Modules et anneaux semi-simples* (French), Springer, Berlin, 2012. Second revised edition of the 1958 edition [MR0098114]. MR3027127

[10] J.W.S. Cassels and A. Fröhlich (eds.), *Algebraic number theory.* Proceedings of an instructional conference organized by the London Mathematical Society (a NATO Advanced Study Institute) with the support of the International Mathematical Union (held at the University of Sussex, Brighton, September 1-17, 1965). Second Edition, including a list of errata, London Mathematical Society, 2010.

[11] Brian Conrad, *Deligne's notes on Nagata compactifications*, J. Ramanujan Math. Soc. **22** (2007), no. 3, 205–257. MR2356346

[12] P. Deligne, *Cohomologie étale* (French), Lecture Notes in Mathematics, vol. 569, Springer-Verlag, Berlin, 1977. Séminaire de géométrie algébrique du Bois-Marie SGA $4\frac{1}{2}$. MR463174

[13] Pierre Deligne, *La conjecture de Weil. II* (French), Inst. Hautes Études Sci. Publ. Math. **52** (1980), 137–252. MR601520

[14] Vladimir Drinfeld, *On a conjecture of Deligne* (English, with English and Russian summaries), Mosc. Math. J. **12** (2012), no. 3, 515–542, 668, DOI 10.17323/1609-4514-2012-12-3-515-542. MR3024821

[15] David Eisenbud, *Commutative algebra*, Graduate Texts in Mathematics, vol. 150, Springer-Verlag, New York, 1995. With a view toward algebraic geometry. MR1322960

[16] Hélène Esnault and Moritz Kerz, *A finiteness theorem for Galois representations of function fields over finite fields (after Deligne)*, Acta Math. Vietnam. **37** (2012), no. 4, 531–562. MR3058662

[17] Lei Fu, *Etale cohomology theory*, Revised edition, Nankai Tracts in Mathematics, vol. 14, World Scientific Publishing Co. Pte. Ltd., Hackensack, NJ, 2015. MR3380806

[18] Philippe Gille, *Le groupe fondamental sauvage d'une courbe affine en caractéristique $p >$ 0* (French), Courbes semi-stables et groupe fondamental en géométrie algébrique (Luminy, 1998), Progr. Math., vol. 187, Birkhäuser, Basel, 2000, pp. 217–231. MR1768103

[19] A. Grothendieck, *Éléments de géométrie algébrique. IV. Étude locale des schémas et des morphismes de schémas IV* (French), Inst. Hautes Études Sci. Publ. Math. **32** (1967), 361. MR238860

[20] A. Grothendieck, L. Illusie, J.-P. Jouanolou, and I. Bucur, *Cohomologie ℓ-adique et fonctions L: Séminaire de Géométrie Algébrique de Bois-Marie 1965/1966*, Lecture Notes in Mathematics, Springer-Verlag, 1977. MR0491704

[21] Alexander Grothendieck and Jacob P. Murre, *The tame fundamental group of a formal neighbourhood of a divisor with normal crossings on a scheme*, Lecture Notes in Mathematics, Vol. 208, Springer-Verlag, Berlin-New York, 1971. MR0316453

[22] *Revêtements étales et groupe fondamental* (French), Lecture Notes in Mathematics, Vol. 224, Springer-Verlag, Berlin-New York, 1971. Séminaire de Géométrie Algébrique du Bois Marie 1960–1961 (SGA 1); Dirigé par Alexandre Grothendieck. Augmenté de deux exposés de M. Raynaud. MR0354651

[23] Robin Hartshorne, *Algebraic geometry*, Springer-Verlag, New York-Heidelberg, 1977. Graduate Texts in Mathematics, No. 52. MR0463157

[24] Kazuya Kato and Takeshi Saito, *Ramification theory for varieties over a perfect field*, Ann. of Math. (2) **168** (2008), no. 1, 33–96, DOI 10.4007/annals.2008.168.33. MR2415398

[25] Nicholas M. Katz, *Wild ramification and some problems of "independence of l"*, Amer. J. Math. **105** (1983), no. 1, 201–227, DOI 10.2307/2374386. MR692111

[26] Nicholas M. Katz, *Local-to-global extensions of representations of fundamental groups* (English, with French summary), Ann. Inst. Fourier (Grenoble) **36** (1986), no. 4, 69–106. MR867916

[27] Nicholas M. Katz, *Gauss sums, Kloosterman sums, and monodromy groups*, Annals of Mathematics Studies, vol. 116, Princeton University Press, Princeton, NJ, 1988. MR955052

[28] Moritz Kerz and Shuji Saito, *Chow group of 0-cycles with modulus and higher-dimensional class field theory*, Duke Math. J. **165** (2016), no. 15, 2811–2897, DOI 10.1215/00127094-3644902. MR3557274

[29] Moritz Kerz and Alexander Schmidt, *On different notions of tameness in arithmetic geometry*, Math. Ann. **346** (2010), no. 3, 641–668, DOI 10.1007/s00208-009-0409-6. MR2578565

[30] János Kollár, *Fundamental groups of rationally connected varieties*, Michigan Math. J. **48** (2000), 359–368, DOI 10.1307/mmj/1030132724. Dedicated to William Fulton on the occasion of his 60th birthday. MR1786496

[31] Laurent Lafforgue, *Chtoucas de Drinfeld et correspondance de Langlands* (French, with English and French summaries), Invent. Math. **147** (2002), no. 1, 1–241, DOI 10.1007/s002220100174. MR1875184

[32] T. Y. Lam, *A first course in noncommutative rings*, 2nd ed., Graduate Texts in Mathematics, vol. 131, Springer-Verlag, New York, 2001. MR1838439

[33] G. Laumon, *Semi-continuité du conducteur de Swan (d'après P. Deligne)* (French), The Euler-Poincaré characteristic (French), Astérisque, vol. 83, Soc. Math. France, Paris, 1981, pp. 173–219. MR629128

[34] G. Laumon, *Transformation de Fourier, constantes d'équations fonctionnelles et conjecture de Weil* (French), Inst. Hautes Études Sci. Publ. Math. **65** (1987), 131–210. MR908218

[35] Qing Liu, *Algebraic geometry and arithmetic curves*, Oxford Graduate Texts in Mathematics, vol. 6, Oxford University Press, Oxford, 2002. Translated from the French by Reinie Erné; Oxford Science Publications. MR1917232

[36] W. Lütkebohmert, *On compactification of schemes*, Manuscripta Math. **80** (1993), no. 1, 95–111, DOI 10.1007/BF03026540. MR1226600

[37] Hideyuki Matsumura, *Commutative algebra*, W. A. Benjamin, Inc., New York, 1970. MR0266911

[38] Hideyuki Matsumura, *Commutative ring theory*, 2nd ed., Cambridge Studies in Advanced Mathematics, vol. 8, Cambridge University Press, Cambridge, 1989. Translated from the Japanese by M. Reid. MR1011461

[39] James S. Milne, *Étale cohomology*, Princeton Mathematical Series, vol. 33, Princeton University Press, Princeton, N.J., 1980. MR559531

[40] James S. Milne, *Fields and galois theory (v4.50)*, 2014, Available at `www.jmilne.org/math/`, p. 138.

[41] Jürgen Neukirch, *Algebraic number theory*, Grundlehren der Mathematischen Wissenschaften [Fundamental Principles of Mathematical Sciences], vol. 322, Springer-Verlag, Berlin, 1999. Translated from the 1992 German original and with a note by Norbert Schappacher; With a foreword by G. Harder. MR1697859

[42] Nikolay Nikolov and Dan Segal, *On finitely generated profinite groups. I. Strong completeness and uniform bounds*, Ann. of Math. (2) **165** (2007), no. 1, 171–238, DOI 10.4007/annals.2007.165.171. MR2276769

[43] Michel Raynaud, *Caractéristique d'Euler-Poincaré d'un faisceau et cohomologie des variétés abéliennes* (French), Séminaire Bourbaki, Vol. 9, Soc. Math. France, Paris, 1995, pp. Exp. No. 286, 129–147. MR1608794

[44] Norbert Roby, *Lois polynomes et lois formelles en théorie des modules* (French), Ann. Sci. École Norm. Sup. (3) **80** (1963), 213–348. MR0161887

[45] Takeshi Saito, *Wild ramification and the characteristic cycle of an l-adic sheaf*, J. Inst. Math. Jussieu **8** (2009), no. 4, 769–829, DOI 10.1017/S1474748008000364. MR2540880

[46] J.-P. Serre, *Espaces fibres algébriques*, Séminaire Claude Chevalley **tome 3** (1958), no. 3, 1–37.

[47] Jean-Pierre Serre, *Sur la rationalité des représentations d'Artin* (French), Ann. of Math. (2) **72** (1960), 405–420, DOI 10.2307/1970142. MR171775

[48] Jean-Pierre Serre, *Zeta and L functions*, Arithmetical Algebraic Geometry (Proc. Conf. Purdue Univ., 1963), Harper & Row, New York, 1965, pp. 82–92. MR0194396

[49] Jean-Pierre Serre, *Linear representations of finite groups*, Springer-Verlag, New York-Heidelberg, 1977. Translated from the second French edition by Leonard L. Scott; Graduate Texts in Mathematics, Vol. 42. MR0450380

[50] Jean-Pierre Serre, *Local fields*, Graduate Texts in Mathematics, vol. 67, Springer-Verlag, New York-Berlin, 1979. Translated from the French by Marvin Jay Greenberg. MR554237

[51] Jean-Pierre Serre, *Galois cohomology*, Corrected reprint of the 1997 English edition, Springer Monographs in Mathematics, Springer-Verlag, Berlin, 2002. Translated from the French by Patrick Ion and revised by the author. MR1867431

[52] Tamás Szamuely, *Galois groups and fundamental groups*, Cambridge Studies in Advanced Mathematics, vol. 117, Cambridge University Press, Cambridge, 2009. MR2548205

[53] P. Webb, *Finite group representations for the pure mathematician*, 2012, course notes available at `http://www.math.umn.edu/~webb/`.

[54] L. Xiao and I. Zhukov, *Ramification of higher local fields, approaches and questions*, Algebra i Analiz **26** (2014), no. 5, 1–63, DOI 10.1090/spmj/1355; English transl., St. Petersburg Math. J. **26** (2015), no. 5, 695–740. MR3408704

Index

Freie Universität Berlin, Mathematisches Institut, Arnimallee 3, D-14195 Berlin, Germany
Email address: kindler@math.fu-berlin.de

Bergische Universität Wuppertal, Fakultät für Mathematik und Naturwissenschaften, Gaußstrasse 20, D-42097 Wuppertal, Germany
Email address: ruelling@uni-wuppertal.de

Published Titles in This Series

21 **K. Ebrahimi-Fard, J. I. Burgos Gil, and D. Manchon, Editors,** Amplitudes, Hodge Theory and Ramification, 2020

20 **David Ellwood, Herwig Hauser, Shigefumi Mori, and Josef Schicho, Editors,** The Resolution of Singular Algebraic Varieties, 2014

19 **James Carlson, Editor,** The Poincaré Conjecture, 2014

18 **Brendan Hassett, James McKernan, Jason Starr, and Ravi Vakil, Editors,** A Celebration of Algebraic Geometry, 2013

17 **David Ellwood, Igor Rodnianski, Gigliola Staffilani, and Jared Wunsch, Editors,** Evolution Equations, 2013

16 **Guillermo Cortiñas, Editor,** Topics in Noncommutative Geometry, 2012

15 **David Ellwood, Charles Newman, Vladas Sidoravicius, and Wendelin Werner, Editors,** Probability and Statistical Physics in Two and More Dimensions, 2012

14 **David A. Ellwood and Emma Previato, Editors,** Grassmannians, Moduli Spaces and Vector Bundles, 2011

13 **James Arthur, James W. Cogdell, Steve Gelbart, David Goldberg, Dinakar Ramakrishnan, and Jiu-Kang Yu, Editors,** On Certain L-Functions, 2011

12 **Alan Carey, David Ellwood, Sylvie Paycha, and Steven Rosenberg, Editors,** Motives, Quantum Field Theory, and Pseudodifferential Operators, 2010

11 **Etienne Blanchard, David Ellwood, Masoud Khalkhali, Matilde Marcolli, Henri Moscovici, and Sorin Popa, Editors,** Quanta of Maths, 2010

10 **Manfred Leopold Einsiedler, David Alexandre Ellwood, Alex Eskin, Dmitry Kleinbock, Elon Lindenstrauss, Gregory Margulis, Stefano Marmi, and Jean-Christophe Yoccoz, Editors,** Homogeneous Flows, Moduli Spaces and Arithmetic, 2010

9 **Reza Akhtar, Patrick Brosnan, and Roy Joshua, Editors,** The Geometry of Algebraic Cycles, 2010

8 **Henri Darmon, David Alexandre Ellwood, Brendan Hassett, and Yuri Tschinkel, Editors,** Arithmetic Geometry, 2009

7 **William Duke and Yuri Tschinkel, Editors,** Analytic Number Theory, 2007

6 **Nigel Higson and John Roe, Editors,** Surveys in Noncommutative Geometry, 2006

5 **David A. Ellwood, Peter S. Ozsváth, András I. Stipsicz, and Zoltán Szabó, Editors,** Floer Homology, Gauge Theory, and Low-Dimensional Topology, 2006

4 **James Arthur, David Ellwood, and Robert Kottwitz, Editors,** Harmonic Analysis, the Trace Formula, and Shimura Varieties, 2005

3 **Michael Douglas, Jerome Gauntlett, and Mark Gross, Editors,** Strings and Geometry, 2004

2 **David Hoffman, Editor,** Global Theory of Minimal Surfaces, 2005

1 **Atish Dabholkar, Sunil Mukhi, and Spenta R. Wadia, Editors,** Strings 2001, 2002